Directory of Power Plants in Canada

UDI-033-92

UDI/McGraw-Hill
1700 K Street, N.W.
Suite 400
Washington, DC 20006
USA

Second Edition
December 1992

C.A.E. Bergesen, Editor

This report has been prepared by Utility Data Institute (UDI), a unit of McGraw-Hill, Inc. The information contained herein has been furnished by sources which UDI believes to be reliable, however, UDI does not guarantee the accuracy, adequacy, timeliness, or completeness of such information.

Neither UDI, McGraw-Hill, Inc., its suppliers of information nor their officers, directors, employees and agents shall be responsible for any errors or omissions or the results obtained from the use of the information in this report.

No part of this publication may be reproduced or utilized in any form or by any means, electronic or mechanical, including photocopying, recording, or by any information storage and retrieval system without permission in writing from the publisher.

Copyright © 1992 by Utility Data Institute, a unit of McGraw-Hill, Inc.
Printed in the U.S.A. All rights reserved.

This report is issued annually. Previous editions may be archived.

UDI Telephone: 202-466-3660
UDI FAX: 202-466-3667

UDI-033-92　　　　　　　　　　　　　　　　　　　　　　　　　　December 1992

PROFILE

The second edition of UDI's *Directory of Power Plants in Canada* lists key design data for every fossil, nuclear, hydroelectric, and gas turbine generating unit in the country. These units are operating, under construction, planned, or shutdown and represent installed electric generating capacity of 110,700 megawatts (MW). They are located at 900 plant sites throughout Canada's 12 provinces and territories, and are operated (or will be operated) by about 240 companies, including the provincially-owned Crown Corporations, investor-owned utilities, municipal power authorities, and a variety of autoproducers. The provincial utilities dominate the industry, controlling 85% of the country's installed capacity.

Over the last two years, the Canadian electric power business has changed substantially. Recent industry issues include utility privatization, the development of a significant non-utility power plant sector, regulatory initiatives related to demand-side management (DSM), and a virtual cessation of the design and development of the gigantic hydroelectric developments characteristic of the Canadian industry for the last two decades. Summary discussions of notable electric power market developments over the past year are included in the report along with basic financial and operating statistics for Canadian utilities.

Unit-level design data shown in the hard-copy of the *Directory of Power Plants in Canada* include operator, size, fuel, age, status, and boiler, turbine, and generator vendor. The source data base contains additional details on turbine type, steam conditions, and air pollution control equipment.

The *Profile Table* on the following pages shows unit counts and capacity totals by unit operating status, unit type, and fuel for all 3,106 units included in UDI's Canadian power plant data base. Note that information on retired Canadian generating units is not printed in the *Directory*, but is contained in the data base sold on diskette.

PROFILE OF 1992 CANADIAN POWER PLANT DATA BASE COVERAGE

PLANT STATUS: IN OPERATION			
PLANT TYPE	*FUEL*	*UNITS*	*CAPACITY (MW)*
GAS TURBINE (SIMPLE-CYCLE)	GAS	36	1,412
	JET FUEL	2	24
	OIL	70	1,275
GAS TURBINE (COMBINED-CYCLE/COGEN)	GAS	12	320
HYDRAULIC	WATER	1,395	62,242
INTERNAL COMBUSTION	GAS	44	55
	OIL	764	576
STEAM TURBINE	COAL	95	18,969
	GAS	80	2,742
	OIL	46	5,452
	URANIUM	20	13,847
	OTHER	52	641
STEAM TURBINE (COMBINED-CYCLE & COGEN)	GAS	18	168
	OIL	25	175
	OTHER	19	271
WIND TURBINE	WIND	4	1
TOTAL IN OPERATION		2,682	108,170

PLANT STATUS: UNDER CONSTRUCTION			
PLANT TYPE	FUEL	UNITS	CAPACITY (MW)
GAS TURBINE (SIMPLE-CYCLE)	GAS/OIL	7	473
GAS TURBINE (COMBINED-CYCLE)	GAS	4	237
HYDRAULIC	WATER	20	2,267
STEAM TURBINE	COAL	3	1,011
	URANIUM	2	1,762
	OTHER	4	183
OTHER	VARIOUS	2	6
TOTAL UNDER CONSTRUCTION		42	5,939

PLANT STATUS: PLANNED			
PLANT TYPE	FUEL	UNITS	CAPACITY (MW)
COMBINED-CYCLE	GAS	11	1,535
GAS TURBINE (SIMPLE-CYCLE)	GAS	6	176
	OIL	4	324
GAS TURBINE (COMBINED-CYCLE)	GAS	3	124
HYDRAULIC	WATER	56	8,506
INTERNAL COMBUSTION	VARIOUS	4	22
STEAM TURBINE	COAL	5	1,705
	REF	4	58
	WOOD	6	114
	OTHER	3	92
STEAM TURBINE (COMBINED-CYCLE & COGEN	WASTE HEAT	2	45
	WOOD OR COAL	3	192
WIND TURBINE	WIND	2	19
TOTAL PLANNED		109	12,912

UDI-033-92   December 1992

PLANT STATUS: SHUTDOWN			
PLANT TYPE	*FUEL*	*UNITS*	*CAPACITY (MW)*
HYDRAULIC	WATER	24	939
INTERNAL COMBUSTION	OIL	10	8
STEAM TURBINE	COAL	8	1,200
	GAS	7	105
TOTAL SHUTDOWN		49	2,252

PLANT STATUS: RETIRED			
PLANT TYPE	*FUEL*	*UNITS*	*CAPACITY (MW)*
GAS TURBINE	GAS/OIL	16	272
IC	GAS/OIL	41	19
HYDRAULIC	WATER	19	31
STEAM TURBINE	VARIOUS	14	68
	URANIUM	3	511
WIND TURBINE	WIND	1	4
TOTAL RETIRED		94	905

PLANT STATUS: UNKNOWN			
PLANT TYPE	*FUEL*	*UNITS*	*CAPACITY (MW)*
INTERNAL COMBUSTION	GAS/OIL	120	61
OTHER	VARIOUS	9	129
TOTAL UNKNOWN		129	190

UDI-033-92   December 1992

TABLE OF CONTENTS

Directory of Power Plants in Canada

	Page
The Electric Power Business in Canada	1
Canadian Electric Utility Profiles	7
BC Hydro	9
Canadian Utilities Ltd.	12
Edmonton Power	14
Hydro-Quebec	15
Manitoba Hydro	20
Maritime Electric Company	23
New Brunswick Power Corp.	25
Newfoundland Light & Power Co.	29
Newfoundland and Labrador Hydro	31
Nova Scotia Power Corp.	32
Ontario Hydro	36
SaskPower	42
TransAlta Utilities	45
1992 Canadian Power Plant Report and Data Base	57
Information Sources and Research Methodology	63
Directory of Power Plants in Canada	65
Appendix A: Electric Power Plant Operators in Canada	A-1
Appendix B: UDI Canadian Power Plant Data (CANADA) File	B-1
Appendix C: Abbreviations Used in the UDI Utility Data Base	C-1
Appendix D: Conversion Factors and Tables for Electric Power Plant Data	D-1

Figures

Figure 1: Electric Generating Capacity Worldwide -- 1990 49

Figure 2: Installed Generating Capacity of the 12 Largest
Utilities in Canada -- 1992 51

Figure 3: Age Distribution of Installed Generating Capacity in
Canada 53

Figure 4: Fuel Distribution of Installed Generating Capacity in
Canada -- 1992 55

UDI-033-92   December 1992

THE ELECTRIC POWER BUSINESS IN CANADA

Canada has an abundance of natural resources, a well-educated population, and a mature market economy. Until recently, the Canadian electric power industry reflected this comfortable situation by being conservative, slow growing, and, essentially, static. However, the large utilities in Canada -- and they include two of the biggest electric companies in the world -- are faced with an onslaught of market forces that will force (and have already forced) some wrenching changes in their business.

"Hydro Faces Dark Days" read the headline in the Toronto *Globe & Mail* last fall: "Quebec Inc(omplete)" said *The Economist* in March reporting on **Hydro-Quebec's** struggle with a **New York Power Authority** decision to cancel a multibillion dollar power-purchase contract. Mine closings, plant deferrals, environmental agitation, complex conservation programs, costly new baseload plants, land claims, capital disallowances, and utility privatization all have been in the news over the last year.

Most Canadian utilities are faced with flat -- or even declining -- electricity sales over the next few years. This comes at a time when many Canadian companies are facing large capital expenditures and a variety of competitive pressures that will force major structural changes within their companies.

A few Canadian utilities continue to experience significant load growth, some of it supported by large, long-term power sales contracts into northern-tier states in the United States. Increasingly, inter-provincial power sales are also an important source of revenue for certain companies.

The Canadian electric power industry has some notable characteristics:

- Canada consumed 4.1% of worldwide electricity in 1990; Canada's electricity consumption is only about 16% of the electricity consumed in the United States.

- At year-end 1991, Canada had 108,000 MW of installed capacity, about 4% of the world total. Canadian capacity represents 15% of that of the U.S. total installed capacity of 677,700 MW (see *Figure 1* for 1990 data).

- Canada is the world's largest producer of hydroelectricity, with 1990 generation of 297 million megawatthours (MWhrs), 14% of the world's total hydroelectric output. The United States and the USSR ranked second and third, respectively.[1]

- Along with hydropower, Canada also produces large amounts of nuclear energy. As a result, Canada may have the lowest-cost power of any comparable economy. In 1989, Canadian residential rates were ¢US[2] 4.8/kilowatthour (kWhr) while industrial rates were ¢US 3.9/kWhr. U.S. rates were 58% and 18% higher, respectively, in 1989.

- In part as a result of these low electricity costs, Canadians also consume more electricity per capita than any other country with the exception of Norway, a far smaller economy. Canadian consumption of electricity per capita in 1990 was 18,149 kWhr, 49% higher than the U.S. value of 12,170 Kwhr.

Although comparative statistics have not been prepared, Canadian electric utilities may be more important to their country's economy than electric utilities in any other comparable economy:

- Canadian electric utilities have increased their contribution to Canadian Gross Domestic Product (GDP) from 2.3% in 1960 to 3.3% in 1991.[3]

- Electric utilities employ 101,000 people directly, representing 1.1% of total Canadian employment.

- The two biggest Canadian utilities (**Ontario Hydro** and **Hydro-Quebec**) are also the second and third biggest companies in the country when ranked by assets.

---

[1]The latest global electricity statistics available are for the year 1990. Sources are the *International Energy Annual*, published by the Energy Information Administration of the U.S. Department of Energy (DOE/EIA-0219) and the *Energy Statistics Yearbook*, published by the United Nations (1992). Although subject to restatement, the UN Yearbook is the *de facto* source for most summary statistics published about electric power worldwide.

[2]Unless otherwise specified, monetary units are Canadian dollars.

[3]These and many other valuable statistics can be found in the publication *Electric Power In Canada 1991* published by Energy Mines and Resources Canada located in Ottawa, Ontario. This publication also has useful reviews of electric power regulation in Canada.

- In 1989, 4.5% of all Canadian electricity generated was exported to the United States. This made Canada the third largest power exporter in the world after France and the USSR. (Canada also imported 2.6% of its consumption in 1989.)

Electric power is only one component of Canada's huge energy resources sector. In 1990, the country was the world's fifth largest primary energy producer, behind Saudi Arabia, and is the sixth largest energy consumer, just behind Germany, and just ahead of the United Kingdom. Coal, oil, natural gas, and uranium are all produced in large quantities. Increasingly, natural gas is being exported to the United States. Uranium has been a major export for years and although production declined in 1991, Canada still provides 30% of the western world's supply.

Setting the Stage

While the Canadian electric power business environment has been relatively stable in the last several years, Canadian electric utilities are now beginning to feel the full impact of the many planning uncertainties now experienced by their cousins in other "advanced" market economies. As in the United States, Japan, and the European Community (EC),[4] these uncertainties can in large measure be traced to:

- Public opposition to the construction of large thermal or hydraulic power plants and transmission facilities.

- Important environmental concerns with coal-fired power plants.

- A complex and changing public perception of nuclear power.

- The strategic importance of maintaining fuel diversity in power generation, complicated by price and availability concerns for fuel oil and natural gas.

Large electric companies worldwide share these common features of their business environment. The power industry's response -- in terms of planning for future system

---

[4]For more information on the electricity business in the EC, refer to UDI's *Directory of Power Plants in the European Community* (UDI-027-91).

development, customer relations, etc. -- is in large measure shaped by the socioeconomic factors characteristic of their country.

In the United States, the relationship between utilities and government is both complex and diverse, reserves of all fuels are large, and the sheer size of the country has led to the development of regionally distinct economies (Sunbelt vs. Rustbelt, etc.). These and other factors have led to the development of hundreds of electric utilities and non-utility power plant operators with thousands of non-standardized generating units. Current U.S. power plant development is focused on small units that can be built rapidly with relatively small capital expenditures. There are regulatory disincentives to the retirement of existing capacity.

In contrast, the electric power industry in Japan is dominated by 10 large utility companies working closely with a federal agency (MITI) which encourages use of a long-term planning horizon.[5] This, combined with the absence of indigenous reserves of fossil fuels, a very high level of engineering and construction ability, a "nation-as-team" concept, and other factors have kept Japanese utilities busy building large nuclear and fossil (coal- and LNG-fired) units. In contrast to the United States, Japanese utilities are actively retiring older generating units.

In Canada, there are similarities to both the American and Japanese situations. Canada shares with the United States a vast amount of land and abundant fossil and nuclear fuel resources. On the other hand, the relationship of the large electric utilities with federal and provincial governments is basically similar to the Japanese model.

---

[5] See UDI's *Directory of Power Plants in Asia and Australia* (UDI-051-92) for more information on the Japanese electric power industry.

The Canadian electric power industry is dominated by the 12 largest electric utilities, which control 92% of all installed[6] capacity. In fact, the three largest utilities -- **Ontario Hydro, Hydro-Quebec,** and **BC Hydro** -- control close to 70% of the installed capacity in the country (see *Figure 2*). The Canadian inventory of power plants is also relatively young, with the substantial majority of installed capacity 20 years of age or less (see *Figure 3*), and dominated by hydraulic power plants as shown in *Figure 4* (although the dominance of hydroelectric plants in the capacity mix has dropped from 80% to about 60% over the last 30 years).

Recent developments in the Canadian electric power business seem to be moving the industry towards what might be termed an American model. This involves continued dependence on large central-station plants for most of the electricity in the country accompanied by a gradual opening of the generation sector to so-called "private" suppliers.

As in the United States, forecasting electricity demand in Canada has gotten steadily more complex. According to Energy Mines & Resources,[7] Canadian GDP will grow at 2.4% per year from 1991-2010, far less than the 4.1% per year over the last 30 years. Electric power demand is estimated to grow at 1.8% per year over that time period, not too different from mid-range estimates for the United States. (The 1990 estimate of Canadian demand growth for the same period was 2.1% per year.) This rate would move total domestic demand from 470,895 gigawatthours (GWhrs) in 1991 to 665,800 GWhrs in 2010 with peak demand moving from 82,863 MW in 1991 to 116,216 MW in 2010.

Although the power generation, transmission, and distribution systems built over the last half century to serve the country's urbanized population will probably not be

---

[6]"Installed" is defined to include generating units that are in operation or shutdown. Shutdown units may be in economic reserve or mothballed, but are still carried on the system as potential generating resources.

[7]Electric Power in Canada 1991 has these and the other forecast statistics cited.

tinkered with -- and this is true in all other OECD countries -- Canada faces some unusual problems due, ironically, to the great success of its utilities in building and operating large, standardized power plants.

As noted in last year's review, it is clear that the construction of the "mega projects" characteristic of the last 20 years of power plant construction in Canada will be increasingly difficult. Much of this may be attributed to gradual centralization of environmental assessment oversight and steady expansion in the number of intervenors in power plant licensing and assessment proceedings. This is a recent development in Canada where the provincial and federal governments jointly develop, permit, manufacture, construct, _and_ operate most of the country's capacity through the provincial utilities. (Note too that many of the mega-projects were built in very lightly inhabited areas in a thinly-populated country.)

Things may get worse as far as the utilities are concerned. The Canadian Environmental Assessment Act, approved by the federal government in June 1992, is to take effect in January 1993. This act for the first time gives specific details of federal government involvement in project reviews. The contentious negotiations between **Hydro-Quebec** and various intervenors over the GRANDE BALEINE project may be only the first example of a steadily more difficult siting and licensing process in Canada. Unfortunately for electric utilities, no near-term developments can be identified in Canada that will make the "large" capacity addition planning process any more certain.

UDI-033-92 December 1992

CANADIAN ELECTRIC UTILITY PROFILES

This report section contains brief, utility-by-utility summaries of electric companies in Canada. These accounts are derived from company reports, from Canadian government statistics, and from the trade and business press. Included are notable events and performance indicators that characterize the electric power business in Canada.

> *New this year are statistical tables showing key utility performance factors, extracted from utility Annual Reports or by direct survey. These numbers reflect the reporting and accounting language used by the companies. However, given the mix of public and private utilities involved -- as well as the great disparity in size and output -- there can be no assurance that the factors are comparable between companies.*

The Canadian electric power industry dates from 1884. Since 1947, electric output in the country increased at an average rate of 5.5% per year. (By way of comparison, real GDP grew by 4.2% per year and population by 1.1% per year in this same period.[8]) This high overall growth rate is by no means equally spread over this huge country for, as shown in the table below, Canadian provinces differ tremendously in size and population density.

To a large extent, the electric power business in Canada is organized along provincial lines since the Canadian constitution put the electric power business primarily within the jurisdiction of the provincial governments. The mandate of the provincial electric utilities is to provide electricity and energy services to provincial industries and residents, either via direct sales or through municipally-owned resellers. The biggest Canadian utilities are still provincially-owned Crown Corporations, although **Nova Scotia Power** was privatized during 1991. The two largest provinces -- Ontario and Quebec -- contain over 60% of Canada's population.

---

[8] *Electric Power In Canada 1991.*

Canadian Provinces & Territories				
	1990 Population (X1000)	Area (sq km)	Area (sq mi)	Pop Density (per sq km)
Alberta	2473	644560	248800	3.84
British Columbia	3139	929977	358971	3.38
Manitoba	1091	548505	211723	1.99
New Brunswick	724	72109	27834	10.04
Newfoundland	573	371788	143510	1.54
Nova Scotia	892	52855	20402	16.88
Ontario	9748	891425	344090	10.94
Prince Edward Island	130	5661	2185	22.97
Quebec	6771	1357148	523859	4.99
Saskatchewan	1000	570850	220348	1.75
Northwest Territory	54	3293891	1271442	0.02
Yukon Territory	26	479096	184931	0.05
Totals	26621	9217865	3558095	2.89

Source: *1992 World Almanac*

The provincial governments have become more important due to recent political developments in Canada and federal efforts to decentralize decision-making and regulation. In two key instances discussed below -- Ontario and British Columbia -- the relatively recent turnover of the provincial government has had a major and rapid impact on the day-to-day business of the provincially-owned utilities.

One result is that the business of the provinces is to a large extent the business of the province's largest businesses. As noted above, the importance of the big Canadian utilities to the overall economy reinforces the business impact of the provinces with the impact of federal government regulations and initiatives. In many cases, this puts

Canada's electric utilities squarely in the socio-political arena to a degree unmatched in any other country.

BC Hydro

This provincially-owned Crown Corporation is the third largest electric utility in Canada. **BC Hydro** was originally formed in 1962 by the merger of several public and private utilities to develop the province's considerable hydroelectric potential. Today, 89% of its 10,500-MW capacity is hydroelectric, 5,995 MW in three large plants at GORDON M SHRUM (2,416 MW), REVELSTOKE (1,843 MW), and MICA (1,736 MW).

British Columbia has been the fastest-growing province in Canada for some time. In fact, over the last 10 years, provincial load has grown by an average of 3.6% per year despite generally difficult economic conditions. Like other Canadian utilities, **BC Hydro** is not contemplating near-term construction of new large power plants, although "SITE C" has been evaluated for a major hydraulic plant. Rather, a combination of aggressive Demand Side Management (DSM) programs coupled with some amount of non-utility generating capacity will, it is hoped, suffice to delay the need for new capacity for at least the rest of the decade.

In April 1992, the BC Utilities Commission rendered a decision supporting utility-proposed changes in **BC Hydro's** rate structure towards a flat-rate structure for residential and commercial customers. This moved from the block-rate structure which had heretofore encouraged energy-intensive industry and residential demand patterns.

In November 1992, the provincial government announced an electricity rate hike cap of one percentage point above inflation for the current year and two points above inflation in future years. This, the first such arrangement in Canada, will obviously require careful balancing *vis a vis* the revenues that **BC Hydro** provides to the government.

The utility has adopted three main strategies to minimize the need for future utility-owned capacity additions.

- **BC Hydro** has been very active in demand side management (DSM) and conservation and has developed the Power Smart scheme, a portfolio of programs designed to encourage electric energy efficiency. The utility has licensed the program to other Canadian utilities. By March 1992, the utility reported savings of 724 Gwhrs due to the impact of the Power Smart program.

- A second initiative termed Resource Smart aims to maximize output from existing plants and T&D facilities. During the year, several older hydraulic plants were refurbished netting savings of about 856 Gwhrs.

- **BC Hydro** is promoting the integration of non-utility generating resources into the utility grid. Two private power plants with a total capacity of 160 MW are expected to connect to the grid in the next few years.

As with **Hydro-Quebec** in the east, **BC Hydro's** planning process is complicated by substantial power exports to the United States. The utility nearly doubled its "power trade" revenue to $170 million in 1991. The provincial government is taking a keen interest in the future of hydropower sales into the Pacific Northwest when a series of existing contracts come up for renewal in 1997/98, not the least because **BC Hydro** pumps hundreds of millions of dollars into the provincial coffers each year.

Much depends on transactions with the United States' giant power marketing agency **Bonneville Power Administration (BPA)**. Already facing the removal of 1,100 MW from the region's power plant pool with the shutdown of the TROJAN nuclear unit, the pending expiration of the 1964 Columbia River Treaty will affect an additional 1,300 MW of capacity **BPA** buys from **BC Hydro**. Beyond this amount, there are various short-term and long-term, firm and interruptible sales that need to be evaluated, and if necessary, replaced with new contracts.

Complicating these ongoing arrangements was a decision in mid-1992 by the recently-elected New Democratic Party (NDP) to reexamine the province's policy regarding power exports. Although **BC Hydro's** sales policies were subsequently reaffirmed later in the year, the utility was put on notice that this area of their business would henceforth be under constant scrutiny.

Another twist is the future of the 540-MW KEMANO project 700 km north of Vancouver, a $1 billion addition to an existing **Alcan Aluminum Ltd** hydroelectric power plant used for a large smelter in Kitimat, BC. **BC Hydro** has a 20-year contract to buy 2,500 GWhr from what will be Canada's largest "independent" power plant (**Alcan** completed an 800-MW hydro plant in 1967). Construction on the four new KEMANO units was stopped in 1991 after two years of work and **Alcan** and the BC government are battling through the courts with environmentalists and a tribal council as to the necessity and/or scope of an environmental assessment.

BC Hydro	1991 Data	1990 Data	% Change
Thermal Generation (million kWhr)	827	1404	-41.1
Hydraulic Generation (million kWhr)	49096	46029	6.7
Electricity Generation (million kWhr)	49923	47433	5.2
Purchased & Net Interchange (million kWhr)	1995	1425	40.0
Line Losses & System Use (million kWhr)	4387	4687	-6.4
Total Energy Sales (million kWhr)	47531	43991	8.0
Electricity Trade (million kWhr)	6969	3480	100.3
Peak Demand (MW)	7458	8122	-8.2
Installed Capacity (MW)	10459	10466	-0.1
Revenue ($C million)	2072.0	1909.0	8.5
Power Trade Revenue ($C million)	170.0	95.0	78.9
Net Income ($C million)	220.0	207.0	6.3
O&M and Admin Costs ($C million)	373.0	339.0	10.0
Fuel Expense ($C million)	11.0	10.0	10.0
Finance Expense ($C million)	617.0	739.0	-16.5

BC Hydro	1991 Data	1990 Data	% Change
Depreciation Expense ($C million)	256.0	229.0	11.8
Taxes ($C million)	108.0	99.0	9.1
Assets-in-Service ($C million)	11319.0	11026.0	2.7
Capital Spending ($C million)	550.0	373.0	47.5
System Energy Cost (cents/kWhr)	3.52	N/A	
Customers (X1000)	1324	1290	2.6
Transmission Line Length (km)	17458	17589	-0.7
Distribution Circuits (km)	51391	51100	0.6
Regular Employees	4834	5006	-3.4
Temporary Employees	1214	1064	14.1

Data for fiscal year ending March 31 of following year. N/A = not available

Canadian Utilities Ltd

Three Canadian electric utilities are units of the Power Division of **Canadian Utilities Ltd.** (CU), an investor-owned utility based in Alberta (which is itself a subsidiary of ATCO Ltd.). These are **Alberta Power Ltd.**, **Yukon Electrical Company Ltd.**, and **Northland Utilities (NWT) Ltd.** **Alberta Power** is the largest electric utility in the group, serving 147,826 customers in Alberta and Saskatchewan. Overall, **CU** had 1991 revenues of $1.15 billion with operating income of $250 million. **CU's** natural gas utility operations account for 57% of the company's overall revenues. Total assets were $3.27 billion.

In general, **CU's** electric power business was healthy despite a depressed economy and warmer than normal weather. With the two-unit SHEERNESS lignite-fired plant now in-service, attention turned to the **Alberta Power** transmission system and three new 144-kV transmission lines were completed along with 14 new substations. **Yukon Electrical Company** and **Northland Utilities (NWT) Ltd.**, 66% owned by **CU**, which serve about 13,000 customers in the Northwest Territories, brought several new diesel generating units into service.

Canadian Utilities Ltd. (Electric)	1991 Data	1990 Data	% Change
Power Sales (million kWhr)	7188	6799	5.7
Peak Demand (MW)	1183	1235	-4.2
Installed Capacity (MW)	1425	1424	0.1
Revenue ($C million)	483.9	452.5	6.9
Operating Income ($C million)	92.6	85.3	8.6
Expenses ($C million)	213.3	189.1	12.8
Depreciation Expense ($C million)	61.3	40.0	53.3
Plant-in-Service ($C million)	2415.3	2244.4	7.6
Capital Spending ($C million)	173.6	207.3	-16.3
Customers (X1000)	159	157	1.3
Transmission Line Length (km)	48	48	1.0
Employees	N/A	N/A	
N/A = not available			

The most interesting developments concern **CU Power International Ltd.**, which has major investments in non-utility power generation projects in British Columbia (BC) and in England. In BC, **CU Power** and **Westcoast Power Inc.** have entered into a 20-year contract to sell power to **BC Hydro** from a $115 million, 120-MW cogeneration plant at Taylor. During 1991, all necessary agreements, approvals, and financing arrangements were satisfactorily concluded and construction started. Completion is scheduled for October 1993.

In England, **CU** has a 45% equity interest in **Thames Power Ltd.** which has a 1,000-MW combined-cycle power plant under development at BARKING in East London. In January 1992, a $751 million turnkey construction contract was announced following the signing of a natural-gas supply contract. The project's total cost is estimated at

$1.4 billion upon completion late in 1995. **CU Power International** is project manager and will operate the plant.

Independent power also figures prominently in Alberta's own grid. One of Canada's largest non-utility power plants may be built at BROOKS about 150 km east of Calgary. The $1.2 billion project is being developed by **Fording Coal Ltd.** a unit of **Canadian Pacific Ltd.**, at the site of large low-sulfur coal deposits. Envisioned as a two-unit, 800-MW plant burning 3 million tons per year (tpy) of coal, the plant might be useful in balancing the installed capacity base in the province, now over-weighted in the north. While a proposal was submitted to the provincial government in July 1991, no decision is expected for two years and construction would take another four years.

Edmonton Power

Edmonton Power is the largest municipally-owned electric utility in Canada and celebrated its 100th year of operations in 1991. The City of Edmonton has owned the utility for the last 90 years, having purchased the first power plant and a six-block distribution system from a group of private citizens for $13,500.

Edmonton Power	1991 Data	1990 Data	% Change
Total Energy Sales (million kWhr)	5753	5659	1.7
Electricity Generation (million kWhr)	4295	4260	0.8
Net Interchange (million kWhr)	1688	1711	-1.3
Total Energy for Dist (million kWhr)	5983	5971	0.2
Peak Demand (MW)	1013	1046	-3.2
Installed Capacity (MW)	1455	1455	0.0
Revenue ($C million)	367.1	354.9	3.4
Net Income ($C million)	83.2	93.3	-10.8

Edmonton Power	1991 Data	1990 Data	% Change
O&M and Admin Costs ($C million)	84.8	77.1	10.0
Fuel & Purchased Power ($C million)	77.3	76.9	0.6
Finance Expense ($C million)	79.7	42.5	87.5
Depreciation Expense ($C million)	44.0	24.8	77.4
Taxes & Local Improvements ($C million)	28.8	28.4	1.4
Plant-in-Service ($C million)	1508.9	1475.1	2.3
Capital Spending ($C million)	91.5	169.9	-46.1
Customers (X1000)	247	245	0.9
Regular Employees	1180	N/A	
N/A/ = not available			

While the utility brought the 406-MW GENESEE 1 into service in 1989, Unit 2 was delayed and rescheduled for service in 1994 by the City Council. Capital costs for the plant are some $1.5 billion, although the investment for Unit 1 was only placed into rate base in 1991. An interesting development was the construction of the necessary apparatus to pipe methane to the CLOVER BAR power plant from a large landfill 2 km from the station. Another major project for the utility is the design and implementation of a district heating scheme based on the downtown ROSSDALE power plant fired by natural gas.

Hydro-Quebec

Hydro-Quebec (H-Q) is the second largest Canadian utility. While a joint-stock company, it has only one shareholder, the province of Quebec. Perhaps more than any other Canadian utility, H-Q is involved in the fierce political debate on the Canadian constitution and the relative political and financial autonomy of the various provinces and ethnic groups. As one of the largest companies in Canada, a major exporter (of electricity), and the source of billions of dollars of construction work, **H-Q** has had to balance the demands of many competing constituencies.

In its annual report, the utility characterized 1991 as "a year of challenges": this is somewhat of an understatement. One of the largest controversies was over the drive to increase electricity exports into New York and New England. Of all the Canadian utilities, **H-Q** has tied its future most closely to the electricity export business and exports account for 7% of **H-Q**'s energy sales. Changes in export expectations can have a notable impact on planning, both for future power plants and near-term operations.

Deferral, or cancellation, of a planned 1,000-MW contract for sales to the **New York Power Authority** received wide attention in the trade and business press on both sides of the border accompanied by many predictions for the future of electricity sales south into the United States. In addition, interruptible power sales in 1991 had to be curtailed due to poor water conditions. This will adversely affect revenue in 1992. Continued weakness in oil and gas prices and adverse economic conditions in New York and New England, kept the market soft for additional power exports.[9]

H-Q has in hand a firm-energy contract with the **New England Power Pool (NEPOOL)** for seven million MWhrs which started in July 1991 and runs through 2001 (with provisions for sales in 2002 and 2003). Overall, New England now imports 5% to 10% of its power from Canada, but things are changing fast in the power procurement business in the region.

---

[9] See *U.S. Electricity Trade with Canada and Mexico*, published by the Energy Information Administration, U.S. Department of Energy, for a detailed review of Canada's power business with the United States. (DOE/EIA-0553, January 1992)

At the U.S. - Canada Energy Outlook Conference held in Boston on November 13, 1992, the Chairman of the Executive Committee of NEPOOL stated that there will be no more long term contracts between his organization and H-Q.[10] Rather, NEPOOL's 92 member utilities, including both large investor-owned companies and many small municipals and cooperatives, will be encouraged to contract independently. Since H-Q's current development plan optimistically calls for 1,500 MW of new power exports even though there are no signed contracts for these sales, H-Q will likely be involved in bidding for much smaller quantities of electricity in the future.

During 1991 and 1992, H-Q continued construction on the gigantic LA GRANDE hydroelectric complex in northern Quebec with three 333-MW units at LA GRANDE 2A brought into service in 1991 and three more in 1992. Next will be the 460-MW BRISAY plant, due in 1993/94, followed by LAFORGE 1 at 820 MW and $1.6 billion in about the same time period. The $2.6 billion, 1,360-MW LA GRANDE 1 plant with 12 units is now scheduled for service in 1994. Reevaluation of load growth forecast caused a delay of at least one year in the schedules for LAFORGE 2 and EASTMAIN 1, neither of which is yet under construction.

Beyond these James Bay projects, H-Q continues planning for the series of dams and powerhouses called GRANDE BALEINE. At present, plans for the complex are being reviewed by no less than five separate environmental review panels empowered after separate agreements with the federal and provincial governments and four organizations representing native groups and authorities. According to press reports, the delay in the 1,000-MW sale into New York (see above) had the effect of deferring the first phase of GRANDE BALEINE at least to the year 2000 with the second and third phases to be on-line within eight years thereafter.

---

[10] See *Northeast Power Report* November 27, 1992 (McGraw-Hill, Inc.).

After GRANDE BALEINE, **Hydro-Quebec** is contemplating an even larger development called NOTTAWAY-BROADBACK-RUPERT (NBR) which might ultimately provide 8,400 MW of generating capacity at four different sites, but this project receives no particular mention in the 1991 Annual Report.

Hydro-Quebec has made a major commitment to evaluate and purchase energy from independent generators. A 1991 call for tenders resulted in over 8,000 MW of proposals. The utility currently expects to sign contracts by June 1993 with six projects totalling 937 MW, all of which are planned to be in service by 1996.[11] One contract for thermal plant generation has been signed, with **Indeck Energy Services** for a 139-MW plant at HULL. Output from three small hydro projects is also under contract.

H-Q also has an active DSM program and hopes to reduce annual demand for electricity in the province by a total of 9.3 million MWhrs by the year 2000. The scope and investment in this DSM program has been in the middle of the capacity-planning struggle between the utility and various Indian rights and environmentalist groups. Some critics contend that **H-Q** can triple the amount of energy saved by DSM -- thereby avoiding the investment in GRANDE BALEINE and NBR -- although this will mean foregoing some portion of planned power exports as well.

Even with the 2.2% growth rate now used by **Hydro Quebec** for the period 1992 to 2010, it is hard to see how new power plants can be avoided indefinitely. Over the next 10 years or so, completion of the current construction projects, along with non-utility power plants and DSM will probably suffice, but what next?

---

[11] See *Independent Power Report* November 20, 1992 (McGraw-Hill, Inc.)

Hydro-Quebec	1991 Data	1990 Data	% Change
In-province Sales (million kWhr)	127200	125900	1.0
Thermal Generation (million kWhr)	4810	N/A	
Hydraulic Generation (million kWhr)	117076	N/A	
Nuclear Generation (million kWhr)	N/A	N/A	
Other Generation (million kWhr)	N/A	N/A	
Electricity Generation (million kWhr)	121900	115200	5.8
Purchased Power (million kWhr)	28137	31679	-11.2
Offsystem Sales (million kWhr)	9815	9203	6.7
Total Energy Sales (million kWhr)	137046	135168	1.4
Peak Demand (MW)	29922	27522	8.7
Installed Capacity (MW)	26839	25682	4.5
Revenue ($C million)	6284.0	5883.0	6.8
Net Income ($C million)	760.0	404.0	88.1
Offsystem Sales Rev ($C million)	304.0	300.0	1.3
Operations Expense ($C million)	1870.0	1704.0	9.7
Fuel Expense ($C million)	N/A	N/A	
Interest Expense ($C million)	2294.0	2339.0	-1.9
Depreciation Expense ($C million)	737.0	677.0	8.9
Purchased Power Exp ($C million)	184.0	303.0	-39.3
Taxes ($C million)	392.0	361.0	8.6
Total Assets ($C million)	41851.0	36684.0	14.1
Plant-in-Service ($C million)	37226.0	34012.0	9.4
Capital Spending ($C million)	4037.0	3133.0	28.9
Unit Revenue (¢C/kWhr)	4.53	4.31	5.1
Unit Expenditure (¢C/kWhr)	4.03	4.05	-0.5
Customers (X1000)	3216	3151	2.1
Transmission Line Length (km)	29684	29309	1.3

Hydro-Quebec	1991 Data	1990 Data	% Change
Distribution Circuits (km)	96921	95402	1.6
Permanent Employees	20755	20067	3.4
Temporary Employees	5985	5222	14.6
N/A = not available			

Manitoba Hydro

Manitoba Hydro is a Crown Corporation with an installed capacity of 4,954 MW. Of this, 92% is hydraulic and 8% is thermal, with a small amount provided by isolated diesel plants. The utility operates as an interconnected system with **Winnipeg Hydro**. **Manitoba Hydro** is the fourth largest utility in the country.

The most notable event of the 1991/92 fiscal year was the official opening of the 10-unit, 1,330-MW LIMESTONE hydroelectric plant on the Nelson River, where the last unit went commercial in September 1992. This run-of-the-river plant was brought in under budget (at $1.448 billion) and featured innovative construction techniques with regard to dam construction. Construction at LIMESTONE started in 1976, stopped in 1978 for seven years, and finally finished up this year making the construction duration 15 years.

In other respects, the year was difficult operationally with two severe storms in July and October and a mechanical failure at GRAND RAPIDS 1 which shut down the entire 472-MW hydroelectric plant for two months. The incident occurred on March 10 when the headcover bolts on Unit 1 failed, resulting in flooding of the lower portion of the powerhouse. The failure, the first of its kind in **Manitoba Hydro's** history, resulted in millions of dollars in damage and the spillage of 41,000 liters of hydraulic and lube oil, virtually all of which was contained in the power house.

The current version of the Manitoba Public Utilities Board development plan, issued in 1990, has four major components:

- A 22-year, 1,000-MW energy sale to **Ontario Hydro**, expected to start in 2000 (but see below);

- Two diversity exchange agreements with American utilities **Northern States Power** and **United Power Association** totalling 300 MW;

- Refurbishment of the BRANDON and SELKIRK thermal plants; and

- A 100-MW reduction in load through DSM programs.

Manitoba Hydro joined other large Canadian utilities in the Power Smart energy conservation program, officially launched in April 1991. The target reduction in demand growth was revised upward during the year from 100 MW and 500 million kWhr to 285 MW and 1,049 million kWhr by 2001. This will require an investment of some $400 million in implementing Power Smart programs.

Partially as a result of increases in the projected DSM savings, Units 1-4 at BRANDON may be put into long-term storage in 1996; some life-extension work has been cancelled as a result. Refurbishment work at existing hydraulic plant continued at a relatively high level in 1991/92.

In June 1991, a six-member Environmental Assessment Review Panel was announced to complete the environmental review of the CONAWAPA site on the Nelson River where the next big hydraulic plant -- 1,390 MW -- will likely be built for service early in the next century. Comprehensive environmental assessments of the power plant and associated transmission facilities, including the Bipole III HVDC portion of an upgraded tie to Ontario, were initiated with public hearings scheduled to start in mid-1992. Construction may start as early as 1994 with full service by 2003.

Like other Canadian companies, **Manitoba Hydro** continues to tie isolated loads to the central system and announced a $110 million program to bring nine new communities onto the grid by 1997. This will reduce the number of utility-owned diesel power plants remaining in the province to six.

Other transmission work during the year included planning for a second 500/230 kV transformer at the Dorsey Converter station for the 500-kV line to Minneapolis, Minnesota, completion of a 47-km, 230-kV line from Reston to a new 230/66 kV substation at Virden, and completion of a third 230-kV line from the Dorsey Converter Station to the LaVerendrye substation in Winnipeg.

Manitoba Hydro	1991 Data	1990 Data	% Change
System Sales (million kWhr)	17700	17600	0.6
Thermal Generation (million kWhr)	321	289	11.1
Hydraulic Generation (million kWhr)	23600	20600	14.6
Other Generation (million kWhr)	23	21	9.5
Purchased Power (million kWhr)	1013	1762	-42.5
Electricity Generation (million kWhr)	23900	20800	14.9
Offsystem Sales (million kWhr)	5742	3872	48.3
Peak Demand (MW)	3480	3586	-3.0
Installed Capacity (MW)	5077	4545	11.7
Revenue ($C million)	757	703	7.7
Net Income ($C million)	18	49	-63.6
Offsystem Sales Rev ($C million)	97	67	45.6
Operating Expenses ($C million)	238	254	-6.5
Finance Expense ($C million)	339	257	32.3
Depreciation Expense ($C million)	115	93	23.1
Fixed Assets ($C million)	5602	4522	23.9
Capital Spending ($C million)	706	440	60.3

Manitoba Hydro	1991 Data	1990 Data	% Change
Customers (X1000)	377	375	0.6
Employees	4522	4461	1.4
Note: Capacity, generation & sales statistics include Winnipeg Hydro. Data for fiscal year ending March 31 of following year.			

Manitoba Hydro has an aggressive power export program. During the fiscal year, sales amounted to 5.7 billion kWhrs and resulted in revenues of $97 million, 13% of overall revenues. Much of the recent transmission system planning and construction in the southern part of the province is to support a 500-MW firm sale to **Northern States Power** which will start in 1993. **Manitoba Hydro** also sells power to **Ontario Hydro** (200 MW for three years starting in November 1990). Overall, **Manitoba Hydro's** extraprovincial sales in FY1991 totalled $66.6 million, or 9.4% of total revenues.

The utility had been seeking approval for a 1,000 MW sale to **Ontario Hydro** for a 22-year period starting in the year 2000, a contract that would be met in part with generation from CONAWAPA and could involve revenues of $13 billion over the life of the contract. On December 17, 1992, **Manitoba Hydro** announced that **Ontario Hydro** had cancelled the sale under the terms of the original agreement dated December 7, 1989. This will likely result in some delay in the construction of CONAWAPA and the associated transmission facilities.

Maritime Electric Company

This small, investor-owned utility serves Prince Edward Island and started service in 1918. Despite a fourth successive annual decline in the rate of increase of energy sales, **Maritime** was able to increase net earnings per common share from $1.61 to $1.79. Doubling of the capacity of the twin 138-kV submarine cables from New

Brunswick from 100 MW to 200 MW allowed **Maritime** to back down generation from the oil-fired CHARLOTTETOWN plant by 11%, leading to overall reductions in fuel cost and allowing a $20 million refurbishment program at the plant to continue. In total, 77% of the electricity distributed in 1991 was purchased from **New Brunswick Power Corp.**

Maritime initiated a demand side management program in 1990 and continued the program in 1991, encouraging customers to purchase more energy-efficient lighting. This effort was coupled with hot water and refrigerator programs. Working with the Island Regulatory and Appeals Commission, the utility has planned additional DSM programs, some of which will be included in rate base.

At mid-year, **Maritime** was granted a rate increase of 5.3% with a further 2.5% increase scheduled for January 1992. A rate of return on average common equity of 13.5% was incorporated into the 1991 rate order. No major power plant construction programs are contemplated at the present time.

Maritime Electric Company Ltd.	1991 Data	1990 Data	% Change
Electricity Generation (million kWhr)	175	200	-12.5
Purchased Power (million kWhr)	585	553	5.8
Total Energy Sales (million kWhr)	686	676	1.5
Revenue ($C million)	74.1	72.2	2.6
Net Income ($C million)	7.3	6.2	17.7
Operating Expenses ($C million)	50.5	51.2	-1.4
Finance Expense ($C million)	4.5	4.7	-4.3
Depreciation Expense ($C million)	5.9	5.0	18.0
Fixed Assets ($C million)	181.4	166.1	9.2
Capital Spending ($C million)	16.7	15.5	7.7
Customers (X1000)	51	51	1.2

Maritime Electric Company Ltd.	1991 Data	1990 Data	% Change
Transmission Line Length (km)	560	N/A	
Distribution Circuits (km)	4400	N/A	
Employees	235	233	7.0
N/A = not available			

New Brunswick Power Corp.

Like many other large state-owned utilities, **NB Power** is looking ahead to privatization. In May 1991, the utility changed its name to **New Brunswick Power Corporation**; the removal of the word "Commission" is designed to reassure major export customers in the United States that the utility is a "commercial business and not a government committee." While the utility's status as a Crown Corporation remains unchanged for the time being, the titles and responsibilities of senior executives were also adjusted. An early retirement program, that resulted in the elimination of 100 positions, was a further sign of change.

NB Power and **Hydro-Quebec** have been the two eastern Canadian utilities most active in the drive for electricity exports into the northeastern United States. Over the last two or three years, the U.S. power export business has fallen off (see the discussion under **Hydro-Quebec**) and there is little hope for near-term improvement due to low fossil-fuel prices, a recessionary dampening in demand, and the growth of the non-utility power sector in the primary target markets. This leaves improvements in sales to neighboring utilities as the major hope for the future.

In fact, **NB Power** saw export sales revenue decline to $22.7 million in 1991/92, a decrease of over 9%. New inter-provincial sales to **Maritime Electric** and **Hydro-Quebec** masked the completion of two American utility participation contracts totalling

125 MW at POINT LEPREAU, which between them resulted in a year-on-year revenue decline of $29.4 million.

Nonetheless, **NB Power** has one of the most active power plant construction programs in Canada. Four 100-MW gas turbines at MILLBANK along with a single 100-MW machine at STE-ROSE were commissioned in November 1991. While the MILLBANK units were installed primarily to serve a 400-MW capacity and energy sale to **Hydro-Quebec**, they will also be available for winter peaking service. STE-ROSE can be started up in complete isolation from the rest of the system and is designed to serve the entire Acadian Peninsula in emergencies.

The big construction project is the 450-MW pulverized-coal unit at BELLEDUNE, one of the largest thermal power plant projects in North America. The first Canadian unit equipped from the start with an FGD scrubber, BELLEDUNE 1 will cost an estimated $965 million when completed in late 1993, $565.5 million of which had been spent through March 31, 1992.

Of the company's currently installed capacity, 44% is oil or coal-fired, 24% is hydraulic, 17% is provided by the 630-MW CANDU reactor at POINT LEPREAU, and the balance is in combustion turbines at four sites. POINT LEPREAU, perennially one of the best-operating reactors in the world, provided 27% of the utility's power in 1991 and a major share of the utility's exports to the United States.

At DALHOUSIE, construction has started on a $338 million conversion to burn Orimulsion (bitumen and water) to be imported from Venezuela. Scrubbers for the two units were ordered from Babcock & Wilcox Canada in February 1992. The $30 million award contained an industrial benefits package aimed at assisting provincial companies. The benefits package was similar to a $60 million package negotiated with ABB, supplier and erector of the MILLBANK combustion turbines.

NB Power has transmission and distribution projects all over the province. A 200-km, 345-kV line from Salisbury to Bathurst is designed to complete a high-voltage "ring" around the province. A second 138-kV line from COLESON COVE into Charlotte County was completed, along with several new substations and transformer upgrades. Finally, a new $9.2 million SCADA control system was completed at the Energy Control Centre.

As with other Canadian utilities, **New Brunswick Power** is starting DSM programs and joined the Power Smart program in July 1991. While plans are in-hand to reduce the need for 110 MW of peaking capacity by 1997, additional generating capacity will likely be required.

The utility is investigating the installation of an integrated gasification combined-cycle plant at GRAND LAKE. Other options are additional units at BELLEDUNE (designed for four units), a downsized 450-MW CANDU unit at POINT LEPREAU, or gas-fired combined-cycle.

While the province cannot be considered a "hot bed" of development activity, non-utility generation has not been ignored. Two 30-year contracts have been signed for power purchases from proposed wood-fired power plants at Sussex and Kedgwick. **NB Power** already purchases electricity from several small hydro plants.

As a medium-size Canadian utility, and one with great interest in and experience with the utility business in the United States, **NB Power** may have a clearer view of the "free-trade" future than most of its Canadian cousins. (It certainly communicates more fully through its Annual Report than most other Canadian utilities.) With existing commitments to completion of several substantial capital projects, the company has no choice but to stick to its core business for the moment and hope for general economic improvements in regions to which it can sell baseload thermal and nuclear power. Long-term, these investments should keep the company competitive in the electric power business over the next few decades.

New Brunswick Power Corp.	1991 Data	1990 Data	% Change
In-province Sales (million kWhrs)	12382	11655	6.2
Thermal Generation (million kWhrs)	8037	6506	23.5
Hydraulic Generation (million kWhrs)	2720	3334	-18.4
Nuclear Generation (million kWhrs)	5793	5858	-1.1
Other Generation (million kWhr)	96	1	9500.0
Electricity Generation (million kWhr)	16646	15699	6.0
Purchased Power (million kWhr)	3574	3620	-1.3
Offsystem Sales (million kWhr)	5901	5813	1.5
Total Energy for Dist (million kWhr)	18664	17790	4.9
In-province Peak Demand (MW)	2728	2566	6.3
Installed Capacity (MW)	3725	3225	15.5
Revenue ($C million)	923.9	908.5	1.7
Net Income ($C million)	25.1	8.4	198.8
Offsystem Sales Rev ($C million)	225.2	248.0	-9.2
O&M and Admin Costs ($C million)	270.0	266.3	1.4
Fuel Expense ($C million)	169.2	191.3	-11.6
Finance Expense ($C million)	199.7	211.6	-5.6
Depreciation Expense ($C million)	106.5	96.1	10.8
Purchased Power Exp ($C million)	123.0	113.8	8.1
Fixed Assets ($C million)	3302.0	2821.3	17.0
Capital Spending ($C million)	544.2	463.6	17.4
Customers (X1000)	320	316	1.4
Transmission Line Length (km)	6451	6329	1.9
Distribution Circuits (km)	25539	25097	1.8
Regular Employees	2753	2729	0.9

New Brunswick Power Corp.	1991 Data	1990 Data	% Change
Temporary Employees	410	358	14.5
Data for fiscal year ending March 31 of the following year.			

Newfoundland Light & Power Company Ltd.

Newfoundland Light & Power Company (NLP) is an investor-owned utility operating as the principal of three subsidiaries of **Fortis Inc.**, formed in 1987. Energy sales at **NLP** increased only 0.5% in real terms in 1991, after a 5.4% increase the previous year, reflecting, according to the annual report, "a disastrous fishery, mine closures, and a general downturn in retail activity." Over 90% of the energy distributed by **NLP** is purchased from **Newfoundland & Labrador Hydro**, with the balance provided by 30 small diesel and hydraulic plants.

A new 25-MW gas turbine is to be installed in 1992 at the end of a radial transmission line serving PORT AUX BASQUES and a 6-MW hydro plant is under construction at ROSE BLANCHE. There is no immediate requirement for new baseload power generating facilities of any size; instead, **NLP** has started the Power Smart program to encourage greater customer awareness of energy efficiency. Additional capital improvements are aimed at reducing the frequency of service interruptions and strengthening the company's computer and telecommunications system.

Newfoundland Light & Power Co. Ltd.	1991 Data	1990 Data	% Change
Energy Sales (million kWhr)	4295	4236	1.4
Hydraulic Generation (million kWhr)	409	409	0.0
Purchased Power (million kWhr)	4129	4055	1.8
Electricity Generation (million kWhr)	409	409	0.0

Newfoundland Light & Power Co. Ltd.	1991 Data	1990 Data	% Change
Peak Demand (MW)	N/A	N/A	
Installed Capacity (MW)	N/A	N/A	
Revenue ($C million)	328.7	305.9	7.5
Net Income ($C million)	27.8	26.2	6.1
Operating Costs ($C million)	56.1	54.3	3.3
Purchased Power Expenses ($C million)	184.1	165.3	11.4
Finance Expense ($C million)	21.5	20.4	5.4
Depreciation Expense ($C million)	24.6	24.0	2.5
Taxes ($C million)	15	15	0.7
Fixed Assets in-Service ($C million)	685.0	649.2	5.5
Capital Spending ($C million)	42.5	67.3	-36.8
Customers (X1000)	196	192	1.7
Transmission Line Length (km)	N/A	N/A	
Distribution Circuits (km)	N/A	N/A	
Employees	957	984	-2.7
N/A = not available			

In December 1991, the company was granted a rate increase of 2.23% based on a return on average equity of 13.25% for the test year 1992. One month earlier, **NLP** intervened in proceedings concerning a rate hike of 3.8% proposed by its major supplier **Newfoundland & Labrador Hydro**. This proposed increase follows an 8% increase in the price of purchased power in July 1990. To the extent that the utility is already private, structural changes in the Canadian utility industry will have a smaller impact on **NLP** than on some of its public cousins.

Newfoundland and Labrador Hydro

Newfoundland and Labrador Hydro (NLH) is a provincially-owned Crown Corporation which controls several other companies, notably **Churchill Falls (Labrador) Corp.**, operator of the 5,400-MW CHURCHILL FALLS hydraulic power plant. In terms of installed capacity -- including CHURCHILL FALLS -- **NLH** is the fourth largest utility in Canada. The utility operates principally as a power wholesaler.

Newfoundland & Labrador Hydro	1991 Data	1990 Data	% Change
Energy Sales (million kWhr)	32539	32242	0.9
Thermal Generation (million kWhr)	1460	1866	-21.8
Hydraulic Generation (million kWhr)	33639	32936	2.1
Other Generation (million kWhr)	73	77	-5.2
Electricity Generation (million kWhr)	35172	34879	0.8
Export Sales (million kWhr)	26330	26131	0.8
Peak Demand (MW)	6900	6938	-0.5
Installed Capacity (MW)	7238	7243	-0.1
Revenue ($C million)	370.0	359.4	2.9
Net Income ($C million)	23.0	21.7	6.0
Operation & Admin Costs ($C million)	97.4	91.4	6.6
Fuel & Purchased Power ($C million)	58.5	54.7	6.9
Finance Expense ($C million)	148.2	148.8	-0.4
Depreciation Expense ($C million)	36.2	34.0	6.5
Fixed Assets in-Service ($C million)	2561.1	2528.0	1.3
Capital Spending ($C million)	50.6	78.5	-35.5
Debt/Equity Ratio	84/16	84/16	
Customers (X1000)	30	N/A	
Transmission Line Length (km)	4537	4537	0.0
Distribution Circuits (km)	N/A	N/A	

Newfoundland & Labrador Hydro	1991 Data	1990 Data	% Change
Permanent Employees	1239	1261	-1.7
Temporary Employees	131	185	-29.2
Note: Includes Churchill Falls. N/A = not available			

Construction during the year was at a relatively low level. As with most other Canadian utilities, refurbishment of existing thermal plant is a priority. In this case, **NLH** completed a $22 million, five-year program at the three-unit HOLYROOD plant that involved upgrading two units from 150 MW to 175 MW, among other improvements. A new 25-MW gas turbine was installed in Happy Valley-Goose Bay, and various transmission plant modifications were also completed.

Earlier plans by **NLH** to build 3,100 MW of new hydraulic plant at GULL ISLAND and MUSKRAT FALLS on the Lower Churchill River are not mentioned in the 1991 Annual Report. This $10 billion project was in large measure dependent on sales to **Hydro-Quebec**, which already purchases much of the power from CHURCHILL FALLS.

Instead, **NLH** is working with **Newfoundland Light & Power** on DSM measures that they hope will save 17 MW of capacity and 30 million kWhrs of energy by the mid-1990s. Another DSM initiative with a major industrial customer is aimed at reducing peak demand by 100 MW. On the supply side, **NLH** hopes to purchase 50 MW of capacity and associated energy from non-utility suppliers. In sum, these measures are designed to put off new power plant construction until the end of the decade.

Nova Scotia Power Corp.

Following the initial announcement on January 9, 1992, **Nova Scotia Power Corp. (NSPC)** became the first large Canadian utility to be privatized with a very successful stock-market placement in July.[12] The original plan was for the provincial government to sell 57% of **NSPC** for about $400 million. Investor interest convinced the government to boost the offering to 75% of the utility at the time that the initial prospectus was filed. Since the issue was still oversubscribed, the government decided to sell their entire ownership of 85.1 million shares. Institutional investors reportedly bought $350 million of the issue. Nova Scotia residents were given an opportunity to buy under a special installment receipt plan.

In the end, the **NSPC** privatization raised about $850 million, most of which will be returned to the provincial government, and is considered the largest equity issue ever in Canada as well as one of the largest privatizations of a government corporation. The successful sale was attributed in part to the utility's 7.5% dividend yield as well as an overall lack of high-yield securities of interest to income-seeking as opposed to capital appreciation seeking investors.

While privatization took place quickly, preparations were started in the mid-1980s. In 1986, all government subsidies were eliminated. Thereafter, 20,000 residents invested more than $430 million in Nova Scotia Power Savings Bonds. In 1991, an early retirement plan was offered that resulted in the elimination of 157 positions. No major changes in the company's operations are expected now that privatization is complete.

NSPC had an active operational year in 1991. A new 160-MW coal-fired unit was completed on time and within budget at TRENTON (Unit 6) while the large fluidized-bed unit at POINT ACONI remained on schedule for service in 1993. Approximately $102 million was spent on the T&D system.

---

[12]The sale was reviewed in a **Wall Street Journal** article on July 30, 1992.

In a Canadian example of the interdependence of the coal and power industries, a fatal explosion at **Curragh's** Westray coal mine in May 1992 caused concern over future plant operations at TRENTON. A deep explosion at the relatively new mine trapped and killed 26 miners and the mine's reopening is uncertain. As summarized below, the tragedy led to a political fight exemplifying many of the tradeoffs between energy development, jobs, and environmental activism.

- About 70% of the low-sulfur coal from the Westray mine was being delivered on a "take or pay" basis to TRENTON 6, an arrangement that saved **NSPS** $100 million in pollution control equipment, according to one analyst quoted in the Toronto *Globe & Mail* on May 12, 1992.

- The Westray mine was the largest employer in the region near Stellarton, NS, with a work force of 240.

- Westray had never reached full output and was unable to provide the 700,000 tpy of coal required for TRENTON 6. Before the disaster, about 200,000 tpy was planned to come from **Curragh's** Wimpey open-pit mine on the same seam where 60 miners would be employed, but...

- The provincial government had not decided whether an environmental review would be needed for Wimpey and, with the closure of Westray, both the miners and the power company are wondering whether Wimpey can get into operation or whether new supplies of low-sulfur coal will have to be imported from the U.S. The provincial government has been under fierce pressure to fast-track the permitting process and get the mine in service.

In August 1992, **NSPS** announced it had signed 14 non-utility power purchase contracts for a total of 60 MW. These include 7 hydro projects, 2 methane (landfill gas) plants, 2 wood plants, and 3 waste-to-energy plants.

One of these contracts is a 33-year power purchase agreement with **Polsky Energy Corp.** which is developing a $54.9 million, 21-MW wood-fired cogeneration plant called the BROOKLYN ENERGY CENTRE. The thermal host is **Bowater Mersey Paper Company**. Energy rates were reported in *Independent Power Report* to be ¢US

4.1/kWhr in the first year rising to an inflation-indexed peak of ¢US 19.53/kWhr. Capacity payments ranged from ¢US 3.81/kWhr to ¢US 4.33/kWhr.

In December 1992, **NSPC** finalized an agreement with **Environmental Technologies, Inc.** of Calgary. This involves power purchased from a 3.8-MW landfill gas project in SACKVILLE. Power costs over the 10-year contract range from ¢6.3-8.2. **NSPC** also agreed to purchase power from a 4-MW plant fueled with methane from the LINGAN coal mine.

The Power Smart DSM program licensed from **BC Hydro** (launched in September 1990) continued with the goal of reducing demand by 175 MW and consumption by 700 GWhrs by 2001. The utility did extensive advertising to make customers aware of the program and of **NSPC's** role. Lightbulb exchange programs were held in shopping malls and a commercial lighting program including rebates was launched in October 1991.

Nova Scotia Power Corp.	1991 Data	1990 Data	% Change
Total Energy Sales (million kWhr)	8681	8674	0.1
Thermal Generation (million kWhr)	8288	N/A	
Hydraulic Generation (million kWhr)	1019	N/A	
Other Generation (million kWhr)	7	N/A	
Electricity Generation (million kWhr)	9314	N/A	
Purchased Power (million kWhr)	136	N/A	
Peak Demand (MW)	1806	1827	-1.1
Installed Capacity (MW)	2129	1964	8.4
Revenue ($C million)	673.1	628.0	7.2
Operating Surplus ($C million)	46.3	24.0	92.9
Operating Costs ($C million)	154.7	140.2	10.3
Fuel Expense ($C million)	224.6	220.8	1.7
Finance Expense ($C million)	166.1	166.5	-0.2

Nova Scotia Power Corp.	1991 Data	1990 Data	% Change
Depreciation Expense ($C million)	73.2	63.2	15.8
Purchased Power Exp ($C million)	3.8	15.9	-76.1
Fixed Assets ($C million)	2430.3	2068.4	17.5
Capital Spending ($C million)	390.1	329.3	18.5
Customers (X1000)	401	396	1.3
Transmission Line Length (km)	5099	5004	1.9
Distribution Circuits (km)	24001	24243	-1.0
Employees	2435	2556	-4.7
Data for fiscal year ending March 31 of following year. N/A = not available			

Ontario Hydro

This Crown Corporation is the largest electric utility in Canada and one of the largest in the world. Like its giant sister utility **Hydro-Quebec**, **Ontario Hydro** has been the focus of intense political debate in recent years, culminating in the summer of 1992 with the abrupt dismissal on September 25, 1992 of its "pro-nuclear" president Alan Holt by the "anti-nuclear" provincial government while he was out of the country on utility business. The confusion surrounding the affair was exacerbated by conflicting statements from various provincial politicians and spokesmen.[13]

Ontario Hydro's Chairman Marc Eliesen subsequently left to run **BC Hydro** and was replaced by Maurice Strong, characterized in the trade press as having a "green" background and reservations about nuclear power. As described below, this change in utility leadership comes at a time when the **Ontario Hydro** Board is reviewing multi-billion dollar spending decisions on the company's nuclear plants.

---

[13] See *Nucleonics Week* October 8, 1992 for a full discussion of the matter (McGraw-Hill, Inc.).

Ontario Hydro has begun a concerted effort to focus on its 3 million customers, most of them served by the dozens of municipal utilities that buy its electricity. The utility needs to communicate and work with this diverse customer base as it faces numerous near-term operational and organizational challenges including:

- Finishing the $14 billion, 3,524-MW nuclear plant at DARLINGTON and bringing the plant into rate base;

- Managing a $4 billion program to refurbish existing fossil, nuclear, and hydraulic power plants, including a large effort to evaluate and repair existing reactors at PICKERING B and BRUCE. An additional $1.4 billion has been committed to upgrade the transmission system;

- Rationalizing and reorganizing a sprawling labor and management staff of over 35,000; and

- Working with customers to implement an 11.8% rate increase in 1992.

Ontario Hydro purchases of electricity from other companies 1991 were down nearly 70% from the previous year. Of the 1991 total, about half was bought from other provinces and the United States, while half was purchased from non-utility generators in the province. (Non-utility power purchases increased by a factor of four from 1990 to 1991.) The large overall decrease in purchases was due to a 19% increase in nuclear generation and a 9% increase in thermal generation, although improved performance by these facilities was partially offset by a 7% decrease in hydraulic generation.

During 1991, **Ontario Hydro** intensified its conservation efforts and spent $179 million on DSM and energy management programs, up almost 80% from the previous year. During the year, customers saved an estimated 250 MW of power and some $28 million in their energy bills according to the Annual Report. In 1992, the target DSM savings goal was revised upward to 9,860 MW by 2014, an increase of 77% from the 5,570 MW goal established in the 25-year Resource Plan released in 1989. This may be the most ambitious single-utility DSM goal in the world.

Altogether, up to $3 billion may be spent by **Ontario Hydro** on DSM by the end of the decade. These programs are designed to defer major plant additions by seven years while reducing borrowing by $9 billion over the next 20 years, but will this happen? Beyond that, how is the province to replace its existing 23,000 MW of fossil and nuclear capacity, and in what time frame?

Of all Canadian utilities, **Ontario Hydro** probably faces the most difficult capacity planning process due to its heavy dependence on nuclear power and its relatively older generating units. In fact, by 2014, the utility had plans to retire one-quarter of its generating capacity, but over the last 10 years, demand has increased by 3% per year.

In an "Update" to the 25-year plan released in January 1992, any decision on the massive investment in new nuclear power stations is deferred, along with other investments in new, large hydraulic plants. (This followed by a month the suspension of planning for six new hydro projects and two additions at existing plants in the Moose River basin. A total of 1,500 MW was involved in these schemes.) Power plant rehabilitation programs, on the other hand, have a high priority.

While the Plan Update was originally described by **Ontario Hydro** as not substantially different than the original, an editorial in the March 1992 *Nuclear Engineering International (NEI)* stated that the five years of utility work on the original 25-year plan "were wasted" and that all the fossil vs. nuclear economic studies have basically been disregarded. Much of this rewrite was at the urging of the New Democratic Party (NDP), which took over the provincial government in November 1990. The NDP is characterized in the trade press as anti-nuclear, and has become centrally involved in budgeting and scheduling of **Ontario Hydro's** refurbishment plans and capacity planning process.

More specifically, the Plan Update moves the need for new capacity out from 2003 to 2014 and puts in place expected lifetimes of over 40 years for **Ontario Hydro's** coal-fired units. According to NEI, previous concerns about CO_2 emissions have been discounted (the plan foresees a doubling of **Hydro's** CO_2 emissions over the next two decades) and a $4 billion investment in new pollution-control equipment at thermal plants is contemplated. Among many other "ripple" effects, NEI is also anxious about the impact of the postponement, deferral, or cancellation (whatever it turns out to be) of a pair of **Ontario Hydro** four-unit plants previously in the Plan of Atomic Energy of Canada (AECL), the supplier of CANDU reactors.

With new, large-plant capacity additions completely in limbo, prolonging operations at existing plants obviously must have high priority. Here again, **Ontario Hydro** has difficulties, most acutely with its 16 older CANDU nuclear units at BRUCE and PICKERING where substantial repairs are considered essential by the utility to maintain its huge investment in nuclear power.

In the most recent write-up of the plant-level oversight now experienced by this *government utility*, **Nucleonics Week** reported in December 1992 that the *provincial government* would allow a $108 million steam generator to go ahead, but put a hold on a $2.2 billion retubing/refurbishment program previously approved. An additional $220 million to replace the steam generators at BRUCE 2 is also subject to increased government review.

A further controversial development was also reported: **Ontario Hydro** has been directed to do an economic viability study of BRUCE A. Apparently, it has been suggested that plant O&M budgets might be cut. According to the utility's analysis, reductions in O&M a decade ago led to a fall-off in plant load factor which has only recently been reversed. Reluctant to initiate another such cycle, utility managers are faced with the substantial task of justifying billions of dollars of capital and O&M expenditures (at BRUCE A alone) in the face of what seems to be determined political opposition.

Where does this leave other supply options? **Ontario Hydro** has been planning a 600-900 MW hydro plant known as NRHD (Niagara River Hydroelectric Development, also known as SIR ADAM BECK III) near the existing SIR ADAM BECK plant. The $2.1 billion power project would be accompanied by a new 230-kV double circuit line to Hamilton along with other transmission system upgrades in the area. Yet, as with GRANDE BALEINE in Quebec, progress on this major plant construction program has been hampered by complicated and time-consuming reviews of various resource and environmental planning activities. The most optimistic timetable does not anticipate service of NRHD before 2000.

Not unexpectedly, **Ontario Hydro** has an aggressive program to stimulate the development of non-utility power plants in the province. The utility purchased 2 million MWhrs from non-utility producers in 1991 and 10 new projects totalling 153 MW went into service. When an additional 16 projects now under construction are complete, some 750 MW of independent capacity will be in place. Within 15 years or so, up to 4,000 MW of non-utility capacity are hoped to be available.

The initial implementation of **Ontario Hydro's** non-utility power purchase program has had its problems. For example, in December 1991, the utility notified potential producers that a temporary freeze was being placed on larger projects due to decreases in projected demand for the rest of the decade, and, in February 1992, the utility deferred action on 53 projects.

Although **Hydro** later reopened discussions with 13 projects (over 5 MW) affected by the freeze, press reports of "6% rate increases" and "$8 billion over the next 20 years" for the purchase of independent power angered representatives of the non-utility power business. As a general matter, small hydro projects are treated differently by the utility's planning staff since they are considered to make "needed" contributions to local supplies.

A 20-year contract was to be signed by **Hydro** with **Tenaska** and a unit of **Northeast Utilities**, both from the U.S., for power from a 100-MW gas-fired cogeneration project in Toronto. Also, a contract was reportedly signed for power from a 102-MW cogeneration plant developed by **Nordic Power** and a unit of **Northstar Energy**. However, as if more problems were needed, some of the largest cogeneration projects in the province are being planned by **Ontario Hydro's** municipal utility customers, including Kingston, Sudbury, and Windsor.

In the scheme of things, transmission line development in Ontario seemed to be less controversial, at least for the moment. Construction continued during the year with the completion of the final leg of the Southwestern Ontario Transmission System, a 500-kV line connecting the 4,000-MW NANTICOKE plant to the Longwood substation. This will help move power from the BRUCE nuclear plant to load centers. New transmission lines are planned between Sudbury and Toronto.

It is difficult not to consider "effects of scale" when reviewing **Ontario Hydro's** recent history. Economies of scale in power plant construction and operation as well as in the traditional business functions such as accounting and administration have long been a feature of strategic planning by electric utilities. **Hydro** seems to be showing some negative effects of size when dealing with big issues such as multi-unit nuclear unit refurbishment, 25-year resource plans (10-year plans are the norm in the U.S.), the world's largest DSM program, etc. The financial and "people" resources involved in just one of these efforts are so large as to be difficult to comprehend, but it all has to be done simultaneously. In large measure, the direction of the whole "big power company" concept may be set in Canada.

Ontario Hydro	1991 Data	1990 Data	% Change
In-province Sales (million kWhr)	130964	130875	0.1
Thermal Generation (million kWhr)	30012	27458	9.3
Hydraulic Generation (million kWhr)	33928	36631	-7.4

Ontario Hydro	1991 Data	1990 Data	% Change
Nuclear Generation (million kWhr)	70773	59469	19.0
Electricity Generation (million kWhr)	134713	123558	9.0
Purchased & Net Interchange (million kWhr)	4376	13764	-68.2
Total Energy for Dist (million kWhr)	139089	137322	1.3
December Peak Demand (MW)	22933	21785	5.3
Installed Capacity (MW)	32333	31350	3.1
Revenue ($C million)	7143.0	6484.0	10.2
Net Income ($C million)	204.0	129.0	58.1
O&M and Admin Costs ($C million)	2037.0	1927.0	5.7
Fuel & Purchased Power ($C million)	1273.0	1497.0	-15.0
Finance Expense ($C million)	2241.0	1788.0	25.3
Depreciation Expense ($C million)	1136.0	908.0	25.1
Provincial Levies ($C million)	252	235	7.2
Purchased Power Exp ($C million)	N/A	N/A	
Fixed Assets ($C million)	38170.0	35139.0	8.6
Capital Spending ($C million)	3934.0	3544.0	11.0
Customers (X1000)	3696	3654	1.1
Transmission Line Length (km)	N/A	N/A	
Distribution Circuits (km)	N/A	N/A	
Regular Employees	28396	26821	5.9
Non-regular Employees	7309	9653	-24.3
N/A = not available			

SaskPower

SaskPower is a Crown Corporation serving approximately 405,000 customers. In 1991, 66% of the utility's sales were from thermal plants, 31% from hydraulic facilities, and 2% was purchased from adjoining provinces and North Dakota. The company recently completed SHAND 1, a 300-MW lignite-fired unit at a total estimated cost of $516 million. A second coal-fired unit may be constructed at the site later in the decade.

SHAND 1 is notable for its SO_2 reduction equipment. A 150-MW LIFAC unit developed by the Finnish boiler maker Tampella is designed to achieve a 75% reduction in SO2 in that portion of the flue-gas stream. The process involves limestone injection into the furnace and SHAND 1 represents the first greenfield LIFAC installation in North America. The emission rate set for the unit, which fires low-sulfur lignite (0.4-0.8% S, 6000 btu/lb), is 0.6 lbs/mmbtu.

Generally strong financial performance allowed the utility to complete a second straight year without a general rate increase. In fact, 1991 was the first full year of a recently enacted rate reduction program for small commercial customers.

With a large and thinly populated service area, T&D projects are a major part of **SaskPower's** business. One interesting project is the Rural Underground Distribution program, which in 1991 resulted in burial of 3,544 km of cable to 2,600 rural customers at a cost of $27 million. Three major transmission projects were initiated or completed during the year.

The utility's projections show a need for additional capacity in the late 1990s. New construction will be delayed to the extent possible by energy conservation programs, including energy audits and promotion of such technologies as ground source heat pumps, compact fluorescent fixtures, farm yard light conversions, and residential timers.

When new capacity is required, **SaskPower** is counting on independents to provide at least part of it. In the spring of 1992, the provincial government -- like BC and Ontario recently converted to the New Democratic Party (NDP) -- announced the creation of a Saskatachewan Energy Conservation and Development Institute to help develop a long-term, energy diversity strategy.[14] This is similar to one started by the Alberta Department of Energy in 1991. One contract was signed by the utility in 1990, but this has not yet been implemented.

The large and politically important Saskatchewan lignite-mining industry based in the southern part of the province will undoubtedly play a role in deciding on the next capacity increments. Currently, 70% of the province's power comes from the 9 to 10 million tpy mined in the lignite belt. In turn, this demand spurs continuing investments in surface-mining equipment, human resources, exploration and development technology, etc.

Nuclear power is also of interest in the province. A joint venture announced in 1990 between **SaskPower** and **Atomic Energy of Canada** created a company to evaluate the construction of a 450-MW CANDU reactor in the province. The venture is designed to assist in the development of Saskatchewan's abundant supplies of uranium, but has been "on again, off again." It was "officially shelved"[15] by the NDP in March 1992, but seemingly revived by Prime Minister Mulroney in a visit to the province in October 1992. This high-level interest in the project is a measure of the importance of AECL's CANDU program in the federal government's estimation.

Saskatchewan Power Corp.	1991 Data	1990 Data	% Change
In-province Sales (million kWhrs)	11861	11631	2.0
Thermal Generation (million kWhrs)	8951	8919	0.4
Hydraulic Generation (million kWhrs)	4214	4219	-0.1
Electricity Generation (million kWhr)	13165	13139	0.2
Purchased Power (million kWhrs)	316	195	62.1

---

[14] See *Independent Power Report* April 10, 1992 (McGraw-Hill, Inc.).

[15] *Ibid.*

Saskatchewan Power Corp.	1991 Data	1990 Data	% Change
Export Sales (million kWhrs)	170	246	-30.9
Total Energy Sales (million kWhrs)	12031	11877	1.3
Peak Demand (MW)	2304	2328	-1.0
Installed Capacity (MW)	2774	2775	0.0
Revenue ($C million)	673.9	674.7	-0.1
Net Income ($C million)	118.1	118.4	-0.3
O&M and Admin Costs ($C million)	212.5	206.4	3.0
Fuel & Purchased Power ($C million)	127.3	122.3	4.1
Finance Expense ($C million)	119.0	134.5	-11.5
Depreciation Expense ($C million)	97.0	93.2	4.1
Plant-in-Service ($C million)	3113.0	2935.0	6.1
Capital Spending ($C million)	365.0	291.0	25.4
Customers (X1000)	405	403	0.5
Transmission Line Length (km)	11656	10975	6.2
Distribution Circuits (km)	132812	131914	0.7
Regular Employees	2294	2201	4.2
Part-time Employees	209	180	16.1
N/A = not available			

TransAlta Utilities

TransAlta is the largest investor-owned utility in Canada and began operations in 1911. Today, the company produces about 70% of the electricity used in Alberta, over 80% of it from fossil-fueled power plants, an unusual fuel mix for a Canadian company. Energy sales were up sharply in 1991. In 1981, **TransAlta** formed a subsidiary **TransAlta Resources** which has become one of the most active non-utility power plant developers in the country.

Plans were announced in October 1992 for **TransAlta Utilities** to reorganize into a holding company. As part of the reorganization, **TransAlta Resources** will be renamed **TransAlta Energy Corp.**, thus allowing easier separation of the regulated from the unregulated components. Construction of two cogeneration projects in Ontario developed by **TransAlta Resources** was completed in 1992. Additional projects are under development in British Columbia, Ontario, and Quebec.

Transalta Utilities Corp.	1991 Data	1990 Data	% Change
In-province Sales (million kWhrs)	24250	22786	6.4
Thermal Generation (million kWhrs)	26102	25584	2.0
Hydraulic Generation (million kWhrs)	2022	2051	-1.4
Electricity Generation (million kWhr)	28124	27635	1.8
Purchased Power (million kWhrs)	-1911	-2809	-32.0
Peak Load (MW)	3844	3808	0.9
Installed Capacity (MW)	4476	4476	0.0
Electric Revenues ($C million)	1160.6	1063.5	9.1
Earnings ($C million)	158.3	145.3	8.9
Operating Expenses ($C million)	211.8	194.1	9.1
Fuel & Purchased Power ($C million)	80.8	63.0	28.3
Finance Expense ($C million)	168.4	176.4	-4.5
Depreciation Expense ($C million)	194.8	180.8	7.7
Taxes ($C million)	180.0	182.0	-1.2
Plant-in-Service ($C million)	5276.0	4989.5	5.7
Capital Spending ($C million)	322.3	280.9	14.7
Unit Revenue (cents C/kWhr)	4.76	4.56	4.4
Customers (X1000)	645	633	1.8
Transmission Line Length (km)	N/A	N/A	
Distribution Circuits (km)	N/A	N/A	
Employees (full-time)	2694	2652	1.6
N/A = not available			

TransAlta's financial position was strengthened by the June 1, 1991 decision of the Public Utilities Board to grant an interim rate increase of 7.15%, the first increase since 1986. This order allowed inclusion of the $340 million investment in SHEERNESS 2, a 280-MW lignite-fired unit jointly-owned with **Alberta Power Ltd.**

TransAlta has two ongoing political struggles bearing on its "private utility" status in generally "public utility" Canada. The first is a complex debate over utility income tax rebates, frozen by the Government of Canada in 1990 and discontinued by the provincial government in the same year. According to the utility this has resulted in a significant inequity between investor-owned utility customers and provincial and municipal utility customers. A more local issue has to do with Alberta's Electric Energy Marketing Agency (EEMA), established 10 years ago to reduce regional rate disparities in the province. As matters now stand, **TransAlta** customers reportedly pay a disproportionately large share of EEMA costs.

In August 1991, **TransAlta** began its first firm-power export program, providing **Bonneville Power Administration** with 100 average MW. Net revenues of approximately $25 million are expected over the four-year contract duration. A long-term coordination agreement with **BC Hydro** was completed in 1991. This is expected to provide $400 million in cost savings over the next 20 years and will facilitate export sales through the existing 500-kV interconnection to **BC Hydro**.

In late 1991, the Alberta Department of Energy made available 125 MW of specially-priced independent power contracts for biomass and other alternate fuels under the Small Power Research and Development Program. Both **TransAlta** and **Alberta Power** are involved in these purchases that include three biomass plants, six hydro plants, a tire-burning plant, and 10 wind-to-energy projects. One of the larger projects at $52.5 million is a 21.5-MW, fluidized-bed, wood-fired unit developed by **Millar Western Power Ltd.** near WHITECOURT.

Two small hydro plants went into service in 1992, the 2.4-MW WATERTON plant and the 2-MW ST MARY plant, both developed by **Canadian Hydro Developers Inc.** at a cost of about $6.5 million. **TransAlta** will buy up to $1.2 million worth of electricity for ¢5.1/kWhr.

The wind energy program in Alberta is the most advanced in the country. Three 65-kW machines went into service in June 1992 near Pincher Creek as the DUTCH VALLEY project. Two larger projects are planned in the area to sell about 10 MW of power to **TransAlta**. In September 1992, the CHINOOK (PINCHER CREEK I) wind project of 10 MW advanced with the hire of **NorWestern Energy Systems** as project manager and **Dutch Industries** as general contractor. Owned by the Peigan Indians of southwest Alberta, CHINOOK will have 44 Danish machines.

On the DSM side, **TransAlta** has licensed the **BC Hydro** Power Smart program. Initial activities include lighting, high-efficiency electric motor, and refrigerator programs.

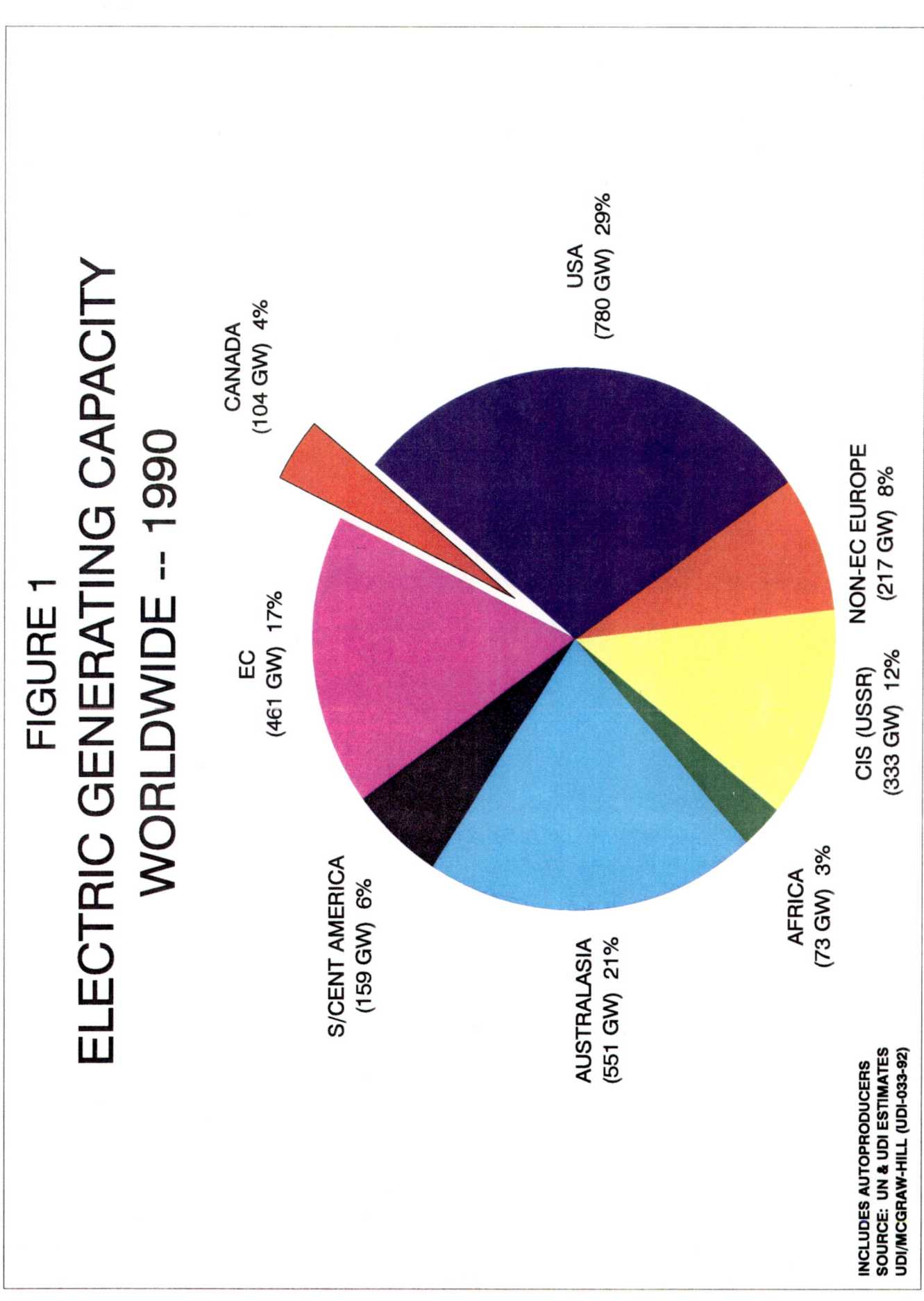

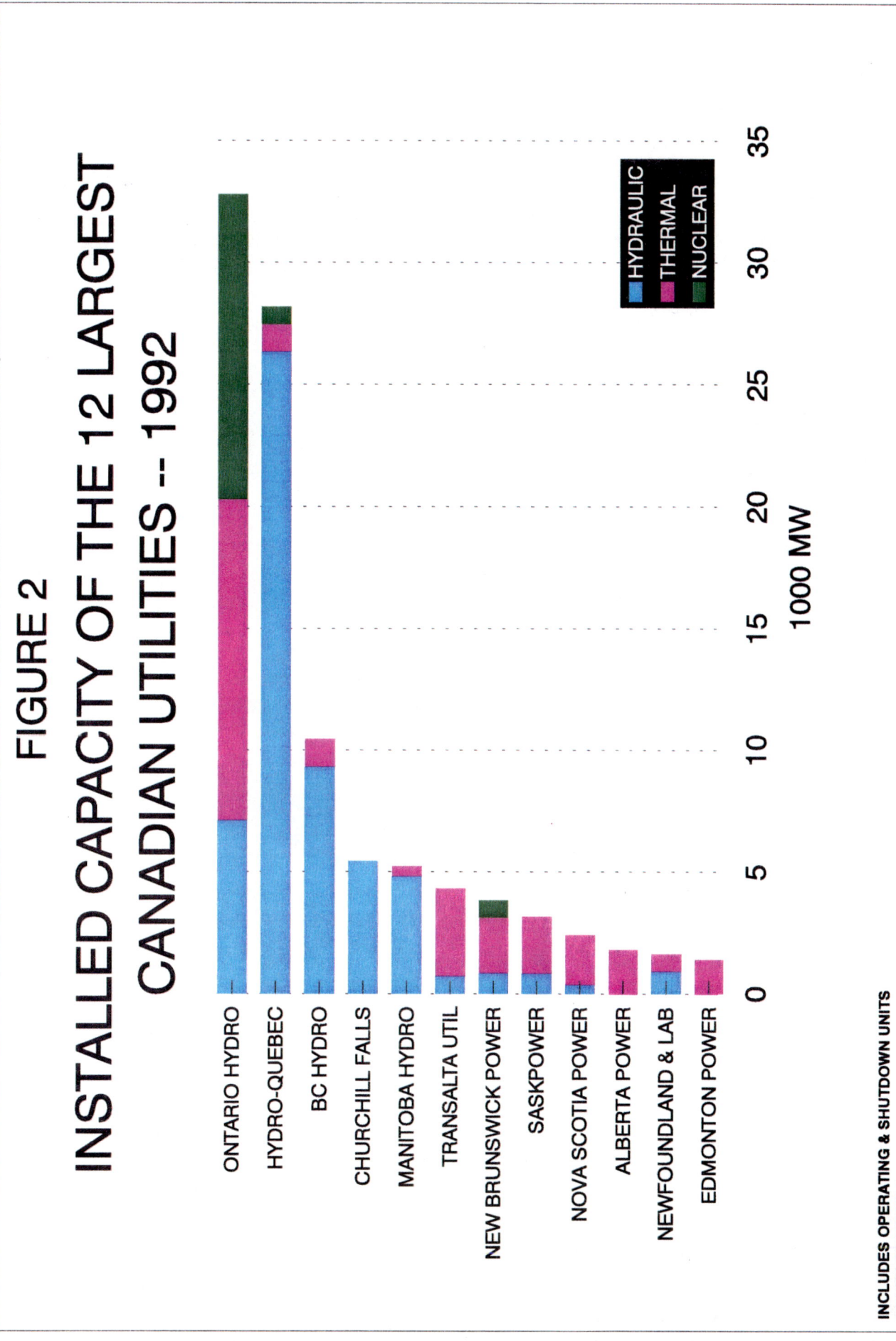

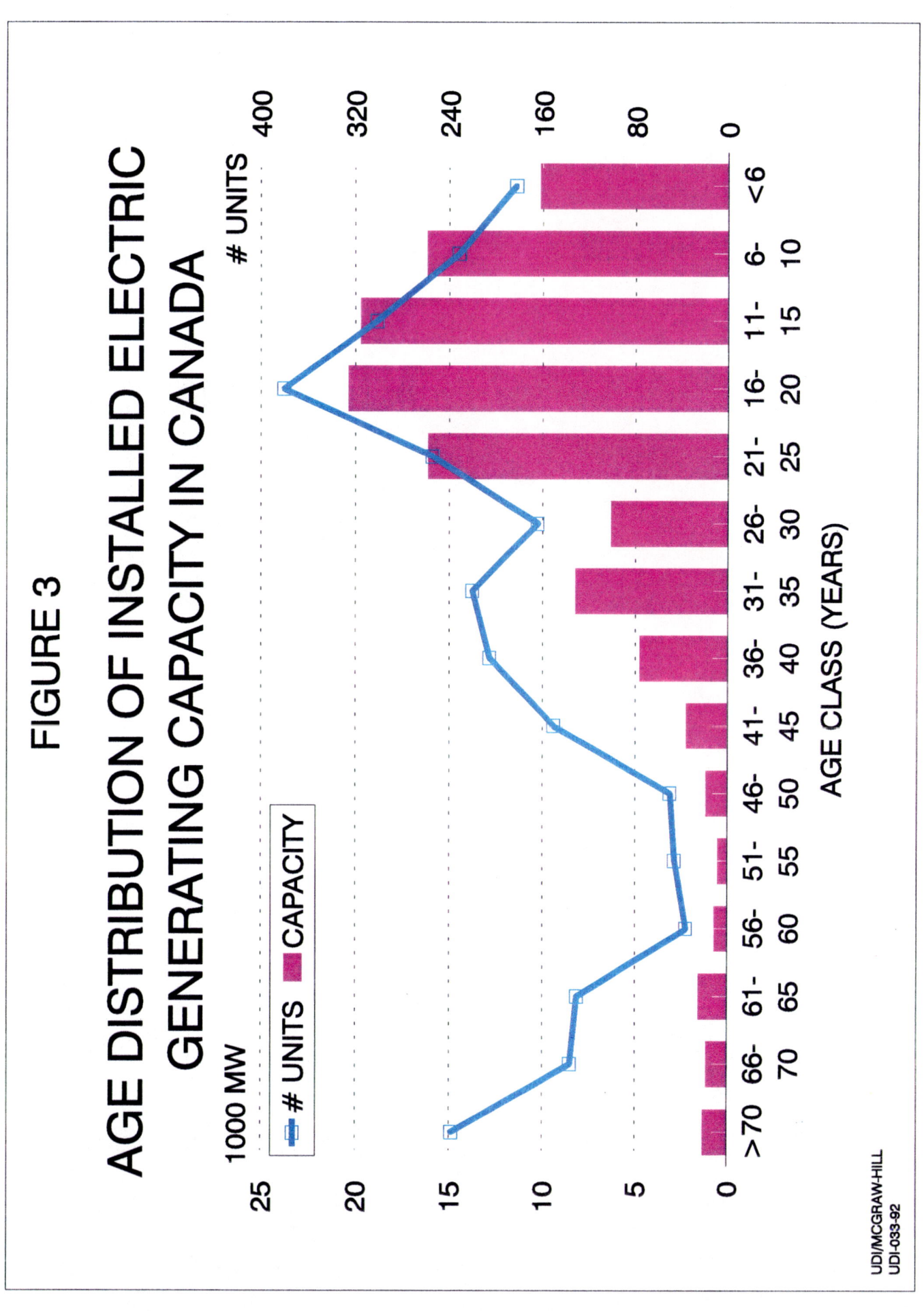

FIGURE 4
FUEL DISTRIBUTION OF INSTALLED GENERATING CAPACITY IN CANADA -- 1992

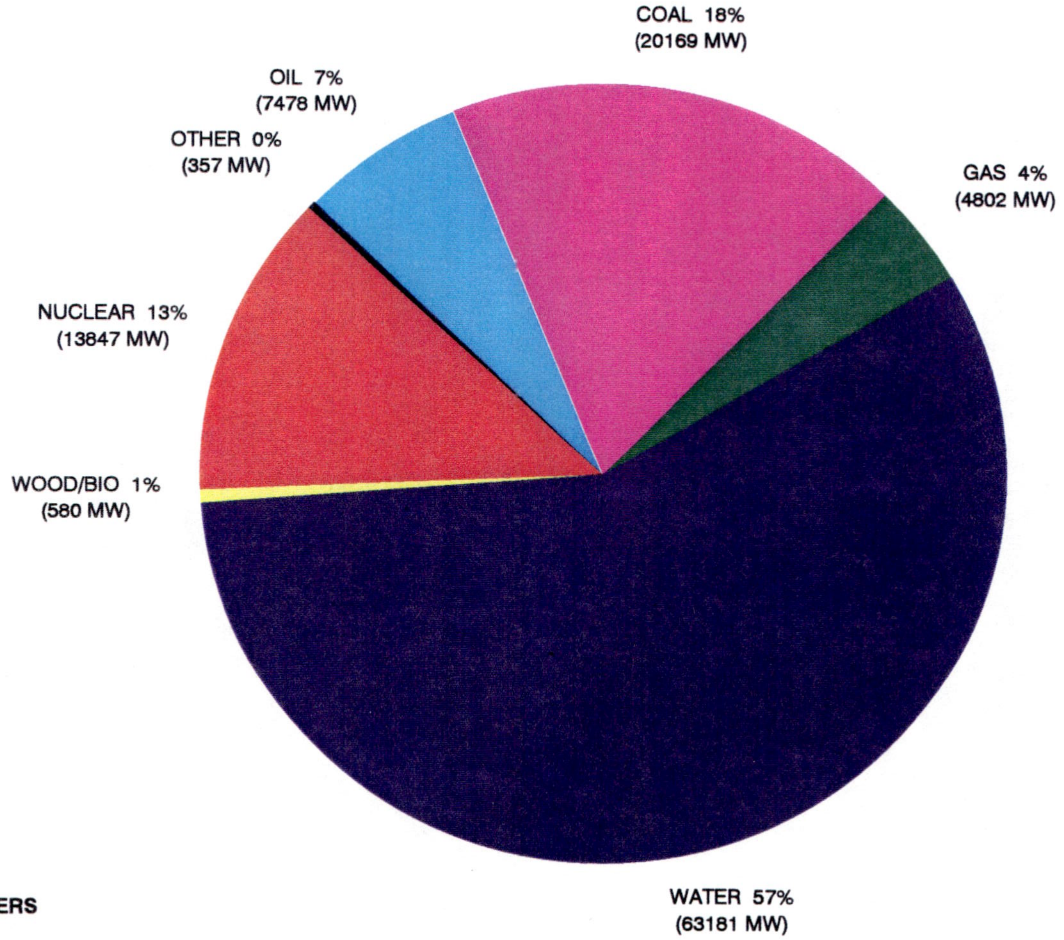

- WATER 57% (63181 MW)
- GAS 4% (4802 MW)
- COAL 18% (20169 MW)
- OIL 7% (7478 MW)
- OTHER 0% (357 MW)
- NUCLEAR 13% (13847 MW)
- WOOD/BIO 1% (580 MW)

INCLUDES AUTOPRODUCERS
UDI/MCGRAW-HILL
UDI-033-92

1991 CANADIAN POWER PLANT REPORT AND DATA BASE

The plant- and unit-specific data for the generating facilities covered in this second edition of UDI's *Directory of Power Plants in Canada* are arranged alphabetically by operating company and plant/unit name. Data are provided on a unitized[16] basis whenever possible and this is indicated by the presence of a unit number designator (Unit 1, 2, etc.).

Absence of such a unit designation indicates that UDI is not currently able to determine whether the capacity shown represents one or more than one individual unit. On occasion, the record may be termed [PLANT NAME] 1&2, indicating that two units are known to be present, but the rated capacity of each unit is unknown.

UDI's Canadian power plant report and data base contain virtually complete coverage for all utility-owned generating units in the country and include all known non-utility and industrial[17] power plants as well. For the newer variety of private power plants, including small hydro, cogeneration, etc., data were obtained from the trade and business press and by direct survey. Typically, non-utility generator (NUG) projects were included only if a power purchase contract has been signed with an electric utility or industrial concern and/or if equipment has been ordered.

The proper nomenclature and description of internal combustion (IC, used synonymously with "diesel") power plants is particularly difficult to obtain and keep up-to-date. Canadian power companies operate 1,000 IC units. Although the total

---

[16] A "unit number" in American usage typically refers to the numbering scheme for an individual power plant turbine/generator. In many countries, the turbine/generator is referred to as a "set". Additional units built at existing power plant sites may be numbered sequentially, or may be given different names such as "[PLANT NAME] NEW" or "[PLANT NAME] TWO", "[PLANT NAME] THREE", etc., with a new numbering scheme started.

[17] In UDI's other international directories and in other international data sources, these non-utility power plants are termed autoproducers.

capacity of these units is approximately 600 MW, remote power generation is very important in the sparsely-settled regions of the country.

In the smaller sizes, IC units can be moved to different locations. If one or more units "disappears" from a plant site, the unit(s) may be retired or they may have been moved. When this occurs, the unit numbering schemes are thrown into disarray. A more complete accounting of the number of IC machines in Canada and the unit numbering schemes must await further research.

Report Format and Coding Conventions

Plant- and unit-level data in this *Directory* are sorted by operating company and plant/unit name. The main table in this report contains the following additional data items (when available):

- Plant/unit capacity.
- Plant/unit status.
- Year-in-service.
- Plant/unit type.
- Primary fuel.
- Primary fuel type.
- Alternate fuel.
- Boiler or reactor manufacturer.
- Turbine manufacturer (also IC engine vendor).
- Generator manufacturer.

UDI's computerized Canadian power plant data base contains additional data on steam conditions, pollution control equipment, and other items, although as yet these fields are not complete. Throughout the report and data base, "N/A" is used to indicate "not applicable" in alphanumeric fields. Blanks in alphanumeric fields indicate data are not available. Blanks in numeric fields may indicate missing data or not applicable.

Much of Canadian utility-owned power plant information was obtained from the Statistics Canada publication *Inventory of Prime Mover and Electric Generating Equipment as of December 31, 1985*.[18] This annual publication continues to be printed, but no longer contains unit-level design information. After the initial data base was completed, additional data were obtained from later editions of the Statistics Canada inventory and from UDI's extensive files of utility and vendor documentation including Annual Reports, other financial and statistical reports, installation lists, press releases, the trade and business press, and direct survey.

The question of data primacy, that is, which of two conflicting sources should be accepted, was resolved on a case-by-case basis. Generally, plant-specific data provided by a utility were given the highest priority, followed by vendor experience lists, and then articles in the trade press.

Plant/Unit Names

UDI's international power plant reports and data bases are designed to provide information on a generating unit basis. For most steam-electric plants, a unit is comprised of a steam generator (boiler or reactor), a steam turbine, and a generator. For gas/combustion turbines,[19] a unit consists of the engine (turbine) and a generator. Combined-cycle units typically add a fired or unfired waste heat boiler to the back end of a simple-cycle gas turbine(s). For hydraulic plants, a unit is the turbine and generator (and SSS vendor is "N/A").

Canadian utilities use an American-style generating unit numbering scheme. By UDI convention, gas turbine unit names include the designation "GT", as in STE-ROSE GT1,

---

[18]*Volume III* of *Electric Power Statistics*, Catalogue 57-206 Annual, published December 1986 at Ottawa, Canada by the Industry Division, Energy Section, Statistics Canada.

[19]"Gas turbine" and "combustion turbine" are used synonymously in UDI publications.

and diesels show "IC" unit number prefixes. Steam turbines at combined-cycle plants use the prefix "SC."

If exact unit names are not known, but a particular number of units are known to be in service, this is shown by listing the plant name with unit numbers, *vis* 1-4. Absence of a unit designation indicates that UDI is not currently able to say whether the capacity shown represents one or more than one individual unit.

Fuel Descriptors, Equipment Vendors, and Other Data

UDI uses the coding and abbreviating conventions established in 10 years of work on power plant data bases. This has resulted in the development of an extensive List of Abbreviations (*Appendix C*).

Plant and Company Addresses

During 1991 and again in 1992, UDI conducted a mailing to all major Canadian utilities to update the collection of Annual Reports and other materials. In 1991, UDI also asked for names and addresses of power plant managers. Some addresses for non-utility power plant operators were obtained from a magnetic tape purchased from Statistics Canada. *Appendix A* is a list of addresses for the major power plant operators and selected power plants in Canada.

Retired Generating Units

To facilitate use of the data base by companies interested in potential repowering schemes, purchase of parts, and comparisons of UDI's statistics with other sources, data for retired generating units are retained in the data base. Format, abbreviations, etc., are consistent with the main data table.

CANADA Data Base

Appendix B is a complete field listing and description of UDI's CANADA Data Base. The computerized data are available for purchase from UDI as a dBASE III file.

Any comments on the report or data base, or specific corrections and updates to any of the information contained therein, are welcome. These should be provided to Christopher Bergesen at UDI (**TEL:** 202-466-3660; **FAX:** 202-466-3667).

This page intentionally left blank.

INFORMATION SOURCES AND RESEARCH METHODOLOGY

Data for UDI's international power plant directories are obtained from four general sources:

- Yearbooks and existing directories.
- Vendor experience lists.
- Utility Annual Reports, utility association reports, and other utility materials.
- Trade and business press.

In the case of UDI's *Canadian Directory*, the primary source of data for existing power plants was the definitive Statistics Canada publication *Inventory of Prime Mover and Electric Generating Equipment* cited above. The 1985 inventory, published in December 1986, contained unit-by-unit information including vendor names. Subsequent editions provided plant-level detail only and omitted vendor names. Data in UDI's Directory were updated to the extent possible with later editions of the Statistics Canada inventories.

General References

General references were used to obtain certain "first-tier" information in the data base (*i.e.*, utility name, plant name, capacity, and fuel). Thereafter, the UDI directory was developed by constantly cross-referencing data from the various sources.

Electrical World Directory of Electric Utilities, 101st Edition, New York: McGraw-Hill, Inc., 1992, 1123 pp.

Electrical World International Directory of Electric Utilities, Third Edition, New York: McGraw-Hill, Inc., 1972, 836 pp.

Energy Statistics 1990, United Nations, New York, NY, 1992.

International Directory of Electric Utilities, New York: McGraw-Hill, Inc., 1991, 326 pp.

"World List of Nuclear Plants," *Nuclear News*, August 1992, La Grange, IL, American Nuclear Society.

World Nuclear Industry Handbook 1989, Nuclear Engineering International, Sutton, Surrey, England, 1989, 356 pp.

Supplier Experience Lists

Experience lists -- also termed "reference" or "installation" -- lists are prepared by heavy electrical equipment manufacturers worldwide. These are an invaluable source of plant design information. Over 60 suppliers have provided their reference lists to UDI (the list of suppliers is available on request).

Utility Annual Reports and Other Materials

Annual Reports, statistical supplements, and public relations materials provided by Canadian utilities were of great value. Where available, these utility reports have been obtained by UDI from 1981 forward. *The assistance of the Canadian utilities, utility organizations, and manufacturers that provided documents and data to UDI is gratefully acknowledged.*

Trade and Business Press

In this category are grouped major articles, lists of contracts awarded, papers presented at symposia and professional meetings, and clippings from magazines and newsletters of orders announced, plants completed, etc. Often, such references provide only one piece of information for a specific power plant or generating unit that might be used in the data base. However, trade press references are considered reliable and timely and are locally important in the data base, particularly for non-utility power plants and projected facilities. Fragmentary data can also be obtained from brochures and other promotional items distributed by equipment vendors.

DIRECTORY

OF

POWER PLANTS IN CANADA

DIRECTORY OF POWER PLANTS IN CANADA
UDI/MCGRAW-HILL

PLANT/UNIT NAME	MW	STA	YEAR	TYPE	FUEL	FUEL TYPE	ALT FUEL	SSS MFR	TURB MFR	GEN MFR
OPERATOR: 146436 CANADA INC										
MORGAN FALLS	0.50	PLN	1993	HY	WAT	CONV	NONE	N/A		
OPERATOR: 156993 CANADA INC										
SHEET HARBOUR	2.20	PLN	1995	HY	WAT	CONV	NONE	N/A		
OPERATOR: 80840 CANADA LTD										
CHUTES A GORRY	2.20	PLN	1996	HY	WAT	CONV	NONE	N/A		
OPERATOR: ABITIBI-PRICE LTD										
BISHOP FALLS 1	1.50	OPR	1908	HY	WAT	CONV	NONE	N/A	SMS	CGE
BISHOP FALLS 2	1.50	OPR	1928	HY	WAT	CONV	NONE	N/A	SMS	WH
BISHOP FALLS 3	2.03	OPR	1933	HY	WAT	CONV	NONE	N/A	SMS	WH
BISHOP FALLS 4	2.03	OPR	1953	HY	WAT	CONV	NONE	N/A	SMS	WH
BISHOP FALLS 5	2.03	OPR	1953	HY	WAT	CONV	NONE	N/A	SMS	WH
BISHOP FALLS 6	2.03	OPR	1953	HY	WAT	CONV	NONE	N/A	SMS	WH
BISHOP FALLS 7	2.03	OPR	1953	HY	WAT	CONV	NONE	N/A	SMS	WH
BISHOP FALLS 8	2.03	OPR	1953	HY	WAT	CONV	NONE	N/A	SMS	WH
BISHOP FALLS 9	2.03	OPR	1953	HY	WAT	CONV	NONE	N/A	SMS	WH
BUCHANS 1	2.17	OPR	1988	HY	WAT	CONV	NONE	N/A		
GRAND FALLS (AP) 1	1.50	OPR	1909	HY	WAT	CONV	NONE	N/A	AGK	BBC
GRAND FALLS (AP) 2	1.50	OPR	1909	HY	WAT	CONV	NONE	N/A	AGK	BBC
GRAND FALLS (AP) 3	1.50	OPR	1911	HY	WAT	CONV	NONE	N/A	AGK	BBC
GRAND FALLS (AP) 4	4.00	OPR	1952	HY	WAT	CONV	NONE	N/A	SMS	WH
GRAND FALLS (AP) 5	4.00	OPR	1952	HY	WAT	CONV	NONE	N/A	SMS	WH
GRAND FALLS (AP) 6	4.00	OPR	1952	HY	WAT	CONV	NONE	N/A	SMS	WH
GRAND FALLS (AP) 7	4.00	OPR	1952	HY	WAT	CONV	NONE	N/A	SMS	WH
GRAND FALLS (AP) 8	26.50	OPR	1955	HY	WAT	CONV	NONE	N/A	DEW	WH
GRAND FALLS (AP) S1	5.00	OPR	1931	ST/S	OIL	HFO	NONE	FW	WH	WH
GRAND FALLS (AP) S2	5.00	OPR	1931	ST/S	OIL	HFO	NONE	FW	WH	WH
IROQUOIS FALLS 01	1.20	OPR	1949	HY	WAT	CONV	NONE	N/A	HOK	WHC
IROQUOIS FALLS 02	1.20	OPR	1949	HY	WAT	CONV	NONE	N/A	HOK	WHC

DIRECTORY OF POWER PLANTS IN CANADA
UDI/MCGRAW-HILL

PLANT/UNIT NAME	MW	STA	YEAR	TYPE	FUEL	FUEL TYPE	ALT FUEL	SSS MFR	TURB MFR	GEN MFR
OPERATOR: ABITIBI-PRICE LTD										
IROQUOIS FALLS 03	2.03	OPR	1949	HY	WAT	CONV	NONE	N/A	SMS	WHC
IROQUOIS FALLS 04	2.03	OPR	1949	HY	WAT	CONV	NONE	N/A	SMS	WHC
IROQUOIS FALLS 05	2.03	OPR	1949	HY	WAT	CONV	NONE	N/A	SMS	WHC
IROQUOIS FALLS 06	2.03	OPR	1949	HY	WAT	CONV	NONE	N/A	SMS	WHC
IROQUOIS FALLS 07	2.03	OPR	1949	HY	WAT	CONV	NONE	N/A	SMS	WHC
IROQUOIS FALLS 08	1.28	OPR	1949	HY	WAT	CONV	NONE	N/A	NOHAB	WHC
IROQUOIS FALLS 09	1.28	OPR	1949	HY	WAT	CONV	NONE	N/A	NOHAB	WHC
IROQUOIS FALLS 10	1.28	OPR	1949	HY	WAT	CONV	NONE	N/A	NOHAB	WHC
IROQUOIS FALLS 11	1.28	OPR	1949	HY	WAT	CONV	NONE	N/A	NOHAB	WHC
IROQUOIS FALLS 12	1.28	OPR	1949	HY	WAT	CONV	NONE	N/A	NOHAB	WHC
IROQUOIS FALLS 13	1.28	OPR	1949	HY	WAT	CONV	NONE	N/A	NOHAB	WHC
IROQUOIS FALLS 14	1.28	OPR	1949	HY	WAT	CONV	NONE	N/A	NOHAB	WHC
ISLAND FALLS (ABL) 1	9.60	OPR	1986	HY	WAT	CONV	NONE	N/A	IPM	CGE
ISLAND FALLS (ABL) 2	9.60	OPR	1979	HY	WAT	CONV	NONE	N/A	DEW	WH
ISLAND FALLS (ABL) 3	9.60	OPR	1981	HY	WAT	CONV	NONE	N/A	DEW	WH
ISLAND FALLS (ABL) 4	9.60	OPR	1982	HY	WAT	CONV	NONE	N/A	DEW	WH
TWIN FALLS (AP) 1	4.05	OPR	1921	HY	WAT	CONV	NONE	N/A	IPM	WHC
TWIN FALLS (AP) 2	4.05	OPR	1921	HY	WAT	CONV	NONE	N/A	IPM	WHC
TWIN FALLS (AP) 3	4.05	OPR	1921	HY	WAT	CONV	NONE	N/A	IPM	WHC
TWIN FALLS (AP) 4	4.05	OPR	1921	HY	WAT	CONV	NONE	N/A	IPM	WHC
TWIN FALLS (AP) 5	4.05	OPR	1927	HY	WAT	CONV	NONE	N/A	IPM	WHC
OPERATOR: ADECON										
PINCHER CREEK II	1.50	CON	1992	WTG	WIND		NONE	N/A		
OPERATOR: AEC POWER LTD										
MILDRED LAKE 1	50.00	OPR	1978	ST	GAS		NONE	BW	GE	GE
MILDRED LAKE 2	50.00	OPR	1978	ST	GAS		NONE	BW	GE	GE
MILDRED LAKE 3	50.00	OPR	1978	ST	GAS		NONE	BW	GE	GE
MILDRED LAKE 4	68.00	OPR	1978	ST	GAS		NONE	BW	GE	GE
MILDRED LAKE GT1	28.00	OPR	1977	GT	GAS		NONE	N/A	GE	GE

UDI-033-92   December 1992

DIRECTORY OF POWER PLANTS IN CANADA
UDI/MCGRAW-HILL

PLANT/UNIT NAME	MW	STA	YEAR	TYPE	FUEL	FUEL TYPE	ALT FUEL	SSS MFR	TURB MFR	GEN MFR
OPERATOR: AEC POWER LTD										
MILDRED LAKE GT2	28.00	OPR	1977	GT	GAS		NONE	N/A	GE	GE
OPERATOR: ALBERTA HOSPITAL										
EDMONTON HOSPITAL 1	2.50	OPR	1971	ST	GAS		NONE	BW/TIW	EW	BBC
PONOKA HOSPITAL 1	0.60	OPR	1961	ST	GAS		NONE	FW	BBC	BBC
PONOKA HOSPITAL 2	0.60	OPR	1961	ST	GAS		NONE	FW	BBC	BBC
PONOKA HOSPITAL 3	0.52	OPR	1984	ST	GAS		NONE	FW	TERR	KATO
PONOKA HOSPITAL IC1	0.20	OPR	1972	IC	OIL	DIST	NONE	N/A	WAU	CANR
OPERATOR: ALBERTA POWER LTD										
ALGAR IC1	0.03	OPR	1977	IC	OIL	DIESEL	NONE	N/A	DEUTZ	STM
BATTLE RIVER 1	30.00	OPR	1956	ST	COAL	SUB	NONE	CE	BBC	BBC
BATTLE RIVER 2	30.00	OPR	1964	ST	COAL	SUB	NONE	CE	BBC	BBC
BATTLE RIVER 3	150.00	OPR	1969	ST	COAL	SUB	NONE	CE	GE	GE
BATTLE RIVER 4	154.00	OPR	1975	ST	COAL	SUB	NONE	CE	GE	GE
BATTLE RIVER 5	376.11	OPR	1981	ST	COAL	SUB	NONE	CE	HT	HT
BERLAND IC1	0.02	OPR	1967	IC	OIL	DIESEL	NONE	N/A	DEUTZ	TAMPER
BUFFALO CREEK IC1	0.50	OPR	1967	IC	GAS		NONE	N/A	W-S	IDEAL
BUFFALO CREEK IC2	0.50	OPR	1967	IC	GAS		NONE	N/A	W-S	IDEAL
BUFFALO CREEK IC3	1.25	OPR	1970	IC	GAS		NONE	N/A	W-S	EP
BUFFALO CREEK IC4	1.25	OPR	1970	IC	GAS		NONE	N/A	W-S	EP
CHIPEWYAN LAKE IC1	0.10	OPR	1984	IC	OIL	DIESEL	NONE	N/A	VOLV	STM
CHIPEWYAN LAKE IC2	0.10	OPR	1984	IC	OIL	DIESEL	NONE	N/A	VOLV	STM
CHIPEWYAN LAKE IC3	0.06	OPR	1986	IC	OIL	DIESEL	NONE	N/A		
CROW LAKE IC1	0.03	OPR	1977	IC	OIL	DIESEL	NONE	N/A	DEUTZ	STM
ECONOMY IC1	0.03	OPR	1977	IC	OIL	DIESEL	NONE	N/A	DEUTZ	STM
FLAT TOP MOUNTAIN IC1	0.01	OPR	1971	IC	OIL	DIESEL	NONE	N/A	DEUTZ	TAMPER
FLAT TOP MOUNTAIN IC2	0.01	OPR	1971	IC	OIL	DIESEL	NONE	N/A	DEUTZ	TAMPER
FOGGY MOUNTAIN IC1	0.01	OPR	1971	IC	OIL	DIESEL	NONE	N/A	DEUTZ	TAMPER
FOGGY MOUNTAIN IC2	0.01	OPR	1971	IC	OIL	DIESEL	NONE	N/A	DEUTZ	TAMPER
FORT CHIPEWYAN IC1	0.50	UNK	1971	IC	OIL	DIESEL	NONE	N/A	CAT	TAMPER

DIRECTORY OF POWER PLANTS IN CANADA
UDI/MCGRAW-HILL

PLANT/UNIT NAME	MW	STA	YEAR	TYPE	FUEL	FUEL TYPE	ALT FUEL	SSS MFR	TURB MFR	GEN MFR
OPERATOR: ALBERTA POWER LTD										
FORT CHIPEWYAN IC2	0.50	OPR	1973	IC	OIL	DIESEL	NONE	N/A	CAT	TAMPER
FORT CHIPEWYAN IC3	0.80	OPR	1974	IC	OIL	DIESEL	NONE	N/A	CAT	TAMPER
FORT CHIPEWYAN IC4	1.09	OPR	1984	IC	OIL	DIESEL	NONE	N/A	MIRR	BRUSH
FORT CHIPEWYAN IC5	1.09	OPR	1984	IC	OIL	DIESEL	NONE	N/A	MIRR	BRUSH
FORT MCMURRAY GT1	3.30	OPR	1975	GT	GAS		NONE	N/A	ALL	IDEAL
FORT MCMURRAY IC1	1.20	UNK	1966	IC	GAS		NONE	N/A	COOP	EE
FORT MCMURRAY IC2	1.20	UNK	1966	IC	GAS		NONE	N/A	COOP	EE
FORT MCMURRAY IC3	2.50	UNK	1968	IC	GAS		NONE	N/A	COOP	EE
FORT MCMURRAY IC4	3.00	UNK	1969	IC	GAS		NONE	N/A	COOP	EE
FORT MCMURRAY IC5	2.07	UNK	1974	IC	GAS		NONE	N/A	FM	FM
FOX LAKE IC1	0.25	UNK	1968	IC	OIL	DIESEL	NONE	N/A	BELLIS	EMC
FOX LAKE IC2	0.15	UNK	1973	IC	OIL	DIESEL	NONE	N/A	CAT	TAMPER
FOX LAKE IC3	0.20	OPR	1984	IC	OIL	DIESEL	NONE	N/A	VOLV	STM
FOX LAKE IC4	0.20	OPR	1985	IC	OIL	DIESEL	NONE	N/A	VOLV	STM
FOX LAKE IC5	0.10	OPR	1985	IC	OIL	DIESEL	NONE	N/A	VOLV	STM
FOX LAKE IC6	0.33	OPR	1987	IC	OIL	DIESEL	NONE	N/A		
FOX LAKE IC7	0.40	OPR	1989	IC	OIL	DIESEL	NONE	N/A		
GARDEN CREEK IC1	0.15	UNK	1983	IC	OIL	DIESEL	NONE	N/A	VOLV	LS
GARDEN CREEK IC2	0.15	OPR	1985	IC	OIL	DIESEL	NONE	N/A	VOLV	LS
GARDEN CREEK IC3	0.15	OPR	1985	IC	OIL	DIESEL	NONE	N/A	VOLV	LS
GARDEN CREEK IC4	0.10	OPR	1980	IC	OIL	DIESEL	NONE	N/A		
GREGOIRE IC1	0.03	UNK	1977	IC	OIL	DIESEL	NONE	N/A	DEUTZ	STM
HR MILNER 1	150.00	OPR	1973	ST	COAL	BIT	NONE	BW	HT	HT
HUNT CREEK IC1	0.13	OPR	1972	IC	OIL	DIESEL	NONE	N/A	CAT	TAMPER
HUNT CREEK IC2	0.13	OPR	1972	IC	OIL	DIESEL	NONE	N/A	CAT	TAMPER
INDIAN CABINS IC1	0.05	OPR	1975	IC	OIL	DIESEL	NONE	N/A	DEUTZ	STM
INDIAN CABINS IC2	0.05	OPR	1975	IC	OIL	DIESEL	NONE	N/A	DEUTZ	STM
INDIAN CABINS IC3	0.03	OPR	1975	IC	OIL	DIESEL	NONE	N/A		
JASPER 1	0.45	OPR	1949	HY	WAT	CONV	NONE	N/A	PWW	CGE
JASPER 2	0.95	OPR	1956	HY	WAT	CONV	NONE	N/A	LF	CGE
JASPER GT1	3.30	OPR	1975	GT	GAS		NONE	N/A	ALL	IDEAL

DIRECTORY OF POWER PLANTS IN CANADA
UDI/MCGRAW-HILL

PLANT/UNIT NAME	MW	STA	YEAR	TYPE	FUEL	FUEL TYPE	ALT FUEL	SSS MFR	TURB MFR	GEN MFR
OPERATOR: ALBERTA POWER LTD										
JASPER GT2	3.30	OPR	1989	GT	GAS		NONE	N/A	ALL	IDEAL
JASPER IC1	3.00	OPR	1959	IC	GAS		NONE	N/A	COOP	EE
JASPER IC2	3.00	OPR	1960	IC	GAS		NONE	N/A	COOP	EE
JASPER IC3	1.20	OPR	1973	IC	GAS		NONE	N/A	WAU	TAMPER
JASPER IC4	1.20	OPR	1974	IC	GAS		NONE	N/A	WAU	TAMPER
JASPER IC5	2.10	OPR	1974	IC	GAS		NONE	N/A	GM	GM
JEAN D'OR PRAIRIE IC1	0.15	UNK	1983	IC	GAS		NONE	N/A	VOLV	STM
JEAN D'OR PRAIRIE IC2	0.11	UNK	1984	IC	GAS		NONE	N/A	VOLV	LS
JEAN D'OR PRAIRIE IC3	0.16	UNK	1984	IC	GAS		NONE	N/A	VOLV	LS
JEAN D'OR PRAIRIE IC4	0.50	OPR	1989	IC	GAS		NONE	N/A		
JEAN D'OR PRAIRIE IC5	0.60	OPR	1989	IC	GAS		NONE	N/A		
JEAN D'OR PRAIRIE IC6	0.15	OPR	1989	IC	GAS		NONE	N/A		
JEAN D'OR PRAIRIE IC7	0.15	OPR	1989	IC	GAS		NONE	N/A		
MARIANNA LAKE IC1	0.10	OPR	1971	IC	OIL	DIESEL	NONE	N/A	CAT	PALM
MARIANNA LAKE IC2	0.13	OPR	1981	IC	OIL	DIESEL	NONE	N/A	CAT	BBC
MARIANNA LAKE IC3	0.13	OPR	1982	IC	OIL	DIESEL	NONE	N/A	CAT	BBC
MARIANNA LAKE IC4	0.13	OPR	1982	IC	OIL	DIESEL	NONE	N/A	CAT	BBC
MAYTOWER IC1	0.03	OPR	1977	IC	OIL	DIESEL	NONE	N/A	DEUTZ	STM
MUSKEG IC1	0.02	UNK	1977	IC	OIL	DIESEL	NONE	N/A	DEUTZ	STM
PANNY RIVER IC1	0.02	OPR	1975	IC	OIL	DIESEL	NONE	N/A	GM	TAMPER
PANNY RIVER IC2	0.80	OPR	1974	IC	OIL	DIESEL	NONE	N/A		
PANNY RIVER IC3	0.50	OPR	1984	IC	OIL	DIESEL	NONE	N/A		
PANNY RIVER IC4	1.03	OPR	1988	IC	OIL	DIESEL	NONE	N/A		
PEACE POINT IC1	0.04	OPR	1961	IC	OIL	DIESEL	NONE	N/A	CAT	CAT
PEACE POINT IC2	0.04	OPR	1970	IC	OIL	DIESEL	NONE	N/A	CAT	CAT
RAINBOW LAKE GT1	27.50	OPR	1968	GT	GAS		NONE	N/A	WHC	WHC
RAINBOW LAKE GT2	46.40	OPR	1970	GT	GAS		NONE	N/A	BBC	BBC
SHEERNESS 1	382.95	OPR	1986	ST	COAL	SUB	NONE	CE	HT	GE
SHEERNESS 2	382.95	OPR	1990	ST	COAL	SUB	NONE	CE	HT	GE
SIMONETTE GT1	18.80	OPR	1966	GT	GAS		NONE	N/A	BBC	BBC
SIMONETTE IC1	0.02	OPR	1977	IC	OIL	DIESEL	NONE	N/A	DEUTZ	STM

DIRECTORY OF POWER PLANTS IN CANADA
UDI/MCGRAW-HILL

PLANT/UNIT NAME	MW	STA	YEAR	TYPE	FUEL	FUEL TYPE	ALT FUEL	SSS MFR	TURB MFR	GEN MFR
OPERATOR: ALBERTA POWER LTD										
SKUNK LAKE IC1	0.17	OPR	1987	IC	OIL	DIESEL	NONE	N/A		
SKUNK LAKE IC2	0.17	OPR	1987	IC	OIL	DIESEL	NONE	N/A		
STEEN RIVER IC1	0.02	OPR	1981	IC	OIL	DIESEL	NONE	N/A	DEUTZ	STM
STEEN RIVER TOWN IC1	0.05	OPR	1975	IC	OIL	DIESEL	NONE	N/A	DEUTZ	STM
STEEN RIVER TOWN IC2	0.05	OPR	1976	IC	OIL	DIESEL	NONE	N/A	DEUTZ	STM
STURGEON GT1	10.00	OPR	1958	GT	GAS		NONE	N/A	BBC	BBC
STURGEON GT2	7.50	OPR	1961	GT	GAS		NONE	N/A	BBC	BBC
THICKWOOD HILLS IC1	0.01	UNK	1977	IC	OIL	DIESEL	NONE	N/A	LIST	STM
THICKWOOD HILLS IC2	0.01	OPR	1977	IC	OIL	DIESEL	NONE	N/A	LIST	STM
THICKWOOD HILLS IC3	0.02	OPR	1988	IC	OIL	DIESEL	NONE	N/A		
TOUCHWOOD IC1	0.01	OPR	1971	IC	OIL	DIESEL	NONE	N/A		
TOUCHWOOD IC2	0.01	OPR	1971	IC	OIL	DIESEL	NONE	N/A		
TROUT LAKE IC1	0.15	OPR	1977	IC	OIL	DIESEL	NONE	N/A	CAT	BBC
TROUT LAKE IC2	0.15	OPR	1977	IC	OIL	DIESEL	NONE	N/A	CAT	BBC
TROUT LAKE IC3	0.15	OPR	1980	IC	OIL	DIESEL	NONE	N/A	CAT	BBC
TROUT LAKE IC4	0.15	OPR	1980	IC	OIL	DIESEL	NONE	N/A	CAT	BBC
OPERATOR: ALBERTA PUBLIC WORKS SUP & SER										
CLARESHOLM 1	0.40	OPR	1960	ST	GAS		NONE	FW/TIW	GE	GE
CLARESHOLM 2	0.13	OPR	1972	ST	GAS		NONE	FW/TIW	WAU	TAMPER
LEGISLATURE BUILDING 1	0.50	OPR	1953	ST	GAS		NONE	FW	BELLIS	LANCS
LEGISLATURE BUILDING 2	0.80	OPR	1959	ST	GAS		NONE	FW	SKIN	GE
LEGISLATURE BUILDING 3	0.80	OPR	1965	ST	GAS		NONE	FW	BELLIS	MTPL
MICHENER CENTRE SOUTH 1	0.10	STN	1926	ST	GAS		NONE	FW/VKE	BELLIS	GE
MICHENER CENTRE SOUTH 2	0.25	STN	1930	ST	GAS		NONE	FW/VKE	BELLIS	MTPL
MICHENER CENTRE SOUTH 3	0.40	OPR	1961	ST	GAS		NONE	FW/VKE	WH	WH
OPERATOR: ALBERTA SUGAR COMPANY										
TABER 1	2.00	OPR	1950	ST	GAS		NONE	BWGM	WH	WH
TABER 2	4.30	OPR	1967	ST	GAS		NONE	BWGM	BBC	BBC

DIRECTORY OF POWER PLANTS IN CANADA
UDI/MCGRAW-HILL

PLANT/UNIT NAME	MW	STA	YEAR	TYPE	FUEL	FUEL TYPE	ALT FUEL	SSS MFR	TURB MFR	GEN MFR
OPERATOR: ALBRIGHT & WILSON AMERIQUE										
BUCKINGHAM 1	1.44	OPR	1914	HY	WAT	CONV	NONE	N/A	SMS	CGE
BUCKINGHAM 2	1.44	OPR	1915	HY	WAT	CONV	NONE	N/A	SMS	CGE
BUCKINGHAM 3	1.44	OPR	1920	HY	WAT	CONV	NONE	N/A	SMS	CGE
BUCKINGHAM 4	1.44	OPR	1928	HY	WAT	CONV	NONE	N/A	SMS	CGE
BUCKINGHAM 5	1.84	OPR	1936	HY	WAT	CONV	NONE	N/A	ACC	CGE
BUCKINGHAM 6	1.98	OPR	1986	HY	WAT	CONV	NONE	N/A		
OPERATOR: ALCAN SMELTERS & CHEMICALS LTD										
CHUTE A CARON 1	45.00	OPR	1931	HY	WAT	CONV	NONE	N/A	SMS	WHC
CHUTE A CARON 2	45.00	OPR	1931	HY	WAT	CONV	NONE	N/A	SMS	WHC
CHUTE A CARON 3	45.00	OPR	1932	HY	WAT	CONV	NONE	N/A	SMS	WHC
CHUTE A CARON 4	45.00	OPR	1934	HY	WAT	CONV	NONE	N/A	SMS	WHC
CHUTE A LA SAVANNE 1	37.45	OPR	1953	HY	WAT	CONV	NONE	N/A	DEW	CGE
CHUTE A LA SAVANNE 2	37.45	OPR	1953	HY	WAT	CONV	NONE	N/A	DEW	CGE
CHUTE A LA SAVANNE 3	37.45	OPR	1953	HY	WAT	CONV	NONE	N/A	DEW	CGE
CHUTE A LA SAVANNE 4	37.45	OPR	1953	HY	WAT	CONV	NONE	N/A	DEW	CGE
CHUTE A LA SAVANNE 5	37.45	OPR	1953	HY	WAT	CONV	NONE	N/A	DEW	CGE
CHUTE DES PASSES 1	148.50	OPR	1959	HY	WAT	CONV	NONE	N/A	EE	CGE
CHUTE DES PASSES 2	148.50	OPR	1959	HY	WAT	CONV	NONE	N/A	EE	CGE
CHUTE DES PASSES 3	148.50	OPR	1959	HY	WAT	CONV	NONE	N/A	EE	CGE
CHUTE DES PASSES 4	148.50	OPR	1960	HY	WAT	CONV	NONE	N/A	EE	CGE
CHUTE DES PASSES 5	148.50	OPR	1960	HY	WAT	CONV	NONE	N/A	EE	CGE
CHUTE DU DIABLE 1	37.45	OPR	1952	HY	WAT	CONV	NONE	N/A	ACC	WHC
CHUTE DU DIABLE 2	37.45	OPR	1952	HY	WAT	CONV	NONE	N/A	ACC	WHC
CHUTE DU DIABLE 3	37.45	OPR	1952	HY	WAT	CONV	NONE	N/A	ACC	WHC
CHUTE DU DIABLE 4	37.45	OPR	1952	HY	WAT	CONV	NONE	N/A	ACC	WHC
CHUTE DU DIABLE 5	37.45	OPR	1952	HY	WAT	CONV	NONE	N/A	ACC	WHC
ISLE MALIGNE 01	28.00	OPR	1925	HY	WAT	CONV	NONE	N/A	ACC	WHC
ISLE MALIGNE 02	28.00	OPR	1925	HY	WAT	CONV	NONE	N/A	ACC	WHC
ISLE MALIGNE 03	28.00	OPR	1925	HY	WAT	CONV	NONE	N/A	ACC	WHC
ISLE MALIGNE 04	28.00	OPR	1925	HY	WAT	CONV	NONE	N/A	ACC	WHC

DIRECTORY OF POWER PLANTS IN CANADA
UDI/MCGRAW-HILL

PLANT/UNIT NAME	MW	STA	YEAR	TYPE	FUEL	FUEL TYPE	ALT FUEL	SSS MFR	TURB MFR	GEN MFR
OPERATOR: **ALCAN SMELTERS & CHEMICALS LTD**										
ISLE MALIGNE 05	28.00	OPR	1925	HY	WAT	CONV	NONE	N/A	ACC	WHC
ISLE MALIGNE 06	28.00	OPR	1925	HY	WAT	CONV	NONE	N/A	ACC	WHC
ISLE MALIGNE 07	28.00	OPR	1925	HY	WAT	CONV	NONE	N/A	ACC	WHC
ISLE MALIGNE 08	28.00	OPR	1925	HY	WAT	CONV	NONE	N/A	ACC	WHC
ISLE MALIGNE 09	28.00	OPR	1926	HY	WAT	CONV	NONE	N/A	ACC	WHC
ISLE MALIGNE 10	28.00	OPR	1926	HY	WAT	CONV	NONE	N/A	ACC	WHC
ISLE MALIGNE 11	28.00	OPR	1928	HY	WAT	CONV	NONE	N/A	ACC	WHC
ISLE MALIGNE 12	28.00	OPR	1937	HY	WAT	CONV	NONE	N/A	ACC	WHC
KEMANO 01	97.60	OPR	1954	HY	WAT	CONV	NONE	N/A	ACC	CGE
KEMANO 02	97.60	OPR	1954	HY	WAT	CONV	NONE	N/A	PWW	WHC
KEMANO 03	97.60	OPR	1954	HY	WAT	CONV	NONE	N/A	DEW	EE
KEMANO 04	105.60	OPR	1956	HY	WAT	CONV	NONE	N/A	PWW	WHC
KEMANO 05	97.60	OPR	1956	HY	WAT	CONV	NONE	N/A	DEW	CGE
KEMANO 06	105.60	OPR	1957	HY	WAT	CONV	NONE	N/A	PWW	EE
KEMANO 07	105.60	OPR	1958	HY	WAT	CONV	NONE	N/A	DEW	CGE
KEMANO 08	105.60	OPR	1967	HY	WAT	CONV	NONE	N/A	DEW	WHC
KEMANO 09	137.10	DEL	1996	HY	WAT	CONV	NONE	N/A	DBSZ	ABB
KEMANO 10	137.10	DEL	1996	HY	WAT	CONV	NONE	N/A	DBSZ	ABB
KEMANO 11	137.10	DEL	1996	HY	WAT	CONV	NONE	N/A	DBSZ	ABB
KEMANO 12	137.10	DEL	1996	HY	WAT	CONV	NONE	N/A	DBSZ	ABB
SHIPSHAW 01	60.00	OPR	1942	HY	WAT	CONV	NONE	N/A	AC	CGE
SHIPSHAW 02	60.00	OPR	1942	HY	WAT	CONV	NONE	N/A	AC	WHC
SHIPSHAW 03	58.50	OPR	1943	HY	WAT	CONV	NONE	N/A	SMS	WHC
SHIPSHAW 04	58.50	OPR	1943	HY	WAT	CONV	NONE	N/A	SMS	WHC
SHIPSHAW 05	60.00	OPR	1943	HY	WAT	CONV	NONE	N/A	AC	CGE
SHIPSHAW 06	60.00	OPR	1943	HY	WAT	CONV	NONE	N/A	AC	WHC
SHIPSHAW 07	60.00	OPR	1943	HY	WAT	CONV	NONE	N/A	AC	CGE
SHIPSHAW 08	60.00	OPR	1943	HY	WAT	CONV	NONE	N/A	AC	WHC
SHIPSHAW 09	60.00	OPR	1943	HY	WAT	CONV	NONE	N/A	AC	CGE
SHIPSHAW 10	60.00	OPR	1943	HY	WAT	CONV	NONE	N/A	AC	WHC
SHIPSHAW 11	60.00	OPR	1943	HY	WAT	CONV	NONE	N/A	SMS	CGE

DIRECTORY OF POWER PLANTS IN CANADA
UDI/MCGRAW-HILL

PLANT/UNIT NAME	MW	STA	YEAR	TYPE	FUEL	FUEL TYPE	ALT FUEL	SSS MFR	TURB MFR	GEN MFR
OPERATOR: ALCAN SMELTERS & CHEMICALS LTD										
SHIPSHAW 12	60.00	OPR	1943	HY	WAT	CONV	NONE	N/A	SMS	WHC
OPERATOR: ALGOMA STEEL CORP LTD										
SAULT STE MARIE 1	0.63	OPR	1942	ST/S	BGAS		NONE	FW	WH	WH
SAULT STE MARIE 2	0.63	OPR	1942	ST/S	BGAS		NONE	FW	WH	WH
SAULT STE MARIE 3	12.50	OPR	1963	ST	BGAS		NONE	FW	WHC	WHC
SAULT STE MARIE 4	12.50	OPR	1963	ST	BGAS		NONE	FW	WHC	WHC
OPERATOR: ALGONQUIN POWER CORP										
CORDOVA LAKE 1	0.78	PLN	9999	HY	WAT	CONV	NONE	N/A	OSS	KATO
OPERATOR: ALLIED CHEMICALS CANADA LTD										
AMHERSTBURG 1	2.50	OPR	1948	ST/S	GAS		NONE	BW/CE	GE	GE
AMHERSTBURG 2	3.75	OPR	1957	ST/S	GAS		NONE	BW/CE	GE	GE
AMHERSTBURG 3	4.70	OPR	1966	ST/S	GAS		NONE	BW/CE	GE	GE
OPERATOR: AMERICAN POWER & WST MGMT										
CHEMAINUS REFUEL GT1	19.75	PLN	1995	GT	GAS		WDGAS	N/A	GE	GE
CHEMAINUS REFUEL GT2	19.75	PLN	1995	GT	GAS		WDGAS	N/A	GE	GE
CHEMAINUS REFUEL GT3	18.00	PLN	1995	GT	GAS		WDGAS	N/A	GE	GE
CHEMAINUS REFUEL GT4	18.00	PLN	1995	GT	GAS		WDGAS	N/A	GE	GE
OPERATOR: AMOCO CANADA PETROLEUM LTD										
BIGSTONE IC1	0.40	OPR	1967	IC	GAS		NONE	N/A	WAU	EMC
BIGSTONE IC2	0.40	OPR	1967	IC	GAS		NONE	N/A	WAU	EMC
BIGSTONE IC3	0.40	OPR	1967	IC	GAS		NONE	N/A	WAU	EMC
BIGSTONE IC4	0.40	OPR	1967	IC	GAS		NONE	N/A	WAU	EMC
EAST CROSSFIELD 1	0.30	OPR	1970	ST	GAS		NONE	TIW		EMC
EAST CROSSFIELD 2	0.30	OPR	1970	ST	GAS		NONE	TIW		EMC
EAST CROSSFIELD IC1	0.40	OPR	1968	IC	GAS		NONE	N/A	WAU	EMC
EAST CROSSFIELD IC2	0.40	OPR	1968	IC	GAS		NONE	N/A	WAU	EMC

DIRECTORY OF POWER PLANTS IN CANADA
UDI/MCGRAW-HILL

PLANT/UNIT NAME	MW	STA	YEAR	TYPE	FUEL	FUEL TYPE	ALT FUEL	SSS MFR	TURB MFR	GEN MFR
OPERATOR: AMOCO CANADA PETROLEUM LTD										
FIR IC1	0.18	OPR	1976	IC	GAS		NONE	N/A	WAU	KATO
FIR IC2	0.18	OPR	1976	IC	GAS		NONE	N/A	WAU	KATO
SOUTH WAPITI IC1	0.45	OPR	1982	IC	GAS		NONE	N/A	WAU	BBC
SOUTH WAPITI IC2	0.45	OPR	1982	IC	GAS		NONE	N/A	WAU	BBC
WHITECOURT (ACP) IC7	0.80	OPR	1965	IC	GAS		NONE	N/A	COOP	GE
WHITECOURT (ACP) IC8	0.80	OPR	1965	IC	GAS		NONE	N/A	COOP	GE
OPERATOR: ATLANTIC COMBUSTION PROD LTD										
DIGBY 1	10.00	PLN	9999	ST	WOOD		NONE			
OPERATOR: ATLANTIC SUGAR LTD										
SAINT JOHN 3	6.00	OPR	1989	ST/S	OIL	HFO	NONE			
OPERATOR: BALOIL LTD/ENERFORCE ENERGY										
ZAMA FIELD 1	2.50	PLN	1993	IC	GAS		NONE	N/A		
OPERATOR: BC FOREST PRODUCTS LTD										
COWICHAN 2	0.80	OPR	1915	ST	WOOD		NONE	VS/CE	AC	AC
COWICHAN 3	2.00	OPR	1918	ST	WOOD		NONE	VS/CE	AC	AC
COWICHAN 4	5.00	OPR	1966	ST	WOOD		NONE	VS/CE	AC	AC
CROFTON 1	38.00	OPR	1981	ST	OIL	HFO	NONE	CE/FW	HT	HT
MACKENZIE 1	20.00	OPR	1979	ST	GAS		NONE	BW/CE	HT	HT
VICTORIA (BCFP) 1	3.00	UNK	1940	ST	WOOD		NONE	BW/VUI	GE	GE
VICTORIA (BCFP) 2	1.50	UNK	1950	ST	WOOD		NONE	BW/VUI	AC	AC
OPERATOR: BC HYDRO										
ABERFELDIE 1	2.50	OPR	1922	HY	WAT	CONV	NONE	N/A	SMS	WHC
ABERFELDIE 2	2.50	OPR	1922	HY	WAT	CONV	NONE	N/A	SMS	WHC
AH-SIN-HEEK IC1	1.00	OPR	1962	IC	OIL	DIESEL	NONE	N/A	GM	GM
AH-SIN-HEEK IC2	1.00	OPR	1962	IC	OIL	DIESEL	NONE	N/A	GM	GM
AH-SIN-HEEK IC3	1.00	OPR	1964	IC	OIL	DIESEL	NONE	N/A	GM	GM

DIRECTORY OF POWER PLANTS IN CANADA
UDI/MCGRAW-HILL

PLANT/UNIT NAME	MW	STA	YEAR	TYPE	FUEL	FUEL TYPE	ALT FUEL	SSS MFR	TURB MFR	GEN MFR
OPERATOR: BC HYDRO										
AH-SIN-HEEK IC4	0.60	OPR	1969	IC	OIL	DIESEL	NONE	N/A	CAT	KATO
AH-SIN-HEEK IC5	0.60	OPR	1969	IC	OIL	DIESEL	NONE	N/A	CAT	KATO
ALOUETTE 1	8.00	OPR	1928	HY	WAT	CONV	NONE	N/A	EE	EE
ANAHIM IC2	0.25	OPR	1972	IC	OIL	DIESEL	NONE	N/A	GM	KATO
ANAHIM IC3	0.25	OPR	1972	IC	OIL	DIESEL	NONE	N/A	GM	KATO
ANAHIM IC4	0.25	OPR	1974	IC	OIL	DIESEL	NONE	N/A	CAT	TAMPER
ANAHIM IC5	0.25	OPR	1974	IC	OIL	DIESEL	NONE	N/A	CAT	TAMPER
ASH RIVER 1	25.20	OPR	1959	HY	WAT	CONV	NONE	N/A	AN	WHC
ATLIN IC1	0.25	UNK	1974	IC	OIL	DIESEL	NONE	N/A	CAT	TAMPER
ATLIN IC2	0.25	UNK	1974	IC	OIL	DIESEL	NONE	N/A	CAT	TAMPER
ATLIN IC3	0.40	OPR	1978	IC	OIL	DIESEL	NONE	N/A	CAT	BBC
ATLIN IC4	0.40	OPR	1978	IC	OIL	DIESEL	NONE	N/A	CAT	BBC
ATLIN IC5	0.40	OPR	1978	IC	OIL	DIESEL	NONE	N/A	CAT	BBC
ATLIN IC6	0.60	OPR	1975	IC	OIL	DIESEL	NONE	N/A	CAT	KATO
ATLIN IC7	0.50	OPR	1966	IC	OIL	DIESEL	NONE	N/A	CAT	KATO
ATLIN IC8	0.60	OPR	1975	IC	OIL	DIESEL	NONE	N/A	CAT	KATO
BAMFIELD IC1	0.25	UNK	1971	IC	OIL	DIESEL	NONE	N/A	CAT	CAT
BAMFIELD IC2	0.30	UNK	1975	IC	OIL	DIESEL	NONE	N/A	CAT	TAMPER
BELLA BELLA IC1	0.60	OPR	1969	IC	OIL	DIESEL	NONE	N/A	CAT	KATO
BELLA BELLA IC2	0.60	OPR	1969	IC	OIL	DIESEL	NONE	N/A	CAT	KATO
BELLA BELLA IC3	0.60	OPR	1970	IC	OIL	DIESEL	NONE	N/A	CAT	KATO
BELLA BELLA IC4	0.60	OPR	1970	IC	OIL	DIESEL	NONE	N/A	CAT	KATO
BELLA BELLA IC5	0.60	UNK	1976	IC	OIL	DIESEL	NONE	N/A	CAT	KATO
BOSTON BAR IC1	0.15	OPR	1951	IC	OIL	DIESEL	NONE	N/A	VENG	EE
BOSTON BAR IC2	0.15	OPR	1951	IC	OIL	DIESEL	NONE	N/A	VENG	EE
BOSTON BAR IC3	0.50	OPR	1955	IC	OIL	DIESEL	NONE	N/A	BENZ	GE
BOSTON BAR IC4	0.50	OPR	1956	IC	OIL	DIESEL	NONE	N/A	BENZ	GE
BOSTON BAR IC5	0.65	OPR	1960	IC	OIL	DIESEL	NONE	N/A	GM	WHC
BRIDGE RIVER I 1	45.00	OPR	1948	HY	WAT	CONV	NONE	N/A	VIW	WHC
BRIDGE RIVER I 2	45.00	OPR	1949	HY	WAT	CONV	NONE	N/A	VIW	WHC
BRIDGE RIVER I 3	45.00	OPR	1949	HY	WAT	CONV	NONE	N/A	VIW	WHC

DIRECTORY OF POWER PLANTS IN CANADA
UDI/MCGRAW-HILL

PLANT/UNIT NAME	MW	STA	YEAR	TYPE	FUEL	FUEL TYPE	ALT FUEL	SSS MFR	TURB MFR	GEN MFR
OPERATOR: BC HYDRO										
BRIDGE RIVER I 4	45.00	OPR	1954	HY	WAT	CONV	NONE	N/A	VIW	WHC
BRIDGE RIVER II 1	62.00	OPR	1959	HY	WAT	CONV	NONE	N/A	VEW	WHC
BRIDGE RIVER II 2	62.00	OPR	1959	HY	WAT	CONV	NONE	N/A	VEW	WHC
BRIDGE RIVER II 3	62.00	OPR	1960	HY	WAT	CONV	NONE	N/A	NEY	WHC
BRIDGE RIVER II 4	62.00	OPR	1960	HY	WAT	CONV	NONE	N/A	NEY	WHC
BURRARD 1	150.00	OPR	1962	ST	GAS		NONE	BESS	AEI	AEI
BURRARD 2	150.00	OPR	1963	ST	GAS		NONE	BESS	AEI	AEI
BURRARD 3	150.00	OPR	1965	ST	GAS		NONE	BESS	AEI	AEI
BURRARD 4	150.00	OPR	1967	ST	GAS		NONE	BESS	AEI	AEI
BURRARD 5	150.00	OPR	1968	ST	GAS		NONE	BESS	AEI/GE	AEI/GE
BURRARD 6	162.50	OPR	1975	ST	GAS		NONE	BESS	EE	EE
CHEAKAMUS 1	70.00	OPR	1957	HY	WAT	CONV	NONE	N/A	VIW	WHC
CHEAKAMUS 2	70.00	OPR	1957	HY	WAT	CONV	NONE	N/A	VIW	WHC
CLAYTON FALLS 1	0.70	OPR	1961	HY	WAT	CONV	NONE	N/A	GGG	CGE
CLOWHOM 1	30.00	OPR	1958	HY	WAT	CONV	NONE	N/A	VIW	WHC
DEASE LAKE IC1	0.50	OPR	1963	IC	OIL	DIESEL	NONE	N/A	CAT	KATO
DEASE LAKE IC2	0.60	OPR	1975	IC	OIL	DIESEL	NONE	N/A	CAT	KATO
DEASE LAKE IC3	0.50	OPR	1978	IC	OIL	DIESEL	NONE	N/A	CAT	COEL
DEASE LAKE IC4	0.50	OPR	1978	IC	OIL	DIESEL	NONE	N/A	CAT	KATO
EDDONTENAJON IC1	0.50	OPR	1966	IC	OIL	DIESEL	NONE	N/A	CAT	KATO
EDDONTENAJON IC2	0.25	OPR	1972	IC	OIL	DIESEL	NONE	N/A	GM	KATO
EDDONTENAJON IC3	0.25	OPR	1972	IC	OIL	DIESEL	NONE	N/A	GM	KATO
EDDONTENAJON IC4	0.15	UNK	1975	IC	OIL	DIESEL	NONE	N/A	GM	KATO
EDDONTENAJON IC5	0.35	OPR	1975	IC	OIL	DIESEL	NONE	N/A	GM	KATO
EDDONTENAJON IC6	0.60	OPR	1975	IC	OIL	DIESEL	NONE	N/A	CAT	KATO
ELKO 1	4.80	OPR	1924	HY	WAT	CONV	NONE	N/A	DEW	CGE
ELKO 2	4.80	OPR	1924	HY	WAT	CONV	NONE	N/A	DEW	CGE
FALLS RIVER 1	4.80	OPR	1930	HY	WAT	CONV	NONE	N/A	DEW	EE
FALLS RIVER 2	4.80	OPR	1930	HY	WAT	CONV	NONE	N/A	DEW	WHC
FORT NELSON GT1	5.00	OPR	1963	GT	GAS		NONE	N/A	OREND	GE
FORT NELSON IC01	3.00	OPR	1957	IC	GAS		NONE	N/A	COOP	WH

DIRECTORY OF POWER PLANTS IN CANADA
UDI/MCGRAW-HILL

PLANT/UNIT NAME	MW	STA	YEAR	TYPE	FUEL	FUEL TYPE	ALT FUEL	SSS MFR	TURB MFR	GEN MFR
OPERATOR: BC HYDRO										
FORT NELSON IC02	3.00	OPR	1957	IC	GAS		NONE	N/A	COOP	WH
FORT NELSON IC03	1.20	OPR	1960	IC	GAS		NONE	N/A	COOP	GE
FORT NELSON IC04	0.60	OPR	1969	IC	GAS		NONE	N/A	COOP	EL
FORT NELSON IC05	0.35	OPR	1963	IC	GAS		NONE	N/A	CAT	COEL
FORT NELSON IC06	3.00	OPR	1974	IC	GAS		NONE	N/A	COOP	WH
FORT NELSON IC07	3.00	OPR	1978	IC	GAS		NONE	N/A	COOP	WH
FORT NELSON IC08	3.00	OPR	1978	IC	GAS		NONE	N/A	COOP	WH
FORT NELSON IC09	5.00	OPR	1963	IC	GAS		NONE	N/A		
FORT NELSON IC10	2.50	OPR	1978	IC	GAS		NONE	N/A		
FORT NELSON IC11	0.88	OPR	1988	IC	GAS		NONE	N/A		
GORDON M SHRUM 01	227.00	OPR	1968	HY	WAT	CONV	NONE	N/A	MHI	CGE
GORDON M SHRUM 02	227.00	OPR	1968	HY	WAT	CONV	NONE	N/A	MHI	CGE
GORDON M SHRUM 03	227.00	OPR	1968	HY	WAT	CONV	NONE	N/A	MHI	CGE
GORDON M SHRUM 04	227.00	OPR	1969	HY	WAT	CONV	NONE	N/A	MHI	CGE
GORDON M SHRUM 05	227.00	OPR	1969	HY	WAT	CONV	NONE	N/A	MHI	CGE
GORDON M SHRUM 06	227.00	OPR	1971	HY	WAT	CONV	NONE	N/A	TS	TS
GORDON M SHRUM 07	227.00	OPR	1972	HY	WAT	CONV	NONE	N/A	TS	TS
GORDON M SHRUM 08	227.00	OPR	1972	HY	WAT	CONV	NONE	N/A	TS	TS
GORDON M SHRUM 09	300.00	OPR	1974	HY	WAT	CONV	NONE	N/A	FUJI	FUJI
GORDON M SHRUM 10	300.00	OPR	1980	HY	WAT	CONV	NONE	N/A	FUJI	FUJI
JOHN HART 1	20.00	OPR	1948	HY	WAT	CONV	NONE	N/A	DEW	WHC
JOHN HART 2	20.00	OPR	1949	HY	WAT	CONV	NONE	N/A	DEW	WHC
JOHN HART 3	20.00	OPR	1949	HY	WAT	CONV	NONE	N/A	DEW	WHC
JOHN HART 4	20.00	OPR	1949	HY	WAT	CONV	NONE	N/A	DEW	WHC
JOHN HART 5	20.00	OPR	1953	HY	WAT	CONV	NONE	N/A	DEW	WHC
JOHN HART 6	20.00	OPR	1953	HY	WAT	CONV	NONE	N/A	DEW	WHC
JORDAN RIVER 1	150.00	OPR	1971	HY	WAT	CONV	NONE	N/A	MHI	MHI
KEOGH GT1	40.50	OPR	1973	GT	OIL	DIST	NONE	N/A	CW	BRUSH
KEOGH GT2	59.20	OPR	1978	GT	OIL	DIST	NONE	N/A	CW	BRUSH
KITKATLA IC1	0.50	OPR	1966	IC	OIL	DIESEL	NONE	N/A		
KITKATLA IC2	0.40	OPR	1984	IC	OIL	DIESEL	NONE	N/A	CAT	KATO

DIRECTORY OF POWER PLANTS IN CANADA
UDI/MCGRAW-HILL

PLANT/UNIT NAME	MW	STA	YEAR	TYPE	FUEL	FUEL TYPE	ALT FUEL	SSS MFR	TURB MFR	GEN MFR
OPERATOR: BC HYDRO										
KITKATLA IC3	0.15	OPR	1984	IC	OIL	DIESEL	NONE	N/A	CAT	EMC
KITKATLA IC4	0.30	OPR	1984	IC	OIL	DIESEL	NONE	N/A	CAT	ELC
KOOTENAY CANAL 1	132.30	OPR	1975	HY	WAT	CONV	NONE	N/A	MHI	CGE
KOOTENAY CANAL 2	132.30	OPR	1975	HY	WAT	CONV	NONE	N/A	MHI	CGE
KOOTENAY CANAL 3	132.30	OPR	1976	HY	WAT	CONV	NONE	N/A	MHI	CGE
KOOTENAY CANAL 4	132.30	OPR	1976	HY	WAT	CONV	NONE	N/A	MHI	CGE
LA JOIE 1	22.00	OPR	1957	HY	WAT	CONV	NONE	N/A	ACC	CGE
LADORE FALLS 1	27.00	OPR	1956	HY	WAT	CONV	NONE	N/A	DEW	CGE
LADORE FALLS 2	27.00	OPR	1957	HY	WAT	CONV	NONE	N/A	DEW	CGE
LAKE BUNTZEN I 1	50.00	OPR	1951	HY	WAT	CONV	NONE	N/A	VIW	WHC
LAKE BUNTZEN II 1	8.90	OPR	1913	HY	WAT	CONV	NONE	N/A	AD	DK
LAKE BUNTZEN II 2	8.90	OPR	1914	HY	WAT	CONV	NONE	N/A	AD	DK
LAKE BUNTZEN II 3	8.90	OPR	1919	HY	WAT	CONV	NONE	N/A	AD	DK
LYTTON IC02	0.35	OPR	1958	IC	OIL	DIESEL	NONE	N/A	CAT	EE
LYTTON IC03	0.28	UNK	1959	IC	OIL	DIESEL	NONE	N/A	CAT	KATO
LYTTON IC04	0.50	UNK	1966	IC	OIL	DIESEL	NONE	N/A	CAT	KATO
LYTTON IC05	0.50	UNK	1966	IC	OIL	DIESEL	NONE	N/A	CAT	KATO
LYTTON IC06	0.50	UNK	1966	IC	OIL	DIESEL	NONE	N/A	CAT	KATO
LYTTON IC07	0.50	UNK	1975	IC	OIL	DIESEL	NONE	N/A	CAT	KATO
LYTTON IC08	1.44	OPR	1989	IC	OIL	DIESEL	NONE	N/A	CAT	KATO
LYTTON IC10	0.83	OPR	1989	IC	OIL	DIESEL	NONE	N/A	CAT	KATO
LYTTON IC11	0.83	OPR	1989	IC	OIL	DIESEL	NONE	N/A	CAT	KATO
MASSET IC1	0.60	OPR	1967	IC	OIL	DIESEL	NONE	N/A	CAT	KATO
MASSET IC2	2.50	OPR	1974	IC	OIL	DIESEL	NONE	N/A	GM	GM
MASSET IC3	2.11	OPR	1978	IC	OIL	DIESEL	NONE	N/A	ALKO	BBC
MASSET IC4	2.11	OPR	1978	IC	OIL	DIESEL	NONE	N/A	ALKO	BBC
MASSET IC5	2.11	OPR	1978	IC	OIL	DIESEL	NONE	N/A	ALKO	BBC
MICA 1	434.00	OPR	1976	HY	WAT	CONV	NONE	N/A	HT	CGE
MICA 2	434.00	OPR	1976	HY	WAT	CONV	NONE	N/A	HT	CGE
MICA 3	434.00	OPR	1976	HY	WAT	CONV	NONE	N/A	LMZ	CGE
MICA 4	434.00	OPR	1977	HY	WAT	CONV	NONE	N/A	LMZ	CGE

DIRECTORY OF POWER PLANTS IN CANADA
UDI/MCGRAW-HILL

PLANT/UNIT NAME	MW	STA	YEAR	TYPE	FUEL	FUEL TYPE	ALT FUEL	SSS MFR	TURB MFR	GEN MFR
OPERATOR: BC HYDRO										
PEACE CANYON 1	175.00	OPR	1980	HY	WAT	CONV	NONE	N/A	LMZ	MHI
PEACE CANYON 2	175.00	OPR	1980	HY	WAT	CONV	NONE	N/A	LMZ	MHI
PEACE CANYON 3	175.00	OPR	1980	HY	WAT	CONV	NONE	N/A	LMZ	MHI
PEACE CANYON 4	175.00	OPR	1980	HY	WAT	CONV	NONE	N/A	LMZ	MHI
PRINCE RUPERT GT1	23.00	OPR	1973	GT	GAS		NONE	N/A	PW	BRUSH
PRINCE RUPERT GT2	23.00	OPR	1973	GT	GAS		NONE	N/A	PW	BRUSH
PUNTLEDGE 1	27.00	OPR	1955	HY	WAT	CONV	NONE	N/A	ACC	WHC
REVELSTOKE 1	460.75	OPR	1984	HY	WAT	CONV	NONE	N/A	FUJI	FUJI
REVELSTOKE 2	460.75	OPR	1984	HY	WAT	CONV	NONE	N/A	FUJI	FUJI
REVELSTOKE 3	460.75	OPR	1984	HY	WAT	CONV	NONE	N/A	FUJI	FUJI
REVELSTOKE 4	460.75	OPR	1984	HY	WAT	CONV	NONE	N/A	FUJI	FUJI
RUSKIN 1	35.20	OPR	1930	HY	WAT	CONV	NONE	N/A	DEW	WHC
RUSKIN 2	35.20	OPR	1938	HY	WAT	CONV	NONE	N/A	DEW	WHC
RUSKIN 3	35.20	OPR	1950	HY	WAT	CONV	NONE	N/A	DEW	WHC
SANDSPIT IC1	0.60	STN	1952	IC	OIL	DIESEL	NONE	N/A	COOP	GE
SANDSPIT IC2	0.60	STN	1952	IC	OIL	DIESEL	NONE	N/A	COOP	GE
SANDSPIT IC3	1.00	STN	1954	IC	OIL	DIESEL	NONE	N/A	COOP	EE
SANDSPIT IC4	1.00	STN	1965	IC	OIL	DIESEL	NONE	N/A	COOP	GE
SANDSPIT IC5	0.50	STN	1966	IC	OIL	DIESEL	NONE	N/A	CAT	COEL
SANDSPIT IC6	0.60	STN	1969	IC	OIL	DIESEL	NONE	N/A	CAT	KATO
SANDSPIT IC7	0.60	STN	1969	IC	OIL	DIESEL	NONE	N/A	CAT	KATO
SANDSPIT IC8	2.50	STN	1975	IC	OIL	DIESEL	NONE	N/A	GM	GM
SANDSPIT IC9	0.50	STN	1966	IC	OIL	DIESEL	NONE	N/A	CAT	KATO
SETON 1	42.00	OPR	1956	HY	WAT	CONV	NONE	N/A	ACC	WHC
SEVEN MILE 1	202.50	OPR	1979	HY	WAT	CONV	NONE	N/A	MHI	HT
SEVEN MILE 2	202.50	OPR	1980	HY	WAT	CONV	NONE	N/A	MHI	HT
SEVEN MILE 3	202.50	OPR	1980	HY	WAT	CONV	NONE	N/A	MHI	HT
SHUSWAP FALLS 1	2.40	OPR	1929	HY	WAT	CONV	NONE	N/A	AC	WH
SHUSWAP FALLS 2	2.80	OPR	1942	HY	WAT	CONV	NONE	N/A	ACC	CGE
SPILLMACHEEN 1	0.90	OPR	1955	HY	WAT	CONV	NONE	N/A	VIW	WHC
SPILLMACHEEN 2	0.90	OPR	1955	HY	WAT	CONV	NONE	N/A	VIW	WHC

DIRECTORY OF POWER PLANTS IN CANADA
UDI/MCGRAW-HILL

PLANT/UNIT NAME	MW	STA	YEAR	TYPE	FUEL	FUEL TYPE	ALT FUEL	SSS MFR	TURB MFR	GEN MFR
OPERATOR: BC HYDRO										
SPILLMACHEEN 3	2.20	OPR	1955	HY	WAT	CONV	NONE	N/A	EE	EE
STAVE FALLS 1	10.50	OPR	1912	HY	WAT	CONV	NONE	N/A	EW	CGE
STAVE FALLS 2	10.50	OPR	1912	HY	WAT	CONV	NONE	N/A	EW	CGE
STAVE FALLS 3	10.50	OPR	1916	HY	WAT	CONV	NONE	N/A	EW	CGE
STAVE FALLS 4	10.50	OPR	1922	HY	WAT	CONV	NONE	N/A	EW	CGE
STAVE FALLS 5	10.50	OPR	1925	HY	WAT	CONV	NONE	N/A	ACC	CGE
STEWART IC1	1.00	OPR	1964	IC	OIL	DIESEL	NONE	N/A	GM	GM
STEWART IC2	0.35	UNK	1964	IC	OIL	DIESEL	NONE	N/A	CAT	COEL
STEWART IC3	0.50	OPR	1965	IC	OIL	DIESEL	NONE	N/A	CAT	KATO
STEWART IC4	0.50	OPR	1965	IC	OIL	DIESEL	NONE	N/A	CAT	COEL
STEWART IC5	1.14	UNK	1968	IC	OIL	DIESEL	NONE	N/A	FM	FM
STEWART IC6	2.50	OPR	1975	IC	OIL	DIESEL	NONE	N/A	GM	GM
STEWART IC7	0.50	OPR	1972	IC	OIL	DIESEL	NONE	N/A		
STRATHCONA 1	33.75	OPR	1958	HY	WAT	CONV	NONE	N/A	ACC	WHC
STRATHCONA 2	33.75	OPR	1968	HY	WAT	CONV	NONE	N/A	TS	WHC
TATLA LAKE IC3	0.15	UNK	1975	IC	OIL	DIESEL	NONE	N/A	DD	KATO
TATLA LAKE IC4	0.15	UNK	1975	IC	OIL	DIESEL	NONE	N/A	DD	KATO
TELEGRAPH CREEK IC1	0.15	UNK	1969	IC	OIL	DIESEL	NONE	N/A	CAT	KATO
TELEGRAPH CREEK IC2	0.15	UNK	1969	IC	OIL	DIESEL	NONE	N/A	CAT	KATO
TELEGRAPH CREEK IC3	0.25	OPR	1972	IC	OIL	DIESEL	NONE	N/A	GM	KATO
TELEGRAPH CREEK IC4	0.25	OPR	1972	IC	OIL	DIESEL	NONE	N/A	GM	KATO
TELEGRAPH CREEK IC5	0.35	OPR	1976	IC	OIL	DIESEL	NONE	N/A	CAT	C-B
WAHLEACH 1	60.00	OPR	1952	HY	WAT	CONV	NONE	N/A	VIW	CGE
WALTER HARDMAN 1	4.00	OPR	1960	HY	WAT	CONV	NONE	N/A	GGG	CGE
WALTER HARDMAN 2	4.00	OPR	1965	HY	WAT	CONV	NONE	N/A	GGG	CGE
WHATSHAN 1	50.00	OPR	1972	HY	WAT	CONV	NONE	N/A	FUJI	HT
OPERATOR: BC PACKERS LTD										
NAMU IC1	0.24	OPR	1962	IC	OIL	DIESEL	NONE	N/A	GM	ENG
NAMU IC2	0.24	OPR	1962	IC	OIL	DIESEL	NONE	N/A	GM	ENG
NAMU IC3	0.24	OPR	1962	IC	OIL	DIESEL	NONE	N/A	GM	ENG

DIRECTORY OF POWER PLANTS IN CANADA
UDI/MCGRAW-HILL

PLANT/UNIT NAME	MW	STA	YEAR	TYPE	FUEL	FUEL TYPE	ALT FUEL	SSS MFR	TURB MFR	GEN MFR
OPERATOR: BC PACKERS LTD										
NAMU IC4	0.24	OPR	1962	IC	OIL	DIESEL	NONE	N/A	GM	ENG
NAMU IC5	0.24	OPR	1963	IC	OIL	DIESEL	NONE	N/A	GM	ENG
NAMU IC6	0.24	OPR	1963	IC	OIL	DIESEL	NONE	N/A	GM	ENG
OPERATOR: BC SUGAR										
VANCOUVER 1	1.25	OPR	1947	ST	GAS		NONE	BWGM	WH	WH
VANCOUVER 2	1.25	OPR	1947	ST	GAS		NONE	BWGM	WH	WH
VANCOUVER 3	3.00	OPR	1974	ST	GAS		NONE	BWGM	PT	GE
OPERATOR: BC SUGAR REFINING COMPANY LTD										
FORT GARRY 1	1.50	OPR	1940	ST/S	GAS		NONE	FW	EL	EL
FORT GARRY 2	2.50	OPR	1953	ST/S	GAS		NONE	FW	BBC	BBC
OPERATOR: BELLETERRE HYDRO ELEC COMM										
WINNEWAY 1	1.17	OPR	1938	HY	WAT	CONV	NONE	N/A	ACC	EE
WINNEWAY 2	1.17	OPR	1942	HY	WAT	CONV	NONE	N/A	ACC	EE
OPERATOR: BJ HARGROVE LTD										
HARGROVE 1	0.15	OPR	1970	HY	WAT	CONV	NONE	N/A	BARBER	WH
HARGROVE 2	0.35	OPR	1978	HY	WAT	CONV	NONE	N/A	BARBER	EE
OPERATOR: BLACK RIVER HYDRO										
MELFORD BROOK	0.23	PLN	1994	HY	WAT	CONV	NONE	N/A		
OPERATOR: BOISE CASCADE CANADA LTD										
CALM LAKE 1	4.68	OPR	1928	HY	WAT	CONV	NONE	N/A	SMS	WHC
CALM LAKE 2	4.68	OPR	1928	HY	WAT	CONV	NONE	N/A	SMS	WHC
FORT FRANCES 1	1.60	OPR	1955	HY	WAT	CONV	NONE	N/A	CVIC	CGE
FORT FRANCES 2	1.60	OPR	1955	HY	WAT	CONV	NONE	N/A	CVIC	CGE
FORT FRANCES 3	1.60	OPR	1955	HY	WAT	CONV	NONE	N/A	CVIC	CGE
FORT FRANCES 4	1.60	OPR	1955	HY	WAT	CONV	NONE	N/A	CVIC	CGE
FORT FRANCES 5	1.60	OPR	1955	HY	WAT	CONV	NONE	N/A	CVIC	CGE

DIRECTORY OF POWER PLANTS IN CANADA
UDI/MCGRAW-HILL

PLANT/UNIT NAME	MW	STA	YEAR	TYPE	FUEL	FUEL TYPE	ALT FUEL	SSS MFR	TURB MFR	GEN MFR
OPERATOR: BOISE CASCADE CANADA LTD										
FORT FRANCES 6	1.60	OPR	1955	HY	WAT	CONV	NONE	N/A	CVIC	CGE
FORT FRANCES 7	1.60	OPR	1955	HY	WAT	CONV	NONE	N/A	CVIC	CGE
FORT FRANCES 8	1.60	OPR	1955	HY	WAT	CONV	NONE	N/A	CVIC	CGE
KENORA 01	1.00	OPR	1923	HY	WAT	CONV	NONE	N/A	SMS	EMC
KENORA 02	1.25	OPR	1923	HY	WAT	CONV	NONE	N/A	SMS	EMC
KENORA 03	1.25	OPR	1923	HY	WAT	CONV	NONE	N/A	SMS	EMC
KENORA 04	1.00	OPR	1923	HY	WAT	CONV	NONE	N/A	SMS	EMC
KENORA 05	1.00	OPR	1923	HY	WAT	CONV	NONE	N/A	SMS	EMC
KENORA 06	1.25	OPR	1923	HY	WAT	CONV	NONE	N/A	SMS	EMC
KENORA 07	1.25	OPR	1924	HY	WAT	CONV	NONE	N/A	SMS	EMC
KENORA 08	1.00	OPR	1924	HY	WAT	CONV	NONE	N/A	SMS	EMC
KENORA 09	1.25	OPR	1924	HY	WAT	CONV	NONE	N/A	SMS	EMC
KENORA 10	1.25	OPR	1924	HY	WAT	CONV	NONE	N/A	SMS	EMC
NORMAN 1	3.30	OPR	1925	HY	WAT	CONV	NONE	N/A	SMS	WHC
NORMAN 2	3.30	OPR	1925	HY	WAT	CONV	NONE	N/A	SMS	WHC
NORMAN 3	3.30	OPR	1925	HY	WAT	CONV	NONE	N/A	SMS	WHC
NORMAN 4	3.30	OPR	1925	HY	WAT	CONV	NONE	N/A	SMS	WHC
NORMAN 5	3.30	OPR	1925	HY	WAT	CONV	NONE	N/A	SMS	WHC
STURGEON FALLS 1	3.83	OPR	1927	HY	WAT	CONV	NONE	N/A	SMS	WHC
STURGEON FALLS 2	3.83	OPR	1927	HY	WAT	CONV	NONE	N/A	SMS	WHC
OPERATOR: BOWATERS MERSEY PAPER CO LTD										
BROOKLYN 1	5.17	OPR	1968	ST/S	OIL	HFO	NONE	BW	FC	GEC
BROOKLYN IC2	1.50	OPR	1988	IC	OIL	DIST	NONE	N/A		
OPERATOR: BPCO INC										
EDMONTON (BPCO) 1	1.13	OPR	1954	ST	GAS		NONE	WWT/TI	GE	GE
OPERATOR: BRACEBRIDGE HYDRO										
BRACEBRIDGE FALLS 1	0.30	OPR	1900	HY	WAT	CONV	NONE	N/A	COOP	CE
BRACEBRIDGE FALLS 2	0.30	OPR	1904	HY	WAT	CONV	NONE	N/A	COOP	CGE

UDI-033-92   December 1992

DIRECTORY OF POWER PLANTS IN CANADA
UDI/MCGRAW-HILL

PLANT/UNIT NAME	MW	STA	YEAR	TYPE	FUEL	FUEL TYPE	ALT FUEL	SSS MFR	TURB MFR	GEN MFR
OPERATOR: BRACEBRIDGE HYDRO										
HIGH FALLS (BH) 1	0.80	OPR	1948	HY	WAT	CONV	NONE	N/A	COOP	CGE
WILSONS FALLS 1	0.60	OPR	1978	HY	WAT	CONV	NONE	N/A	WK	CGE
OPERATOR: BRITISH GAS/ATLANTIC PACKAGING										
ATLANTIC PACKAGING GT1	40.00	PLN	1995	GT/C	GAS		NONE	N/A	GE	GE
ATLANTIC PACKAGING SC1	20.00	PLN	1995	ST/C	WSTH		NONE			
OPERATOR: CABOT ENERGY CORP										
PORT HAWKESBURY (CE)	150.00	PLN	1996	ST/H	COAL		NONE			
OPERATOR: CALGARY ELECTRIC SYSTEM										
CALGARY IC1	2.75	OPR	1967	IC	OIL	DIST	NONE	N/A	GM	GM
CALGARY IC2	2.75	OPR	1967	IC	OIL	DIST	NONE	N/A	GM	GM
OPERATOR: CAMERON FALLS POWER CORP										
CAMERON FALLS 1	4.35	OPR	1991	HY	WAT	CONV	NONE	N/A	VA	IDEAL
OPERATOR: CAMPBELLFORD HYDRO										
CROW BAY 1	0.90	OPR	1981	HY	WAT	CONV	NONE	N/A	BARBER	AC
CROW BAY 2	1.18	OPR	1912	HY	WAT	CONV	NONE	N/A	GE	CGE
OPERATOR: CANADA TUNGSTEN MINING CORP										
TUNGSTEN IC01	0.50	OPR	1962	IC	OIL	DIESEL	NONE	N/A	CAT	EMC
TUNGSTEN IC02	0.50	OPR	1962	IC	OIL	DIESEL	NONE	N/A	CAT	EMC
TUNGSTEN IC03	0.50	OPR	1962	IC	OIL	DIESEL	NONE	N/A	CAT	EMC
TUNGSTEN IC04	0.60	OPR	1971	IC	OIL	DIESEL	NONE	N/A	CAT	EMC
TUNGSTEN IC05	0.80	OPR	1973	IC	OIL	DIESEL	NONE	N/A	CAT	TAMPER
TUNGSTEN IC06	0.60	OPR	1974	IC	OIL	DIESEL	NONE	N/A	CAT	GE
TUNGSTEN IC07	0.60	OPR	1974	IC	OIL	DIESEL	NONE	N/A	CAT	GE
TUNGSTEN IC08	0.60	OPR	1975	IC	OIL	DIESEL	NONE	N/A	CAT	TAMPER
TUNGSTEN IC09	2.50	OPR	1979	IC	OIL	DIESEL	NONE	N/A	HAWK	BRUSH
TUNGSTEN IC10	2.50	OPR	1979	IC	OIL	DIESEL	NONE	N/A	HAWK	BRUSH

DIRECTORY OF POWER PLANTS IN CANADA
UDI/MCGRAW-HILL

PLANT/UNIT NAME	MW	STA	YEAR	TYPE	FUEL	FUEL TYPE	ALT FUEL	SSS MFR	TURB MFR	GEN MFR
OPERATOR: CANADIAN CREW ENERGY DEVELOP										
MEAGER CREEK 1	60.00	PLN	1995	ST	GEO		NONE	N/A		
OPERATOR: CANADIAN FOREST PRODUCTS LTD										
ENGLEWOOD IC1	0.25	OPR	1969	IC	OIL	DIESEL	NONE	N/A	CAT	KATO
ENGLEWOOD IC2	0.60	OPR	1973	IC	OIL	DIESEL	NONE	N/A	FINN	KATO
ENGLEWOOD IC3	0.50	OPR	1975	IC	OIL	DIESEL	NONE	N/A	GM	CANR
ENGLEWOOD IC4	0.25	OPR	1976	IC	OIL	DIESEL	NONE	N/A	CAT	KATO
ENGLEWOOD IC5	0.25	OPR	1977	IC	OIL	DIESEL	NONE	N/A	CAT	WORT
OPERATOR: CANADIAN GENERAL ELECTRIC CO										
PETERBOROUGH (GE) 1	2.00	OPR	1931	ST	GAS		NONE	CE	GE	GE
OPERATOR: CANADIAN HYDRO DEVELOPERS INC										
BELLY RIVER 1	3.00	OPR	1991	HY	WAT	CONV	NONE	N/A	HYDROL	HYDROL
ST MARY DAM 1	2.30	OPR	1992	HY	WAT	CONV	NONE	N/A	NEY	KATO
WATERTON 1	2.80	OPR	1992	HY	WAT	CONV	NONE	N/A	NEY	KATO
OPERATOR: CANADIAN NIAGARA POWER CO LTD										
WB RANKINE 01	7.50	OPR	1904	HY	WAT	CONV	NONE	N/A	GE	CGE
WB RANKINE 02	7.50	OPR	1904	HY	WAT	CONV	NONE	N/A	GE	CGE
WB RANKINE 03	7.50	OPR	1905	HY	WAT	CONV	NONE	N/A	GE	CGE
WB RANKINE 04	7.50	OPR	1906	HY	WAT	CONV	NONE	N/A	GE	CGE
WB RANKINE 05	7.50	OPR	1906	HY	WAT	CONV	NONE	N/A	GE	CGE
WB RANKINE 06	9.38	OPR	1910	HY	WAT	CONV	NONE	N/A	WHC	WHC
WB RANKINE 07	9.38	OPR	1913	HY	WAT	CONV	NONE	N/A	WHC	WHC
WB RANKINE 08	9.38	OPR	1916	HY	WAT	CONV	NONE	N/A	WHC	WHC
WB RANKINE 09	9.38	OPR	1916	HY	WAT	CONV	NONE	N/A	WHC	WHC
WB RANKINE 10	9.38	OPR	1917	HY	WAT	CONV	NONE	N/A	WHC	WHC
WB RANKINE 11	9.38	OPR	1924	HY	WAT	CONV	NONE	N/A	WHC	WHC

DIRECTORY OF POWER PLANTS IN CANADA
UDI/MCGRAW-HILL

PLANT/UNIT NAME	MW	STA	YEAR	TYPE	FUEL	FUEL TYPE	ALT FUEL	SSS MFR	TURB MFR	GEN MFR
OPERATOR: CANADIAN SALT COMPANY LTD										
LINDBERGH 1	0.96	OPR	1958	ST	GAS		NONE	FW	BBC	BBC
LINDBERGH 2	0.60	OPR	1964	ST	GAS		NONE	FW	GE	GE
OPERATOR: CAPE BRETON COUNTY										
SYDNEY WTE	0.48	PLN	1993	ST	REF		NONE			
OPERATOR: CARIBOO PULP & PAPER CO										
QUESNEL 1	28.00	OPR	1972	ST	LIQ		NONE	BW/FW	TS	TS
OPERATOR: CARMICHAEL POWER COMPANY										
CARMICHAEL FALLS II 1	18.70	OPR	1991	HY	WAT	CONV	NONE	N/A		
OPERATOR: CASSIAR MINING CORP										
CASSIAR IC01	0.98	OPR	1964	IC	OIL	DIESEL	NONE	N/A	MIRR	GE
CASSIAR IC02	1.40	OPR	1971	IC	OIL	DIESEL	NONE	N/A	RHN	BRUSH
CASSIAR IC03	1.40	OPR	1972	IC	OIL	DIESEL	NONE	N/A	RHN	BRUSH
CASSIAR IC04	1.40	OPR	1973	IC	OIL	DIESEL	NONE	N/A	RHN	BRUSH
CASSIAR IC05	1.40	OPR	1974	IC	OIL	DIESEL	NONE	N/A	RHN	BRUSH
CASSIAR IC06	1.40	OPR	1975	IC	OIL	DIESEL	NONE	N/A	RHN	BRUSH
CASSIAR IC07	1.40	OPR	1976	IC	OIL	DIESEL	NONE	N/A	RHN	BRUSH
CASSIAR IC08	1.40	OPR	1978	IC	OIL	DIESEL	NONE	N/A	RHN	BRUSH
CASSIAR IC09	1.40	OPR	1979	IC	OIL	DIESEL	NONE	N/A	RHN	BRUSH
CASSIAR IC10	1.40	OPR	1979	IC	OIL	DIESEL	NONE	N/A	RHN	BRUSH
CASSIAR IC11	0.60	OPR	1981	IC	OIL	DIESEL	NONE	N/A	CAT	CANR
CASSIAR IC12	1.50	OPR	1985	IC	OIL	DIESEL	NONE	N/A	CAT	IDEAL
OPERATOR: CELANESE CANADA LTD										
CLOVER BAR PLANT 1	6.60	OPR	1953	ST	GAS		NONE	FW/BW	WH	WH
CLOVER BAR PLANT 2	6.60	OPR	1953	ST	GAS		NONE	FW/BW	WH	WH
CLOVER BAR PLANT 3	6.60	OPR	1953	ST	GAS		NONE	FW/BW	WH	WH
DRUMMONDVILLE (CCL) 1	1.50	OPR	1935	ST/S	OIL	HFO	NONE	BW/CE	PAR	PAR

DIRECTORY OF POWER PLANTS IN CANADA
UDI/MCGRAW-HILL

PLANT/UNIT NAME	MW	STA	YEAR	TYPE	FUEL	FUEL TYPE	ALT FUEL	SSS MFR	TURB MFR	GEN MFR
OPERATOR: CELANESE CANADA LTD										
DRUMMONDVILLE (CCL) 2	2.50	OPR	1950	ST/S	OIL	HFO	NONE	FW	GE	GE
DRUMMONDVILLE (CCL) 3	3.50	OPR	1953	ST/S	OIL	HFO	NONE	CE	GE	GE
OPERATOR: CENTRAL COAST POWER CORP										
OCEAN FALLS 1	1.90	OPR	1917	HY	WAT	CONV	NONE	N/A	PWW	CGE
OCEAN FALLS 2	1.90	OPR	1917	HY	WAT	CONV	NONE	N/A	PWW	CGE
OCEAN FALLS 3	4.20	OPR	1923	HY	WAT	CONV	NONE	N/A	PWW	CGE
OCEAN FALLS 4	4.20	OPR	1932	HY	WAT	CONV	NONE	N/A	PWW	CGE
OPERATOR: CENTRALE SPC INC										
CHICOUTIMI (SPC) 1	32.00	OPR	1956	HY	WAT	CONV	NONE	N/A	SMS	CGE
OPERATOR: CHAPLEAU COGENERATION LTD										
CHAPLEAU	7.17	OPR	1987	ST	WOOD		NONE	FW	DET	DET
OPERATOR: CHINOOK PROJECT INC										
SOUTHERN ALBERTA	9.90	PLN		WTG	WIND	HAWT	NONE	N/A		
OPERATOR: CHURCHILL FALLS LABRADOR CORP										
CHURCHILL FALLS 01	500.00	OPR	1971	HY	WAT	CONV	NONE	N/A	DEW	CGE
CHURCHILL FALLS 02	475.00	OPR	1971	HY	WAT	CONV	NONE	N/A	MIL	MIL
CHURCHILL FALLS 03	500.00	OPR	1972	HY	WAT	CONV	NONE	N/A	DEW	CGE
CHURCHILL FALLS 04	500.00	OPR	1972	HY	WAT	CONV	NONE	N/A	MIL	MIL
CHURCHILL FALLS 05	500.00	OPR	1973	HY	WAT	CONV	NONE	N/A	DEW	CGE
CHURCHILL FALLS 06	503.50	OPR	1973	HY	WAT	CONV	NONE	N/A	MIL	MIL
CHURCHILL FALLS 07	500.00	OPR	1973	HY	WAT	CONV	NONE	N/A	DEW	CGE
CHURCHILL FALLS 08	500.00	OPR	1973	HY	WAT	CONV	NONE	N/A	MIL	MIL
CHURCHILL FALLS 09	500.00	OPR	1974	HY	WAT	CONV	NONE	N/A	DEW	CGE
CHURCHILL FALLS 10	475.00	OPR	1974	HY	WAT	CONV	NONE	N/A	MIL	MIL
CHURCHILL FALLS 11	475.00	OPR	1974	HY	WAT	CONV	NONE	N/A	MIL	MIL

DIRECTORY OF POWER PLANTS IN CANADA
UDI/MCGRAW-HILL

PLANT/UNIT NAME	MW	STA	YEAR	TYPE	FUEL	FUEL TYPE	ALT FUEL	SSS MFR	TURB MFR	GEN MFR
OPERATOR: *CIP INC*										
GOLD RIVER PULP MILL 1	1.50	OPR	1966	ST	BIO		NONE	FW/CE	PAR	PAR
GOLD RIVER PULP MILL 2	27.96	OPR	1982	ST	BIO		NONE	FW/CE	STAL	STAL
OPERATOR: *CLYDE COATS*										
GABRIOLA ISLAND 1	0.13	OPR	1987	HY	WAT	CONV	NONE	N/A	GILK	WH
OPERATOR: *COATICOOK ELECTRIC DEPT*										
COATICOOK 1	0.72	OPR	1927	HY	WAT	CONV	NONE	N/A	WH	EE
COATICOOK 2	0.72	OPR	1927	HY	WAT	CONV	NONE	N/A	WH	EE
PENMAN 1	0.55	OPR	1985	HY	WAT	CONV	NONE	N/A		
SAINT PAUL 1	0.45	OPR	1985	HY	WAT	CONV	NONE	N/A		
OPERATOR: *COBALT POWER COMPANY*										
RAGGED CHUTE 1	3.20	OPR	1991	HY	WAT	CONV	NONE	N/A	DBSZ	
OPERATOR: *COMINCO LTD*										
ARSENIC PLANT IC1	0.12	OPR	1981	IC	OIL	DIESEL	NONE	N/A	DD	BBC
BRILLIANT 1	27.20	OPR	1944	HY	WAT	CONV	NONE	N/A	DEW	WHC
BRILLIANT 2	27.20	OPR	1944	HY	WAT	CONV	NONE	N/A	DEW	WHC
BRILLIANT 3	27.20	OPR	1949	HY	WAT	CONV	NONE	N/A	DEW	WHC
BRILLIANT 4	27.20	OPR	1968	HY	WAT	CONV	NONE	N/A	DEW	WHC
C-1 POWERHOUSE IC1	0.50	OPR	1980	IC	OIL	DIESEL	NONE	N/A	DD	BBC
C-1 POWERHOUSE IC2	0.50	OPR	1980	IC	OIL	DIESEL	NONE	N/A	DD	BBC
C-1 POWERHOUSE IC3	0.50	OPR	1980	IC	OIL	DIESEL	NONE	N/A	DD	BBC
ROBERTSON SHAFT IC1	0.50	OPR	1975	IC	OIL	DIESEL	NONE	N/A	CAT	GE
WANETA 1	72.00	OPR	1954	HY	WAT	CONV	NONE	N/A	DEW	WHC
WANETA 2	72.00	OPR	1954	HY	WAT	CONV	NONE	N/A	DEW	WHC
WANETA 3	72.00	OPR	1963	HY	WAT	CONV	NONE	N/A	DEW	WHC
WANETA 4	76.50	OPR	1966	HY	WAT	CONV	NONE	N/A	ACC	CGE

DIRECTORY OF POWER PLANTS IN CANADA
UDI/MCGRAW-HILL

PLANT/UNIT NAME	MW	STA	YEAR	TYPE	FUEL	FUEL TYPE	ALT FUEL	SSS MFR	TURB MFR	GEN MFR
OPERATOR: CONSOLIDATED-BATHURST LTD										
BATHURST 1	6.00	OPR	1937	ST/S	OIL	HFO	NONE	CE	BBC	BBC
BATHURST 2	7.61	OPR	1946	ST/S	OIL	HFO	NONE	BW	BBC	BBC
BATHURST 3	7.00	OPR	1958	ST/S	OIL	HFO	NONE	BW	GE	GE
GRAND BAIE I 1	0.83	OPR	1917	HY	WAT	CONV	NONE	N/A	SMS	WH
GRAND BAIE II 1	0.46	OPR	1918	HY	WAT	CONV	NONE	N/A	SMS	CGE
GREAT FALLS 1	3.60	OPR	1921	HY	WAT	CONV	NONE	N/A	BV	CGE
GREAT FALLS 2	3.60	OPR	1921	HY	WAT	CONV	NONE	N/A	BV	CGE
GREAT FALLS 3	3.60	OPR	1930	HY	WAT	CONV	NONE	N/A	AC	CGE
OPERATOR: CONWEST EXPLORATION CO LTD										
BLACK RIVER (MARATHON) 1	4.50	OPR	1992	HY	WAT	CONV	NONE	N/A	HYDROL	G-A
BLACK RIVER (MARATHON) 2	4.50	OPR	1992	HY	WAT	CONV	NONE	N/A	HYDROL	G-A
BLACK RIVER (MARATHON) 3	4.50	OPR	1992	HY	WAT	CONV	NONE	N/A	HYDROL	G-A
DRYDEN FALLS	2.60	PLN	9999	HY	WAT	CONV	NONE	N/A		
SALMON INLET	15.00	PLN	9999	HY	WAT	CONV	NONE	N/A		
OPERATOR: COPPER BEACH ESTATES LTD										
BEACH 1	2.00	OPR	1916	HY	WAT	CONV	NONE	N/A	PWW	WHC
BEACH 2	2.00	OPR	1917	HY	WAT	CONV	NONE	N/A	PWW	WHC
OPERATOR: CORNER BROOK PULP & PAPER LTD										
CORNER BROOK 1	6.60	OPR	1956	ST	OIL	HFO	NONE	FW	PAR	PAR
OPERATOR: CRESTBROOK PULP & PAPER LTD										
ALBERTA MILL 1	41.40	OPR	1990	ST/S	UNK				MHI	
SKOOKUMCHUCK 1	15.00	OPR	1968	ST	GAS		NONE	MHI	MHI	MHI
OPERATOR: CROWN FOREST INDUSTRIES LTD										
CAMPBELL RIVER (CFI) 1	0.80	UNK	1964	ST	WOOD		NONE	CE/BW	WH	GE
CAMPBELL RIVER (CFI) 2	3.26	UNK	1965	ST	WOOD		NONE	CE/BW	GE	GE
CAMPBELL RIVER (CFI) 3	25.00	OPR	1981	ST	OIL	HFO	NONE	CE/BW	MHI	MHI

DIRECTORY OF POWER PLANTS IN CANADA
UDI/MCGRAW-HILL

PLANT/UNIT NAME	MW	STA	YEAR	TYPE	FUEL	FUEL TYPE	ALT FUEL	SSS MFR	TURB MFR	GEN MFR
OPERATOR: CROWN FOREST INDUSTRIES LTD										
KELOWNA 1	2.00	OPR	1954	ST	WOOD		NONE	BW	GE	GE
KELOWNA 2	3.50	OPR	1961	ST	WOOD		NONE	BW	AC	AC
KELOWNA 3	1.00	OPR	1963	ST	WOOD		NONE	BW	GE	GE
OPERATOR: DAISHOWA INC										
FORESTVILLE 1	1.00	OPR	1954	HY	WAT	CONV	NONE	N/A	BARBER	EE
OPERATOR: DEER LAKE POWER CO										
DEER LAKE 1	11.28	OPR	1925	HY	WAT	CONV	NONE	N/A	ARG	BTH
DEER LAKE 2	11.31	OPR	1925	HY	WAT	CONV	NONE	N/A	ARG	BTH
DEER LAKE 3	11.31	OPR	1925	HY	WAT	CONV	NONE	N/A	ARG	BTH
DEER LAKE 4	11.28	OPR	1925	HY	WAT	CONV	NONE	N/A	ARG	BTH
DEER LAKE 5	11.31	OPR	1925	HY	WAT	CONV	NONE	N/A	ARG	BTH
DEER LAKE 6	11.28	OPR	1925	HY	WAT	CONV	NONE	N/A	ARG	BTH
DEER LAKE 7	11.28	OPR	1925	HY	WAT	CONV	NONE	N/A	ARG	BTH
DEER LAKE 8	22.80	OPR	1929	HY	WAT	CONV	NONE	N/A	NN	CGE
DEER LAKE 9	22.80	OPR	1929	HY	WAT	CONV	NONE	N/A	NN	CGE
WATSONS BROOK 1	4.60	OPR	1958	HY	WAT	CONV	NONE	N/A	EE	EE
WATSONS BROOK 2	4.60	OPR	1958	HY	WAT	CONV	NONE	N/A	EE	EE
OPERATOR: DESTEC ENERGY/LONG LAKE ENERGY										
KINGSTON	226.00	PLN	1995	CC	GAS		NONE			
OPERATOR: DOMINION TEXTILE INC										
MAGOG (DT) 1	1.00	OPR	1920	HY	WAT	CONV	NONE	N/A	WH	CGE
MAGOG (DT) 2	1.00	OPR	1920	HY	WAT	CONV	NONE	N/A	WH	CGE
OPERATOR: DOMTAR CHEMICALS GROUP										
AMHERST 1	0.70	OPR	1946	ST/S	OIL	HFO	NONE	DB	WORT	EMC
UNITY 1	1.15	OPR	1948	ST	GAS		NONE	FW/MV	WORT	EE

DIRECTORY OF POWER PLANTS IN CANADA
UDI/MCGRAW-HILL

PLANT/UNIT NAME	MW	STA	YEAR	TYPE	FUEL	FUEL TYPE	ALT FUEL	SSS MFR	TURB MFR	GEN MFR
OPERATOR: DOW CHEMICAL OF CANADA LTD										
DOW ALBERTA GT1	99.50	OPR	1979	GT	GAS		NONE	N/A	GE	GE
DOW ALBERTA GT2	99.50	OPR	1979	GT	GAS		NONE	N/A	GE	GE
SARNIA 1	28.80	OPR	1963	ST	GAS		NONE	FW/BW	WHC	WHC
SARNIA 2	28.80	OPR	1963	ST	GAS		NONE	FW/BW	WHC	WHC
SARNIA GT1	54.40	OPR	1972	GT	GAS		NONE	N/A	GE	GE
SARNIA GT2	54.40	OPR	1972	GT	GAS		NONE	N/A	GE	GE
SARNIA GT3	72.25	OPR	1977	GT	GAS		NONE	N/A	BBC	EMC
OPERATOR: DUPONT CANADA INC										
MAITLAND GT1	38.30	CON	1992	GT/S	GAS		NONE	N/A	GE	GE
OPERATOR: DUTCH VALLEY PRODUCE LTD										
DUTCH VALLEY 1	0.07	OPR	1992	WTG	WIND		NONE	N/A		
DUTCH VALLEY 2	0.07	OPR	1992	WTG	WIND		NONE	N/A		
DUTCH VALLEY 3	0.07	OPR	1992	WTG	WIND		NONE	N/A		
OPERATOR: EAST TWIN CREEK HYDRO LTD										
EAST TWIN CREEK 1	1.50	OPR	1991	HY	WAT	CONV	NONE	N/A		
OPERATOR: EASTERN CANADIAN COAL GAS VENT										
LINGAN-PHELAN	4.00	PLN	1993	IC	MGAS		NONE	N/A		
OPERATOR: EASTERN POWER DEVELOPERS INC										
BROCK WEST 1	11.50	OPR	1991	ST	METH		NONE	NB	GE	GE
BROCK WEST 2	11.50	OPR	1991	ST	METH		NONE	NB	GE	GE
OPERATOR: EB EDDY FOREST PRODUCTS LTD										
CHAUDIERE FALLS 1	4.00	OPR	1955	HY	WAT	CONV	NONE	N/A	ACC	CGE
CHAUDIERE FALLS 2	4.00	OPR	1955	HY	WAT	CONV	NONE	N/A	ACC	CGE
CHAUDIERE FALLS 3	3.75	OPR	1955	HY	WAT	CONV	NONE	N/A	ACC	CGE
EDDY 1	3.00	OPR	1909	HY	WAT	CONV	NONE	N/A	SMS	ACB
EDDY 2	3.00	OPR	1909	HY	WAT	CONV	NONE	N/A	SMS	ACB

DIRECTORY OF POWER PLANTS IN CANADA
UDI/MCGRAW-HILL

PLANT/UNIT NAME	MW	STA	YEAR	TYPE	FUEL	FUEL TYPE	ALT FUEL	SSS MFR	TURB MFR	GEN MFR
OPERATOR: EB EDDY FOREST PRODUCTS LTD										
EDDY 3	3.00	OPR	1909	HY	WAT	CONV	NONE	N/A	SMS	ACB
ESPANOLA NEW 1	8.00	PLN	1994	HY	WAT	CONV	NONE	N/A		
OPERATOR: ECHO BAY MINES LTD										
LUPIN MINE IC01	0.60	OPR	1982	IC	OIL	DIESEL	NONE	N/A	CAT	BBC
LUPIN MINE IC02	0.60	OPR	1982	IC	OIL	DIESEL	NONE	N/A	CAT	BBC
LUPIN MINE IC03	0.60	OPR	1982	IC	OIL	DIESEL	NONE	N/A	CAT	BBC
LUPIN MINE IC04	0.60	OPR	1982	IC	OIL	DIESEL	NONE	N/A	CAT	BBC
LUPIN MINE IC05	0.60	OPR	1982	IC	OIL	DIESEL	NONE	N/A	CAT	BBC
LUPIN MINE IC06	0.13	OPR	1982	IC	OIL	DIESEL	NONE	N/A	CAT	STM
LUPIN MINE IC07	0.30	OPR	1982	IC	OIL	DIESEL	NONE	N/A	CUM	BBC
LUPIN MINE IC08	0.60	OPR	1982	IC	OIL	DIESEL	NONE	N/A	CAT	TAMPER
LUPIN MINE IC09	0.60	OPR	1982	IC	OIL	DIESEL	NONE	N/A	CAT	TAMPER
LUPIN MINE IC10	1.87	OPR	1982	IC	OIL	DIESEL	NONE	N/A	RHN	LA
LUPIN MINE IC11	1.87	OPR	1982	IC	OIL	DIESEL	NONE	N/A	RHN	LA
LUPIN MINE IC12	1.87	OPR	1982	IC	OIL	DIESEL	NONE	N/A	RHN	LA
LUPIN MINE IC13	2.50	OPR	1982	IC	OIL	DIESEL	NONE	N/A	GM	GM
OPERATOR: EDMONTON POWER										
CLOVER BAR 1	165.00	OPR	1970	ST	GAS		METH	BW	EW	OERL
CLOVER BAR 2	165.00	OPR	1973	ST	GAS		METH	BW	EW	OERL
CLOVER BAR 3	165.00	OPR	1977	ST	GAS		METH	BW	HT	HT
CLOVER BAR 4	165.00	OPR	1979	ST	GAS		METH	BW	HT	HT
GENESEE 1	406.00	OPR	1989	ST	COAL	SUB	NONE	CE	GEC	GEC
GENESEE 2	406.00	CON	1994	ST	COAL	SUB	NONE	CE	GEC	GEC
ROSSDALE 1	15.00	OPR	1944	ST	GAS		NONE	BW	PAR	PAR
ROSSDALE 2	30.00	OPR	1949	ST	GAS		NONE	BW	PAR	PAR
ROSSDALE 3	30.00	OPR	1953	ST	GAS		NONE	BW	PAR	PAR
ROSSDALE 4	30.00	OPR	1955	ST	GAS		NONE	BW	BBC	BBC
ROSSDALE 5	75.00	OPR	1960	ST	GAS		NONE	BW	BBC	BBC
ROSSDALE 6	75.00	OPR	1963	ST	GAS		NONE	BW	PAR	PAR

DIRECTORY OF POWER PLANTS IN CANADA
UDI/MCGRAW-HILL

PLANT/UNIT NAME	MW	STA	YEAR	TYPE	FUEL	FUEL TYPE	ALT FUEL	SSS MFR	TURB MFR	GEN MFR
OPERATOR: EDMONTON POWER										
ROSSDALE 7	75.00	OPR	1966	ST	GAS		NONE	BW	PAR	PAR
OPERATOR: EDMUNSTON ELECTRIC DEPT										
GREEN RIVER 1	0.90	OPR	1930	HY	WAT	CONV	NONE	N/A	ACC	WH
GREEN RIVER 2	1.00	OPR	1984	HY	WAT	CONV	NONE	N/A	AC	WH
GREEN RIVER 3	1.00	OPR	1984	HY	WAT	CONV	NONE	N/A	AC	WH
OPERATOR: ELDORADO RESOURCES LTD										
ELDOR MINES 1	7.50	UNK	1961	HY	WAT	CONV	NONE	N/A	AC	WH
OPERATOR: ENERFOR CORP										
KEDGWICK 1	25.00	PLN	1996	ST	WOOD		NONE			
SUSSEX 1	25.00	PLN	1996	ST	WOOD		NONE			
OPERATOR: ENVIRONMENTAL TECHNOLOGIES INC										
KIRKLAND LANDFILL	3.70	PLN		IC	METH		NONE	N/A		
SACKVILLE	3.80	CON	1992	IC	METH		NONE	N/A		
OPERATOR: ESSO RESOURCES CANADA LTD										
NORMAN WELLS (ERC) IC1	6.50	OPR	1984	GT	GAS		NONE	N/A		
NORMAN WELLS (ERC) IC2	6.50	OPR	1984	GT	GAS		NONE	N/A		
NORMAN WELLS (ERC) IC3	6.50	OPR	1984	GT	GAS		NONE	N/A		
OPERATOR: EVANS PRODUCTS COMPANY LTD										
GOLDEN 1	7.50	OPR	1946	ST	WOOD		NONE	BW	PAR	PAR
OPERATOR: FER ET TITANE DU QUEBEC INC										
HAVRE ST PIERRE IC1	1.00	OPR	1963	IC	OIL	DIST	NONE	N/A	GM	GM
HAVRE ST PIERRE IC2	1.00	OPR	1963	IC	OIL	DIST	NONE	N/A	GM	GM
HAVRE ST PIERRE IC3	0.50	OPR	1975	IC	OIL	DIST	NONE	N/A	CAT	CAT
HAVRE ST PIERRE IC4	0.50	OPR	1975	IC	OIL	DIST	NONE	N/A	CAT	CAT
HAVRE ST PIERRE IC5	0.35	OPR	1979	IC	OIL	DIST	NONE	N/A	CAT	BBC

DIRECTORY OF POWER PLANTS IN CANADA
UDI/MCGRAW-HILL

PLANT/UNIT NAME	MW	STA	YEAR	TYPE	FUEL	FUEL TYPE	ALT FUEL	SSS MFR	TURB MFR	GEN MFR
OPERATOR: FOOTHILLS HOSPITAL										
FOOTHILLS HOSPITAL 1	1.00	OPR	1966	ST	GAS		NONE	FW	WH	WH
FOOTHILLS HOSPITAL 2	1.00	OPR	1966	ST	GAS		NONE	FW	WH	WH
FOOTHILLS HOSPITAL 3	6.00	OPR	1971	ST	GAS		NONE	BW	STAL	ASEA
FOOTHILLS HOSPITAL 4	10.00	OPR	1980	ST	GAS		NONE	TIW	STAL	ASEA
OPERATOR: FORCES MOTRICES ST-FRANCOIS										
EAST-ANGUS	1.70	PLN	1996	HY	WAT	CONV	NONE	N/A		
OPERATOR: FORDING COAL LTD										
BROOKS 1	400.00	PLN	2000	ST	COAL	SUB	NONE			
BROOKS 2	400.00	PLN	2000	ST	COAL	SUB	NONE			
OPERATOR: FRASER LTD										
ATHOLVILLE 1	5.00	OPR	1956	ST/S	OIL	HFO	NONE	BW	BBC	BBC
ATHOLVILLE 2	19.20	OPR	1983	ST/S	OIL	HFO	NONE	BW	ASEA	ASEA
EDMUNSTON 1	3.80	OPR	1947	ST/S	OIL	HFO	NONE	CE	BBC	BBC
EDMUNSTON 2	12.50	OPR	1958	ST/S	OIL	HFO	NONE	CE	WH	WH
EDMUNSTON H1	1.00	OPR	1918	HY	WAT	CONV	NONE	N/A	WH	CGE
EDMUNSTON H2	1.00	OPR	1918	HY	WAT	CONV	NONE	N/A	WH	CGE
OPERATOR: FW TAYLOR LUMBER CO										
MIDDLE MUSQUODOBOIT	2.00	PLN	1994	ST	WOOD		NONE			
OPERATOR: GANANOQUE LIGHT & POWER LTD										
BREWERS MILLS 1	0.30	OPR	1940	HY	WAT	CONV	NONE	N/A	WH	CGE
BREWERS MILLS 2	0.30	OPR	1940	HY	WAT	CONV	NONE	N/A	WH	CGE
BREWERS MILLS 3	0.30	OPR	1940	HY	WAT	CONV	NONE	N/A	WH	CGE
GANANOQUE 1	0.60	OPR	1939	HY	WAT	CONV	NONE	N/A	WH	CGE
JONES FALLS 1	0.18	OPR	1948	HY	WAT	CONV	NONE	N/A	ACC	CGE
JONES FALLS 2	0.80	OPR	1948	HY	WAT	CONV	NONE	N/A	BARBER	CGE
JONES FALLS 3	0.80	OPR	1950	HY	WAT	CONV	NONE	N/A	BARBER	CGE

DIRECTORY OF POWER PLANTS IN CANADA
UDI/MCGRAW-HILL

PLANT/UNIT NAME	MW	STA	YEAR	TYPE	FUEL	FUEL TYPE	ALT FUEL	SSS MFR	TURB MFR	GEN MFR
OPERATOR: GANANOQUE LIGHT & POWER LTD										
JONES FALLS 4	0.80	OPR	1950	HY	WAT	CONV	NONE	N/A	BARBER	CGE
KINGSTON MILLS 1	0.60	OPR	1914	HY	WAT	CONV	NONE	N/A	ACC	CGE
KINGSTON MILLS 2	0.80	OPR	1926	HY	WAT	CONV	NONE	N/A	BV	CGE
KINGSTON MILLS 3	0.50	OPR	1977	HY	WAT	CONV	NONE	N/A	LF	WH
STATION SIX IC1	1.36	OPR	1959	IC	GAS		NONE	N/A	MBD	BRUSH
STATION SIX IC2	1.36	OPR	1959	IC	GAS		NONE	N/A	MBD	BRUSH
STATION SIX IC3	1.25	OPR	1967	IC	GAS		NONE	N/A	NBG	WH
STATION SIX IC4	1.20	OPR	1967	IC	GAS		NONE	N/A	COOP	EE
STATION SIX IC5	0.60	OPR	1978	IC	GAS		NONE	N/A	CAT	GE
STATION SIX IC6	2.25	OPR	1989	IC	GAS		NONE	N/A		
WASHBURN 1	0.15	OPR	1984	HY	WAT	CONV	NONE	N/A	BV	CGE
OPERATOR: GRAND RIVER CONSERVATION AUTH										
CONESTOGO DAM 1	0.45	OPR	1991	HY	WAT	CONV	NONE	N/A		
OPERATOR: GREAT LAKES FOREST PRODUCTS										
FORT WILLIAM 2	17.10	OPR	1963	ST	GAS		NONE	CE	SS	SS
FORT WILLIAM 3	25.47	OPR	1974	ST	GAS		NONE	CE	STAL	ASEA
FORT WILLIAM 4	34.00	OPR	1975	ST	GAS		NONE	CE	STAL	ASEA
OPERATOR: GREAT LAKES POWER INC										
ANDREWS FALLS 1	8.10	OPR	1938	HY	WAT	CONV	NONE	N/A	SMS	CGE
ANDREWS FALLS 2	8.10	OPR	1942	HY	WAT	CONV	NONE	N/A	SMS	CGE
ANDREWS FALLS 3	22.50	OPR	1975	HY	WAT	CONV	NONE	N/A	DEW	CGE
CARMICHAEL FALLS I 1	3.20	OPR	1991	HY	WAT	CONV	NONE	N/A	DBSZ	
CARMICHAEL FALLS I 2	3.20	OPR	1991	HY	WAT	CONV	NONE	N/A	DBSZ	
CLERGUE 1	18.20	OPR	1982	HY	WAT	CONV	NONE	N/A	AC	CGE
CLERGUE 2	18.20	OPR	1982	HY	WAT	CONV	NONE	N/A	AC	CGE
CLERGUE 3	18.20	OPR	1982	HY	WAT	CONV	NONE	N/A	AC	CGE
GARTSHORE FALLS 1	20.00	OPR	1958	HY	WAT	CONV	NONE	N/A	DEW	WHC
HIGH FALLS (GLP) 1	6.75	OPR	1929	HY	WAT	CONV	NONE	N/A	SMS	CGE

DIRECTORY OF POWER PLANTS IN CANADA
UDI/MCGRAW-HILL

PLANT/UNIT NAME	MW	STA	YEAR	TYPE	FUEL	FUEL TYPE	ALT FUEL	SSS MFR	TURB MFR	GEN MFR
OPERATOR: _GREAT LAKES POWER INC_										
HIGH FALLS (GLP) 2	6.75	OPR	1930	HY	WAT	CONV	NONE	N/A	SMS	CGE
HIGH FALLS (GLP) 3	9.68	OPR	1950	HY	WAT	CONV	NONE	N/A	SMS	CGE
HOGG 1	15.00	OPR	1964	HY	WAT	CONV	NONE	N/A	ACC	CGE
HOLLINGSWORTH FALLS 1	20.00	OPR	1959	HY	WAT	CONV	NONE	N/A	DEW	CGE
MACKAY 1	9.00	OPR	1937	HY	WAT	CONV	NONE	N/A	SMS	CGE
MACKAY 2	9.00	OPR	1941	HY	WAT	CONV	NONE	N/A	SMS	CGE
MACKAY 3	22.50	OPR	1957	HY	WAT	CONV	NONE	N/A	SMS	CGE
MAGPIE (GLP) 1	2.65	OPR	1990	HY	WAT	CONV	NONE	N/A	DBSZ	
MCPHAIL FALLS 1	5.00	OPR	1954	HY	WAT	CONV	NONE	N/A	SMS	CGE
MCPHAIL FALLS 2	5.00	OPR	1954	HY	WAT	CONV	NONE	N/A	SMS	CGE
MISSION FALLS 1	2.65	OPR	1990	HY	WAT	CONV	NONE	N/A	DBSZ	
SCOTT FALLS 1	6.80	OPR	1952	HY	WAT	CONV	NONE	N/A	SMS	CGE
SCOTT FALLS 2	6.80	OPR	1952	HY	WAT	CONV	NONE	N/A	SMS	CGE
STEEPHILL FALLS 1	2.65	OPR	1990	HY	WAT	CONV	NONE	N/A	DBSZ	
OPERATOR: _GULF CANADA RESOURCES INC_										
RIMBEY 1	1.00	OPR	1961	ST	GAS		NONE	CE	WHC	WHC
RIMBEY 2	1.00	OPR	1961	ST	GAS		NONE	CE	WHC	WHC
RIMBEY 3	1.00	OPR	1961	ST	GAS		NONE	CE	WHC	WHC
RIMBEY 4	1.00	OPR	1963	ST	GAS		NONE	BW	WHC	WHC
OPERATOR: _HART JAUNE POWER CO_										
FIFTY FOOT FALLS 1	16.15	OPR	1960	HY	WAT	CONV	NONE	N/A	EE	WHC
FIFTY FOOT FALLS 2	16.15	OPR	1960	HY	WAT	CONV	NONE	N/A	EE	WHC
FIFTY FOOT FALLS 3	16.15	OPR	1960	HY	WAT	CONV	NONE	N/A	EE	WHC
OPERATOR: _HEINZ CORP_										
LEAMINGTON GT1	3.71	OPR	1990	GT/S	GAS		NONE	N/A	ALL	
LEAMINGTON GT2	3.71	OPR	1990	GT/S	GAS		NONE	N/A	ALL	

UDI-033-92   December 1992

DIRECTORY OF POWER PLANTS IN CANADA
UDI/MCGRAW-HILL

PLANT/UNIT NAME	MW	STA	YEAR	TYPE	FUEL	FUEL TYPE	ALT FUEL	SSS MFR	TURB MFR	GEN MFR
OPERATOR: HIRAM WALKER & SON LTD										
WALKERVILLE 3	2.50	OPR	1956	ST	GAS		NONE	BW	GE	GE
WALKERVILLE 4	5.00	OPR	1970	ST	GAS		NONE	BW	GE	GE
OPERATOR: HOWE SOUND PULP & PAPER LTD										
PORT MELLON 2	3.00	OPR	1947	ST	LIQ		NONE	CE/BW	WH	WH
PORT MELLON B1	62.50	OPR	1990	ST	LIQ		WOOD		MHI	
PORT MELLON B2	50.00	CON	1991	ST	LIQ		WOOD			
OPERATOR: HUDSON BAY MINING & SMELTING										
FLIN FLON 1	6.00	OPR	1951	ST	OIL	HFO	NONE	BW	GE	GE
FLIN FLON 2	15.00	OPR	1976	ST	OIL	HFO	NONE	BW	AC	AC
SNOW LAKE IC1	0.93	OPR	1980	IC	OIL	DIESEL	NONE	N/A	CANR	CANR
SNOW LAKE IC2	0.08	OPR	1980	IC	OIL	DIESEL	NONE	N/A	GM	BBC
SNOW LAKE IC3	0.08	OPR	1980	IC	OIL	DIESEL	NONE	N/A	GM	BBC
SPRUCE POINT IC1	0.60	OPR	1980	IC	OIL	DIESEL	NONE	N/A	EE	TAMPER
SPRUCE POINT IC2	0.60	OPR	1980	IC	OIL	DIESEL	NONE	N/A	EE	TAMPER
SPRUCE POINT IC3	0.93	OPR	1980	IC	OIL	DIESEL	NONE	N/A	EE	TAMPER
SPRUCE POINT IC4	0.93	OPR	1983	IC	OIL	DIESEL	NONE	N/A	EE	TAMPER
OPERATOR: HYDRO-FRASER INC										
RIVIERE-DU-LOUP	1.50	PLN	1996	HY	WAT	CONV	NONE	N/A		
OPERATOR: HYDRO-QUEBEC										
AKUILVIK IC1	0.18	UNK	1981	IC	OIL	DIESEL	NONE	N/A	CAT	BBC
AKUILVIK IC2	0.18	UNK	1981	IC	OIL	DIESEL	NONE	N/A	CAT	BBC
AKUILVIK IC3	0.25	OPR	1984	IC	OIL	DIESEL	NONE	N/A	CAT	LIMA
AKUILVIK IC4	0.30	OPR	1988	IC	OIL	DIESEL	NONE	N/A		
AKUILVIK IC5	0.30	OPR	1988	IC	OIL	DIESEL	NONE	N/A		
ANSE ST JEAN 1	0.40	OPR	1957	HY	WAT	CONV	NONE	N/A	GGG	EE
APALUK IC1	0.15	OPR	1981	IC	OIL	DIESEL	NONE	N/A	CAT	BBC
APALUK IC2	0.15	OPR	1981	IC	OIL	DIESEL	NONE	N/A	CAT	BBC

DIRECTORY OF POWER PLANTS IN CANADA
UDI/MCGRAW-HILL

PLANT/UNIT NAME	MW	STA	YEAR	TYPE	FUEL	FUEL TYPE	ALT FUEL	SSS MFR	TURB MFR	GEN MFR
OPERATOR: HYDRO-QUEBEC										
APALUK IC3	0.25	OPR	1985	IC	OIL	DIESEL	NONE	N/A	CAT	LIMA
BEAUHARNOIS 01	37.30	STN	1932	HY	WAT	CONV	NONE	N/A	DEW	CGE
BEAUHARNOIS 02	40.00	OPR	1932	HY	WAT	CONV	NONE	N/A	DEW	CGE
BEAUHARNOIS 03	40.00	OPR	1932	HY	WAT	CONV	NONE	N/A	DEW	CGE
BEAUHARNOIS 04	40.00	OPR	1934	HY	WAT	CONV	NONE	N/A	DEW	OERL
BEAUHARNOIS 05	40.00	OPR	1935	HY	WAT	CONV	NONE	N/A	DEW	CGE
BEAUHARNOIS 06	40.00	OPR	1935	HY	WAT	CONV	NONE	N/A	DEW	CGE
BEAUHARNOIS 07	37.30	OPR	1939	HY	WAT	CONV	NONE	N/A	DEW	CGE
BEAUHARNOIS 08	37.30	OPR	1941	HY	WAT	CONV	NONE	N/A	DEW	CGE
BEAUHARNOIS 09	37.30	OPR	1941	HY	WAT	CONV	NONE	N/A	DEW	CGE
BEAUHARNOIS 10	37.30	OPR	1948	HY	WAT	CONV	NONE	N/A	DEW	CGE
BEAUHARNOIS 11	40.00	OPR	1950	HY	WAT	CONV	NONE	N/A	DEW	WHC
BEAUHARNOIS 12	41.12	OPR	1950	HY	WAT	CONV	NONE	N/A	DEW	CGE
BEAUHARNOIS 13	40.00	STN	1951	HY	WAT	CONV	NONE	N/A	ACC	WHC
BEAUHARNOIS 14	41.12	OPR	1951	HY	WAT	CONV	NONE	N/A	DEW	CGE
BEAUHARNOIS 15	41.12	OPR	1951	HY	WAT	CONV	NONE	N/A	ACC	CGE
BEAUHARNOIS 16	40.00	OPR	1952	HY	WAT	CONV	NONE	N/A	ACC	CGE
BEAUHARNOIS 17	40.00	STN	1952	HY	WAT	CONV	NONE	N/A	DEW	CGE
BEAUHARNOIS 18	40.00	OPR	1953	HY	WAT	CONV	NONE	N/A	ACC	CGE
BEAUHARNOIS 19	40.00	OPR	1953	HY	WAT	CONV	NONE	N/A	DEW	CGE
BEAUHARNOIS 20	40.00	OPR	1953	HY	WAT	CONV	NONE	N/A	ACC	WHC
BEAUHARNOIS 21	55.25	OPR	1959	HY	WAT	CONV	NONE	N/A	EE	WHC
BEAUHARNOIS 22	55.25	OPR	1959	HY	WAT	CONV	NONE	N/A	EE	WHC
BEAUHARNOIS 23	55.25	OPR	1959	HY	WAT	CONV	NONE	N/A	EE	WHC
BEAUHARNOIS 24	55.25	OPR	1959	HY	WAT	CONV	NONE	N/A	EE	WHC
BEAUHARNOIS 25	55.25	OPR	1959	HY	WAT	CONV	NONE	N/A	EE	WHC
BEAUHARNOIS 26	55.25	OPR	1960	HY	WAT	CONV	NONE	N/A	EE	WHC
BEAUHARNOIS 27	55.25	OPR	1960	HY	WAT	CONV	NONE	N/A	EE	WHC
BEAUHARNOIS 28	55.25	OPR	1960	HY	WAT	CONV	NONE	N/A	EE	WHC
BEAUHARNOIS 29	55.25	OPR	1961	HY	WAT	CONV	NONE	N/A	EE	WHC
BEAUHARNOIS 30	55.25	OPR	1961	HY	WAT	CONV	NONE	N/A	EE	WHC

DIRECTORY OF POWER PLANTS IN CANADA
UDI/MCGRAW-HILL

PLANT/UNIT NAME	MW	STA	YEAR	TYPE	FUEL	FUEL TYPE	ALT FUEL	SSS MFR	TURB MFR	GEN MFR
OPERATOR: HYDRO-QUEBEC										
BEAUHARNOIS 31	46.75	OPR	1981	HY	WAT	CONV	NONE	N/A	DEW	CGE
BEAUHARNOIS 32	46.75	OPR	1982	HY	WAT	CONV	NONE	N/A	DEW	CGE
BEAUHARNOIS 33	46.75	OPR	1983	HY	WAT	CONV	NONE	N/A	DEW	CGE
BEAUHARNOIS 34	46.75	OPR	1983	HY	WAT	CONV	NONE	N/A	DEW	CGE
BEAUHARNOIS 35	46.75	OPR	1983	HY	WAT	CONV	NONE	N/A	DEW	CGE
BEAUHARNOIS 36	46.75	OPR	1984	HY	WAT	CONV	NONE	N/A	DEW	CGE
BEAUHARNOIS 37	46.75	OPR	1986	HY	WAT	CONV	NONE	N/A	DEW	CGE
BEAUHARNOIS 38	46.75	OPR	1987	HY	WAT	CONV	NONE	N/A	DEW	CGE
BEAUHARNOIS 39	46.75	OPR	1984	HY	WAT	CONV	NONE	N/A	DEW	CGE
BEAUMONT 1	40.50	OPR	1958	HY	WAT	CONV	NONE	N/A	ACC	CGE
BEAUMONT 2	40.50	OPR	1958	HY	WAT	CONV	NONE	N/A	ACC	CGE
BEAUMONT 3	40.50	OPR	1958	HY	WAT	CONV	NONE	N/A	ACC	CGE
BEAUMONT 4	40.50	OPR	1958	HY	WAT	CONV	NONE	N/A	ACC	CGE
BEAUMONT 5	40.50	OPR	1959	HY	WAT	CONV	NONE	N/A	ACC	CGE
BEAUMONT 6	40.50	OPR	1959	HY	WAT	CONV	NONE	N/A	ACC	CGE
BECANCOUR GT1	95.00	CON	1993	GT	GAS		OIL	N/A		
BECANCOUR GT2	95.00	CON	1993	GT	GAS		OIL	N/A		
BECANCOUR GT3	95.00	CON	1993	GT	GAS		OIL	N/A		
BECANCOUR GT4	95.00	CON	1993	GT	GAS		OIL	N/A		
BERSIMIS I 01	114.00	STN	1956	HY	WAT	CONV	NONE	N/A	EE	MV
BERSIMIS I 02	114.00	STN	1956	HY	WAT	CONV	NONE	N/A	EE	MV
BERSIMIS I 03	114.00	STN	1957	HY	WAT	CONV	NONE	N/A	NEY	CGE
BERSIMIS I 04	114.00	OPR	1957	HY	WAT	CONV	NONE	N/A	EE	MV
BERSIMIS I 05	114.00	OPR	1957	HY	WAT	CONV	NONE	N/A	EE	MV
BERSIMIS I 06	114.00	OPR	1958	HY	WAT	CONV	NONE	N/A	NEY	CGE
BERSIMIS I 07	114.00	OPR	1958	HY	WAT	CONV	NONE	N/A	NEY	CGE
BERSIMIS I 08	114.00	OPR	1959	HY	WAT	CONV	NONE	N/A	NEY	CGE
BERSIMIS I 09	120.00	OPR	1987	HY	WAT	CONV	NONE	N/A		
BERSIMIS I 10	120.00	OPR	1987	HY	WAT	CONV	NONE	N/A		
BERSIMIS I 11	120.00	OPR	1987	HY	WAT	CONV	NONE	N/A		
BERSIMIS II 1	131.00	OPR	1959	HY	WAT	CONV	NONE	N/A	DEW	CGE

DIRECTORY OF POWER PLANTS IN CANADA
UDI/MCGRAW-HILL

PLANT/UNIT NAME	MW	STA	YEAR	TYPE	FUEL	FUEL TYPE	ALT FUEL	SSS MFR	TURB MFR	GEN MFR
OPERATOR: HYDRO-QUEBEC										
BERSIMIS II 2	131.00	OPR	1959	HY	WAT	CONV	NONE	N/A	DEW	CGE
BERSIMIS II 3	131.00	OPR	1959	HY	WAT	CONV	NONE	N/A	DEW	CGE
BERSIMIS II 4	131.00	OPR	1960	HY	WAT	CONV	NONE	N/A	DEW	CGE
BERSIMIS II 5	131.00	OPR	1960	HY	WAT	CONV	NONE	N/A	DEW	CGE
BERSIMIS II 6	159.60	OPR	1987	HY	WAT	CONV	NONE	N/A		
BERSIMIS II 7	159.60	OPR	1987	HY	WAT	CONV	NONE	N/A		
BLANC SABLON IC1	0.80	OPR	1973	IC	OIL	DIESEL	NONE	N/A	CAT	TAMPER
BLANC SABLON IC2	0.80	OPR	1973	IC	OIL	DIESEL	NONE	N/A	CAT	KATO
BLANC SABLON IC3	0.80	OPR	1974	IC	OIL	DIESEL	NONE	N/A	CAT	TAMPER
BLANC SABLON IC4	0.80	OPR	1977	IC	OIL	DIESEL	NONE	N/A	CAT	TAMPER
BLANC SABLON IC5	0.80	OPR	1980	IC	OIL	DIESEL	NONE	N/A	CAT	BBC
BLANC SABLON IC6	0.80	OPR	1980	IC	OIL	DIESEL	NONE	N/A	CAT	BBC
BLANC SABLON IC7	0.80	OPR	1981	IC	OIL	DIESEL	NONE	N/A	CAT	TAMPER
BLANC SABLON IC8	0.80	OPR	1984	IC	OIL	DIESEL	NONE	N/A	CAT	BBC
BRISAY 1	191.00	CON	1993	HY	WAT	CONV	NONE	N/A		
BRISAY 2	191.00	CON	1993	HY	WAT	CONV	NONE	N/A		
BRYSON 1	18.00	OPR	1925	HY	WAT	CONV	NONE	N/A	AEI	WHC
BRYSON 2	18.00	OPR	1929	HY	WAT	CONV	NONE	N/A	SMS	WHC
BRYSON 3	25.00	OPR	1981	HY	WAT	CONV	NONE	N/A	DEW	CGE
CADILLAC GT1	54.00	OPR	1976	GT	OIL	DIST	NONE	N/A	RR	BRUSH
CADILLAC GT2	54.00	OPR	1977	GT	OIL	DIST	NONE	N/A	RR	BRUSH
CADILLAC GT3	54.00	OPR	1977	GT	OIL	DIST	NONE	N/A	RR	BRUSH
CARILLON 01	46.75	OPR	1962	HY	WAT	CONV	NONE	N/A	DEW	CGE
CARILLON 02	46.75	OPR	1962	HY	WAT	CONV	NONE	N/A	DEW	CGE
CARILLON 03	46.75	OPR	1962	HY	WAT	CONV	NONE	N/A	DEW	CGE
CARILLON 04	46.75	OPR	1962	HY	WAT	CONV	NONE	N/A	DEW	CGE
CARILLON 05	46.75	OPR	1963	HY	WAT	CONV	NONE	N/A	DEW	CGE
CARILLON 06	46.75	OPR	1963	HY	WAT	CONV	NONE	N/A	DEW	CGE
CARILLON 07	46.75	OPR	1963	HY	WAT	CONV	NONE	N/A	DEW	CGE
CARILLON 08	46.75	OPR	1963	HY	WAT	CONV	NONE	N/A	DEW	CGE
CARILLON 09	46.75	OPR	1963	HY	WAT	CONV	NONE	N/A	DEW	CGE

DIRECTORY OF POWER PLANTS IN CANADA
UDI/MCGRAW-HILL

PLANT/UNIT NAME	MW	STA	YEAR	TYPE	FUEL	FUEL TYPE	ALT FUEL	SSS MFR	TURB MFR	GEN MFR
OPERATOR: HYDRO-QUEBEC										
CARILLON 10	46.75	OPR	1963	HY	WAT	CONV	NONE	N/A	DEW	CGE
CARILLON 11	46.75	OPR	1964	HY	WAT	CONV	NONE	N/A	DEW	CGE
CARILLON 12	46.75	OPR	1964	HY	WAT	CONV	NONE	N/A	DEW	CGE
CARILLON 13	46.75	OPR	1964	HY	WAT	CONV	NONE	N/A	DEW	CGE
CARILLON 14	46.75	OPR	1964	HY	WAT	CONV	NONE	N/A	DEW	CGE
CHELSEA 1	28.80	OPR	1927	HY	WAT	CONV	NONE	N/A	DEW	WHC
CHELSEA 2	28.80	OPR	1927	HY	WAT	CONV	NONE	N/A	DEW	WHC
CHELSEA 3	28.80	OPR	1927	HY	WAT	CONV	NONE	N/A	DEW	WHC
CHELSEA 4	28.80	OPR	1929	HY	WAT	CONV	NONE	N/A	DEW	WHC
CHELSEA 5	28.80	OPR	1939	HY	WAT	CONV	NONE	N/A	DEW	WHC
CHUTE BELL 1	1.60	OPR	1915	HY	WAT	CONV	NONE	N/A	AC	CGE
CHUTE BELL 2	1.60	OPR	1915	HY	WAT	CONV	NONE	N/A	AC	CGE
CHUTE BELL 3	1.60	OPR	1920	HY	WAT	CONV	NONE	N/A	AC	CGE
CHUTE BURROUGHS 1	1.60	OPR	1929	HY	WAT	CONV	NONE	N/A	SMS	CGE
CHUTE GARNEAU 1	2.24	OPR	1925	HY	WAT	CONV	NONE	N/A	EW	WHC
CHUTE HEMMINGS 1	4.80	OPR	1925	HY	WAT	CONV	NONE	N/A	DEW	CGE
CHUTE HEMMINGS 2	4.80	OPR	1925	HY	WAT	CONV	NONE	N/A	DEW	CGE
CHUTE HEMMINGS 3	4.80	OPR	1925	HY	WAT	CONV	NONE	N/A	DEW	CGE
CHUTE HEMMINGS 4	4.80	OPR	1925	HY	WAT	CONV	NONE	N/A	DEW	CGE
CHUTE HEMMINGS 5	4.80	OPR	1925	HY	WAT	CONV	NONE	N/A	DEW	CGE
CHUTE HEMMINGS 6	4.80	OPR	1925	HY	WAT	CONV	NONE	N/A	DEW	CGE
CHUTE-DES-CHATS 1	22.33	OPR	1931	HY	WAT	CONV	NONE	N/A	DEW	WHC
CHUTE-DES-CHATS 2	22.33	OPR	1931	HY	WAT	CONV	NONE	N/A	DEW	WHC
CHUTE-DES-CHATS 3	22.33	OPR	1931	HY	WAT	CONV	NONE	N/A	DEW	WHC
CHUTE-DES-CHATS 4	22.33	OPR	1931	HY	WAT	CONV	NONE	N/A	DEW	WHC
CITIERE GT1	50.22	OPR	1979	GT	OIL	DIST	NONE	N/A	PW	BBC
CITIERE GT2	50.22	OPR	1979	GT	OIL	DIST	NONE	N/A	PW	BBC
CITIERE GT3	50.22	OPR	1979	GT	OIL	DIST	NONE	N/A	PW	BBC
CITIERE GT4	50.22	OPR	1980	GT	OIL	DIST	NONE	N/A	PW	BBC
CORBEAU 1	1.00	OPR	1926	HY	WAT	CONV	NONE	N/A	MV	EMC
CORBEAU 2	1.00	OPR	1926	HY	WAT	CONV	NONE	N/A	MV	EMC

DIRECTORY OF POWER PLANTS IN CANADA
UDI/MCGRAW-HILL

PLANT/UNIT NAME	MW	STA	YEAR	TYPE	FUEL	FUEL TYPE	ALT FUEL	SSS MFR	TURB MFR	GEN MFR
OPERATOR: HYDRO-QUEBEC										
DRUMMONDVILLE 1	2.50	OPR	1910	HY	WAT	CONV	NONE	N/A	BV	WHC
DRUMMONDVILLE 2	2.50	OPR	1910	HY	WAT	CONV	NONE	N/A	BV	WHC
DRUMMONDVILLE 3	4.80	OPR	1925	HY	WAT	CONV	NONE	N/A	DEW	WHC
DRUMMONDVILLE 4	4.80	OPR	1925	HY	WAT	CONV	NONE	N/A	DEW	WHC
EASTMAIN I 1	170.00	PLN	9999	HY	WAT	CONV	NONE	N/A		
EASTMAIN I 2	170.00	PLN	9999	HY	WAT	CONV	NONE	N/A		
EASTMAIN I 3	170.00	PLN	9999	HY	WAT	CONV	NONE	N/A		
GENTILLY 2	685.00	OPR	1980	ST	UR	CANDU	NONE	AECL	GE	GE
GRAND-MERE 1	15.73	OPR	1915	HY	WAT	CONV	NONE	N/A	IPM	WHC
GRAND-MERE 2	18.00	OPR	1915	HY	WAT	CONV	NONE	N/A	IPM	WHC
GRAND-MERE 3	15.73	OPR	1915	HY	WAT	CONV	NONE	N/A	IPM	WHC
GRAND-MERE 4	15.73	OPR	1916	HY	WAT	CONV	NONE	N/A	IPM	WHC
GRAND-MERE 5	15.73	OPR	1916	HY	WAT	CONV	NONE	N/A	IPM	WHC
GRAND-MERE 6	15.73	OPR	1916	HY	WAT	CONV	NONE	N/A	IPM	WHC
GRAND-MERE 7	15.73	OPR	1921	HY	WAT	CONV	NONE	N/A	IPM	WHC
GRAND-MERE 8	15.73	OPR	1922	HY	WAT	CONV	NONE	N/A	IPM	WHC
GRAND-MERE 9	21.50	OPR	1984	HY	WAT	CONV	NONE	N/A	DEW	WHC
GRANDE BALEINE I 1	392.00	PLN	9999	HY	WAT	CONV	NONE	N/A		
GRANDE BALEINE I 2	392.00	PLN	9999	HY	WAT	CONV	NONE	N/A		
GRANDE BALEINE I 3	392.00	PLN	9999	HY	WAT	CONV	NONE	N/A		
GRANDE BALEINE I 4	392.00	PLN	9999	HY	WAT	CONV	NONE	N/A		
GRANDE BALEINE I 5	392.00	PLN	9999	HY	WAT	CONV	NONE	N/A		
GRANDE BALEINE II 1	180.00	PLN	9999	HY	WAT	CONV	NONE	N/A		
GRANDE BALEINE II 2	180.00	PLN	9999	HY	WAT	CONV	NONE	N/A		
GRANDE BALEINE II 3	180.00	PLN	9999	HY	WAT	CONV	NONE	N/A		
GRANDE BALEINE III 1	187.00	PLN	9999	HY	WAT	CONV	NONE	N/A		
GRANDE BALEINE III 2	187.00	PLN	9999	HY	WAT	CONV	NONE	N/A		
GRANDE BALEINE III 3	187.00	PLN	9999	HY	WAT	CONV	NONE	N/A		
HART JAUNE 1	16.15	OPR	1960	HY	WAT	CONV	NONE	N/A		
HART JAUNE 2	16.15	OPR	1960	HY	WAT	CONV	NONE	N/A		
HART JAUNE 3	16.15	OPR	1960	HY	WAT	CONV	NONE	N/A		

DIRECTORY OF POWER PLANTS IN CANADA
UDI/MCGRAW-HILL

PLANT/UNIT NAME	MW	STA	YEAR	TYPE	FUEL	FUEL TYPE	ALT FUEL	SSS MFR	TURB MFR	GEN MFR
OPERATOR: HYDRO-QUEBEC										
HULL II 1	5.76	OPR	1920	HY	WAT	CONV	NONE	N/A	BV	MAW
HULL II 2	5.76	OPR	1920	HY	WAT	CONV	NONE	N/A	VOI	MAW
HULL II 3	5.76	OPR	1923	HY	WAT	CONV	NONE	N/A	VOI	MAW
HULL II 4	10.00	OPR	1969	HY	WAT	CONV	NONE	N/A	AC	CGE
ILE D'ENTREE IC1	0.18	UNK	1974	IC	OIL	DIESEL	NONE	N/A	CAT	GE
ILE D'ENTREE IC2	0.12	UNK	1975	IC	OIL	DIESEL	NONE	N/A	CAT	WHC
ILE D'ENTREE IC3	0.20	UNK	1977	IC	OIL	DIESEL	NONE	N/A	GM	BBC
ILE D'ENTREE IC4	0.40	UNK	1979	IC	OIL	DIESEL	NONE	N/A	CAT	GE
ILE D'ENTREE IC5	0.50	OPR	1979	IC	OIL	DIESEL	NONE	N/A	GM	STEN
ILE D'ENTREE IC6	0.35	UNK	1980	IC	OIL	DIESEL	NONE	N/A	CAT	WH
ILE D'ENTREE IC7	0.32	OPR	1988	IC	OIL	DIESEL	NONE	N/A		
ILES-DE-LA-MADELEINE IC01	2.27	OPR	1968	IC	OIL	DIESEL	NONE	N/A	DEUTZ	SS
ILES-DE-LA-MADELEINE IC02	2.27	UNK	1968	IC	OIL	DIESEL	NONE	N/A	DEUTZ	SS
ILES-DE-LA-MADELEINE IC03	3.07	OPR	1970	IC	OIL	DIESEL	NONE	N/A	MAN	SS
ILES-DE-LA-MADELEINE IC04	3.07	OPR	1971	IC	OIL	DIESEL	NONE	N/A	MAN	SS
ILES-DE-LA-MADELEINE IC05	3.07	OPR	1973	IC	OIL	DIESEL	NONE	N/A	MAN	SS
ILES-DE-LA-MADELEINE IC06	3.07	OPR	1974	IC	OIL	DIESEL	NONE	N/A	MAN	SS
ILES-DE-LA-MADELEINE IC07	2.04	OPR	1974	IC	OIL	DIESEL	NONE	N/A	MLW	CANR
ILES-DE-LA-MADELEINE IC08	2.04	OPR	1974	IC	OIL	DIESEL	NONE	N/A	MLW	CANR
ILES-DE-LA-MADELEINE IC09	2.04	OPR	1975	IC	OIL	DIESEL	NONE	N/A	MLW	CANR
ILES-DE-LA-MADELEINE IC10	2.04	OPR	1975	IC	OIL	DIESEL	NONE	N/A	MLW	CANR
ILES-DE-LA-MADELEINE IC11	2.04	OPR	1975	IC	OIL	DIESEL	NONE	N/A	MLW	CANR
ILES-DE-LA-MADELEINE IC12	5.97	OPR	1977	IC	OIL	DIESEL	NONE	N/A	MAN	SS
ILES-DE-LA-MADELEINE IC13	5.97	OPR	1977	IC	OIL	DIESEL	NONE	N/A	MAN	SS
ILES-DE-LA-MADELEINE IC14	6.80	UNK	1979	IC	OIL	DIESEL	NONE	N/A	GMT	SS
ILES-DE-LA-MADELEINE IC15	6.80	OPR	1979	IC	OIL	DIESEL	NONE	N/A	GMT	SS
ILES-DE-LA-MADELEINE IC16	6.80	OPR	1980	IC	OIL	DIESEL	NONE	N/A	GMT	IDEAL
ILES-DE-LA-MADELEINE IC17	2.04	OPR	1988	IC	OIL	DIESEL	NONE	N/A		
ILES-DE-LA-MADELEINE IC18	2.04	OPR	1988	IC	OIL	DIESEL	NONE	N/A		
ILES-DE-LA-MADELEINE IC19	2.04	OPR	1988	IC	OIL	DIESEL	NONE	N/A		
ILES-DE-LA-MADELEINE IC21	2.04	OPR	1988	IC	OIL	DIESEL	NONE	N/A		

DIRECTORY OF POWER PLANTS IN CANADA
UDI/MCGRAW-HILL

PLANT/UNIT NAME	MW	STA	YEAR	TYPE	FUEL	FUEL TYPE	ALT FUEL	SSS MFR	TURB MFR	GEN MFR
OPERATOR: HYDRO-QUEBEC										
ILES-DE-LA-MADELEINE IC22	2.04	OPR	1988	IC	OIL	DIESEL	NONE	N/A		
ILES-DE-LA-MADELEINE IC23	2.04	OPR	1988	IC	OIL	DIESEL	NONE	N/A		
ILES-DE-LA-MADELEINE IC24	2.04	OPR	1989	IC	OIL	DIESEL	NONE	N/A		
ILES-DE-LA-MADELEINE IC25	11.52	OPR	1991	IC	OIL	DIESEL	NONE	N/A	NSZD	SIE
ILES-DE-LA-MADELEINE IC26	11.52	OPR	1991	IC	OIL	DIESEL	NONE	N/A	NSZD	SIE
ILES-DE-LA-MADELEINE IC27	11.52	OPR	1991	IC	OIL	DIESEL	NONE	N/A	NSZD	SIE
ILES-DE-LA-MADELEINE IC28	11.52	OPR	1991	IC	OIL	DIESEL	NONE	N/A	NSZD	SIE
ILES-DE-LA-MADELEINE IC29	11.52	OPR	1991	IC	OIL	DIESEL	NONE	N/A	NSZD	SIE
ILES-DE-LA-MADELEINE IC30	11.52	OPR	1991	IC	OIL	DIESEL	NONE	N/A	NSZD	SIE
INUKJUAK IC1	0.40	OPR	1981	IC	OIL	DIESEL	NONE	N/A	CAT	LIST
INUKJUAK IC2	0.60	OPR	1981	IC	OIL	DIESEL	NONE	N/A	CAT	BBC
INUKJUAK IC3	0.60	OPR	1984	IC	OIL	DIESEL	NONE	N/A	CAT	LIMA
IVUJIVIK IC1	0.18	OPR	1985	IC	OIL	DIESEL	NONE	N/A	GM	LAW
IVUJIVIK IC2	0.40	OPR	1985	IC	OIL	DIESEL	NONE	N/A	GM	LAW
IVUJIVIK IC3	0.40	OPR	1985	IC	OIL	DIESEL	NONE	N/A	GM	LAW
JOHAN-BEETZ IC1	0.16	UNK	1967	IC	OIL	DIESEL	NONE	N/A	GM	TAMPER
JOHAN-BEETZ IC2	0.25	UNK	1974	IC	OIL	DIESEL	NONE	N/A	GM	TAMPER
JOHAN-BEETZ IC3	0.20	UNK	1974	IC	OIL	DIESEL	NONE	N/A	CAT	TAMPER
KANGIQSUALUJJUAQ IC1	0.13	UNK	1984	IC	OIL	DIESEL	NONE	N/A	CAT	CAT
KANGIQSUALUJJUAQ IC2	0.25	OPR	1984	IC	OIL	DIESEL	NONE	N/A	CAT	BBC
KANGIQSUALUJJUAQ IC3	0.25	OPR	1984	IC	OIL	DIESEL	NONE	N/A	CAT	BBC
KANGIQSUALUJJUAQ IC4	0.40	OPR	1986	IC	OIL	DIESEL	NONE	N/A		
KANGIQSUJUAQ IC1	0.21	OPR	1981	IC	OIL	DIESEL	NONE	N/A	CAT	LS
KANGIQSUJUAQ IC2	0.21	OPR	1981	IC	OIL	DIESEL	NONE	N/A	CAT	LS
KANGIQSUJUAQ IC3	0.40	OPR	1982	IC	OIL	DIESEL	NONE	N/A	CAT	LS
KANGIRSUK IC1	0.25	OPR	1981	IC	OIL	DIESEL	NONE	N/A	GM	BBC
KANGIRSUK IC2	0.25	UNK	1981	IC	OIL	DIESEL	NONE	N/A	CAT	BBC
KANGIRSUK IC3	0.25	UNK	1984	IC	OIL	DIESEL	NONE	N/A	CAT	LIMA
KANGIRSUK IC4	0.40	UNK	1987	IC	OIL	DIESEL	NONE	N/A	CAT	LIMA
KANGIRSUK IC5	0.40	UNK	1987	IC	OIL	DIESEL	NONE	N/A	CAT	LIMA
KUUJJUAQ IC1	0.80	UNK	1975	IC	OIL	DIESEL	NONE	N/A	CAT	TAMPER

DIRECTORY OF POWER PLANTS IN CANADA
UDI/MCGRAW-HILL

PLANT/UNIT NAME	MW	STA	YEAR	TYPE	FUEL	FUEL TYPE	ALT FUEL	SSS MFR	TURB MFR	GEN MFR
OPERATOR: *HYDRO-QUEBEC*										
KUUJJUAQ IC2	0.80	OPR	1978	IC	OIL	DIESEL	NONE	N/A	CAT	BBC
KUUJJUAQ IC3	0.80	UNK	1980	IC	OIL	DIESEL	NONE	N/A	CAT	BBC
KUUJJUAQ IC4	0.40	UNK	1989	IC	OIL	DIESEL	NONE	N/A		
KUUJJUAQ IC5	0.80	UNK	1989	IC	OIL	DIESEL	NONE	N/A		
KUUJJUAQ IC6	0.80	UNK	1989	IC	OIL	DIESEL	NONE	N/A		
KUUJJUARAPIK IC1	0.80	OPR	1973	IC	OIL	DIESEL	NONE	N/A		
KUUJJUARAPIK IC2	0.80	OPR	1974	IC	OIL	DIESEL	NONE	N/A		
KUUJJUARAPIK IC3	0.80	OPR	1989	IC	OIL	DIESEL	NONE	N/A		
LA FORGE I 1	142.00	CON	1995	HY	WAT	CONV	NONE	N/A	MIL	MIL
LA FORGE I 2	142.00	CON	1995	HY	WAT	CONV	NONE	N/A	MIL	MIL
LA FORGE I 3	142.00	CON	1995	HY	WAT	CONV	NONE	N/A	MIL	MIL
LA FORGE I 4	142.00	CON	1995	HY	WAT	CONV	NONE	N/A	MIL	MIL
LA FORGE I 5	142.00	CON	1995	HY	WAT	CONV	NONE	N/A	MIL	MIL
LA FORGE I 6	142.00	CON	1995	HY	WAT	CONV	NONE	N/A	MIL	MIL
LA FORGE II 1	135.00	PLN	9999	HY	WAT	CONV	NONE	N/A		
LA FORGE II 2	135.00	PLN	9999	HY	WAT	CONV	NONE	N/A		
LA GABELLE 1	27.36	OPR	1970	HY	WAT	CONV	NONE	N/A	DEW	WHC
LA GABELLE 2	27.73	OPR	1971	HY	WAT	CONV	NONE	N/A	DEW	WHC
LA GABELLE 3	27.36	OPR	1972	HY	WAT	CONV	NONE	N/A	DEW	WHC
LA GABELLE 4	27.36	OPR	1973	HY	WAT	CONV	NONE	N/A	DEW	WHC
LA GABELLE 5	26.78	OPR	1975	HY	WAT	CONV	NONE	N/A	DEW	WHC
LA GRANDE I 1	116.30	CON	1994	HY	WAT	CONV	NONE	N/A	MIL	MIL
LA GRANDE I 2	116.30	CON	1994	HY	WAT	CONV	NONE	N/A	MIL	MIL
LA GRANDE I 3	116.30	CON	1994	HY	WAT	CONV	NONE	N/A	MIL	MIL
LA GRANDE I 4	116.30	CON	1994	HY	WAT	CONV	NONE	N/A	MIL	MIL
LA GRANDE II 01	333.00	OPR	1979	HY	WAT	CONV	NONE	N/A	DEW	CGE
LA GRANDE II 02	333.00	OPR	1979	HY	WAT	CONV	NONE	N/A	MIL	G-A
LA GRANDE II 03	333.00	OPR	1979	HY	WAT	CONV	NONE	N/A	DEW	CGE
LA GRANDE II 04	333.00	OPR	1979	HY	WAT	CONV	NONE	N/A	MIL	G-A
LA GRANDE II 05	333.00	OPR	1980	HY	WAT	CONV	NONE	N/A	DEW	CGE
LA GRANDE II 06	333.00	OPR	1980	HY	WAT	CONV	NONE	N/A	MIL	G-A

DIRECTORY OF POWER PLANTS IN CANADA
UDI/MCGRAW-HILL

PLANT/UNIT NAME	MW	STA	YEAR	TYPE	FUEL	FUEL TYPE	ALT FUEL	SSS MFR	TURB MFR	GEN MFR
OPERATOR: HYDRO-QUEBEC										
LA GRANDE II 07	333.00	OPR	1980	HY	WAT	CONV	NONE	N/A	DEW	CGE
LA GRANDE II 08	333.00	OPR	1980	HY	WAT	CONV	NONE	N/A	MIL	G-A
LA GRANDE II 09	333.00	OPR	1980	HY	WAT	CONV	NONE	N/A	DEW	CGE
LA GRANDE II 10	333.00	OPR	1980	HY	WAT	CONV	NONE	N/A	MIL	G-A
LA GRANDE II 11	333.00	OPR	1981	HY	WAT	CONV	NONE	N/A	DEW	CGE
LA GRANDE II 12	333.00	OPR	1981	HY	WAT	CONV	NONE	N/A	MIL	G-A
LA GRANDE II 13	333.00	OPR	1981	HY	WAT	CONV	NONE	N/A	DEW	CGE
LA GRANDE II 14	333.00	OPR	1981	HY	WAT	CONV	NONE	N/A	MIL	G-A
LA GRANDE II 15	333.00	OPR	1981	HY	WAT	CONV	NONE	N/A	DEW	CGE
LA GRANDE II 16	333.00	OPR	1981	HY	WAT	CONV	NONE	N/A	MIL	G-A
LA GRANDE II A1	338.00	OPR	1991	HY	WAT	CONV	NONE	N/A	MIL	MIL
LA GRANDE II A2	338.00	OPR	1991	HY	WAT	CONV	NONE	N/A	MIL	MIL
LA GRANDE II A3	338.00	OPR	1991	HY	WAT	CONV	NONE	N/A	MIL	MIL
LA GRANDE II A4	338.00	OPR	1992	HY	WAT	CONV	NONE	N/A	MIL	MIL
LA GRANDE II A5	338.00	OPR	1992	HY	WAT	CONV	NONE	N/A	MIL	MIL
LA GRANDE II A6	338.00	OPR	1992	HY	WAT	CONV	NONE	N/A	MIL	MIL
LA GRANDE III 01	192.00	OPR	1982	HY	WAT	CONV	NONE	N/A	DEW	MIL
LA GRANDE III 02	192.00	OPR	1982	HY	WAT	CONV	NONE	N/A	DEW	MIL
LA GRANDE III 03	192.00	OPR	1982	HY	WAT	CONV	NONE	N/A	DEW	MIL
LA GRANDE III 04	192.00	OPR	1983	HY	WAT	CONV	NONE	N/A	DEW	MIL
LA GRANDE III 05	192.00	OPR	1983	HY	WAT	CONV	NONE	N/A	DEW	MIL
LA GRANDE III 06	192.00	OPR	1983	HY	WAT	CONV	NONE	N/A	DEW	MIL
LA GRANDE III 07	192.00	OPR	1983	HY	WAT	CONV	NONE	N/A	DEW	MIL
LA GRANDE III 08	192.00	OPR	1983	HY	WAT	CONV	NONE	N/A	DEW	MIL
LA GRANDE III 09	192.00	OPR	1983	HY	WAT	CONV	NONE	N/A	DEW	MIL
LA GRANDE III 10	192.00	OPR	1983	HY	WAT	CONV	NONE	N/A	DEW	MIL
LA GRANDE III 11	192.00	OPR	1984	HY	WAT	CONV	NONE	N/A	DEW	MIL
LA GRANDE III 12	192.00	OPR	1984	HY	WAT	CONV	NONE	N/A	DEW	MIL
LA GRANDE IV 1	294.50	OPR	1984	HY	WAT	CONV	NONE	N/A	MIL	MIL
LA GRANDE IV 2	294.50	OPR	1984	HY	WAT	CONV	NONE	N/A	MIL	MIL
LA GRANDE IV 3	294.50	OPR	1984	HY	WAT	CONV	NONE	N/A	MIL	MIL

DIRECTORY OF POWER PLANTS IN CANADA
UDI/MCGRAW-HILL

PLANT/UNIT NAME	MW	STA	YEAR	TYPE	FUEL	FUEL TYPE	ALT FUEL	SSS MFR	TURB MFR	GEN MFR
OPERATOR: HYDRO-QUEBEC										
LA GRANDE IV 4	294.50	OPR	1984	HY	WAT	CONV	NONE	N/A	MIL	MIL
LA GRANDE IV 5	294.50	OPR	1984	HY	WAT	CONV	NONE	N/A	MIL	MIL
LA GRANDE IV 6	294.50	OPR	1984	HY	WAT	CONV	NONE	N/A	MIL	MIL
LA GRANDE IV 7	294.50	OPR	1986	HY	WAT	CONV	NONE	N/A	MIL	MIL
LA GRANDE IV 8	294.50	OPR	1986	HY	WAT	CONV	NONE	N/A	MIL	MIL
LA GRANDE IV 9	294.50	OPR	1986	HY	WAT	CONV	NONE	N/A	MIL	MIL
LA ROMAINE IC1	0.60	OPR	1979	IC	OIL	DIESEL	NONE	N/A	CAT	BBC
LA ROMAINE IC2	0.60	OPR	1979	IC	OIL	DIESEL	NONE	N/A	CAT	TAMPER
LA ROMAINE IC3	0.80	OPR	1982	IC	OIL	DIESEL	NONE	N/A	CAT	KATO
LA ROMAINE IC4	0.80	OPR	1985	IC	OIL	DIESEL	NONE	N/A	CAT	BBC
LA TABATIERE IC1	0.80	UNK	1975	IC	OIL	DIESEL	NONE	N/A	CAT	TAMPER
LA TABATIERE IC2	0.80	OPR	1978	IC	OIL	DIESEL	NONE	N/A	CAT	BBC
LA TABATIERE IC3	0.80	OPR	1978	IC	OIL	DIESEL	NONE	N/A	CAT	BBC
LA TABATIERE IC4	0.80	OPR	1980	IC	OIL	DIESEL	NONE	N/A	CAT	BBC
LA TABATIERE IC5	0.80	OPR	1980	IC	OIL	DIESEL	NONE	N/A	CAT	BBC
LA TABATIERE IC6	0.70	OPR	1982	IC	OIL	DIESEL	NONE	N/A	CAT	COEL
LA TABATIERE IC7	0.80	OPR	1988	IC	OIL	DIESEL	NONE	N/A		
LA TABATIERE IC8	1.10	OPR	1989	IC	OIL	DIESEL	NONE	N/A		
LA TUQUE 1	36.00	OPR	1940	HY	WAT	CONV	NONE	N/A	DEW	CGE
LA TUQUE 2	36.00	OPR	1940	HY	WAT	CONV	NONE	N/A	DEW	CGE
LA TUQUE 3	36.00	OPR	1943	HY	WAT	CONV	NONE	N/A	DEW	CGE
LA TUQUE 4	36.00	OPR	1955	HY	WAT	CONV	NONE	N/A	DEW	CGE
LA TUQUE 5	36.00	OPR	1984	HY	WAT	CONV	NONE	N/A	DEW	CGE
LA TUQUE 6	36.00	OPR	1985	HY	WAT	CONV	NONE	N/A	DEW	CGE
LES CEDRES 01	9.00	OPR	1914	HY	WAT	CONV	NONE	N/A	IPM	CGE
LES CEDRES 02	9.00	OPR	1914	HY	WAT	CONV	NONE	N/A	IPM	CGE
LES CEDRES 03	9.00	OPR	1914	HY	WAT	CONV	NONE	N/A	IPM	CGE
LES CEDRES 04	9.00	OPR	1914	HY	WAT	CONV	NONE	N/A	IPM	CGE
LES CEDRES 05	9.00	OPR	1914	HY	WAT	CONV	NONE	N/A	WSM	CGE
LES CEDRES 06	9.00	OPR	1914	HY	WAT	CONV	NONE	N/A	WSM	CGE
LES CEDRES 07	9.00	OPR	1914	HY	WAT	CONV	NONE	N/A	WSM	CGE

DIRECTORY OF POWER PLANTS IN CANADA
UDI/MCGRAW-HILL

PLANT/UNIT NAME	MW	STA	YEAR	TYPE	FUEL	FUEL TYPE	ALT FUEL	SSS MFR	TURB MFR	GEN MFR
OPERATOR: HYDRO-QUEBEC										
LES CEDRES 08	9.00	OPR	1914	HY	WAT	CONV	NONE	N/A	IPM	CGE
LES CEDRES 09	9.00	OPR	1914	HY	WAT	CONV	NONE	N/A	IPM	CGE
LES CEDRES 10	9.00	OPR	1916	HY	WAT	CONV	NONE	N/A	IPM	CGE
LES CEDRES 11	9.00	OPR	1918	HY	WAT	CONV	NONE	N/A	IPM	CGE
LES CEDRES 12	9.00	OPR	1918	HY	WAT	CONV	NONE	N/A	IPM	CGE
LES CEDRES 13	9.00	OPR	1922	HY	WAT	CONV	NONE	N/A	DEW	CGE
LES CEDRES 14	9.00	OPR	1922	HY	WAT	CONV	NONE	N/A	DEW	CGE
LES CEDRES 15	9.00	OPR	1923	HY	WAT	CONV	NONE	N/A	DEW	CGE
LES CEDRES 16	9.00	OPR	1924	HY	WAT	CONV	NONE	N/A	DEW	CGE
LES CEDRES 17	9.00	OPR	1924	HY	WAT	CONV	NONE	N/A	DEW	CGE
LES CEDRES 18	9.00	OPR	1924	HY	WAT	CONV	NONE	N/A	DEW	CGE
MAGPIE 1	0.90	OPR	1961	HY	WAT	CONV	NONE	N/A	LF	CGE
MAGPIE 2	0.90	OPR	1961	HY	WAT	CONV	NONE	N/A	LF	CGE
MANIC I 1	61.47	OPR	1966	HY	WAT	CONV	NONE	N/A	ACC	A-J
MANIC I 2	61.47	OPR	1966	HY	WAT	CONV	NONE	N/A	ACC	A-J
MANIC I 3	61.47	OPR	1966	HY	WAT	CONV	NONE	N/A	ACC	A-J
MANIC II 1	126.90	OPR	1965	HY	WAT	CONV	NONE	N/A	DEW	CGE
MANIC II 2	126.90	OPR	1965	HY	WAT	CONV	NONE	N/A	DEW	CGE
MANIC II 3	126.90	OPR	1965	HY	WAT	CONV	NONE	N/A	DEW	CGE
MANIC II 4	126.90	OPR	1965	HY	WAT	CONV	NONE	N/A	DEW	CGE
MANIC II 5	126.90	OPR	1965	HY	WAT	CONV	NONE	N/A	DEW	CGE
MANIC II 6	126.90	OPR	1966	HY	WAT	CONV	NONE	N/A	DEW	CGE
MANIC II 7	126.90	OPR	1966	HY	WAT	CONV	NONE	N/A	DEW	CGE
MANIC II 8	126.90	OPR	1967	HY	WAT	CONV	NONE	N/A	DEW	CGE
MANIC III 1	197.20	OPR	1975	HY	WAT	CONV	NONE	N/A	DEW	MIL
MANIC III 2	197.20	OPR	1975	HY	WAT	CONV	NONE	N/A	DEW	MIL
MANIC III 3	197.20	OPR	1976	HY	WAT	CONV	NONE	N/A	DEW	MIL
MANIC III 4	197.20	OPR	1976	HY	WAT	CONV	NONE	N/A	DEW	MIL
MANIC III 5	197.20	OPR	1976	HY	WAT	CONV	NONE	N/A	DEW	MIL
MANIC III 6	197.20	OPR	1976	HY	WAT	CONV	NONE	N/A	DEW	MIL
MANIC V A1	161.50	OPR	1970	HY	WAT	CONV	NONE	N/A	MIL	MIL

DIRECTORY OF POWER PLANTS IN CANADA
UDI/MCGRAW-HILL

PLANT/UNIT NAME	MW	STA	YEAR	TYPE	FUEL	FUEL TYPE	ALT FUEL	SSS MFR	TURB MFR	GEN MFR
OPERATOR: HYDRO-QUEBEC										
MANIC V A2	161.50	OPR	1970	HY	WAT	CONV	NONE	N/A	MIL	MIL
MANIC V A3	161.50	OPR	1970	HY	WAT	CONV	NONE	N/A	MIL	MIL
MANIC V A4	161.50	OPR	1970	HY	WAT	CONV	NONE	N/A	MIL	MIL
MANIC V A5	161.50	OPR	1970	HY	WAT	CONV	NONE	N/A	MIL	MIL
MANIC V A6	161.50	OPR	1971	HY	WAT	CONV	NONE	N/A	MIL	MIL
MANIC V A7	161.50	OPR	1971	HY	WAT	CONV	NONE	N/A	MIL	MIL
MANIC V A8	161.50	OPR	1971	HY	WAT	CONV	NONE	N/A	MIL	MIL
MANIC V B1	266.00	OPR	1989	HY	WAT	CONV	NONE	N/A		
MANIC V B2	266.00	OPR	1989	HY	WAT	CONV	NONE	N/A		
MITIS I 1	2.40	OPR	1922	HY	WAT	CONV	NONE	N/A	SMS	WHC
MITIS I 2	4.00	OPR	1929	HY	WAT	CONV	NONE	N/A	SMS	WHC
MITIS II 1	4.25	OPR	1947	HY	WAT	CONV	NONE	N/A	SMS	WHC
NATASHQUAN IC1	0.50	OPR	1969	IC	OIL	DIESEL	NONE	N/A	CAT	TAMPER
NATASHQUAN IC2	0.80	OPR	1971	IC	OIL	DIESEL	NONE	N/A	CAT	KATO
NATASHQUAN IC3	0.80	OPR	1973	IC	OIL	DIESEL	NONE	N/A	CAT	TAMPER
OUTARDES II 1	151.30	OPR	1978	HY	WAT	CONV	NONE	N/A	MIL	MIL
OUTARDES II 2	151.30	OPR	1978	HY	WAT	CONV	NONE	N/A	MIL	MIL
OUTARDES II 3	151.30	OPR	1978	HY	WAT	CONV	NONE	N/A	MIL	MIL
OUTARDES III 1	189.05	OPR	1969	HY	WAT	CONV	NONE	N/A	DEW	CGE
OUTARDES III 2	189.05	OPR	1969	HY	WAT	CONV	NONE	N/A	DEW	CGE
OUTARDES III 3	189.05	OPR	1969	HY	WAT	CONV	NONE	N/A	DEW	CGE
OUTARDES III 4	189.05	OPR	1969	HY	WAT	CONV	NONE	N/A	DEW	CGE
OUTARDES IV 1	158.00	OPR	1969	HY	WAT	CONV	NONE	N/A	NEY	CGE
OUTARDES IV 2	158.00	OPR	1969	HY	WAT	CONV	NONE	N/A	NEY	CGE
OUTARDES IV 3	158.00	OPR	1969	HY	WAT	CONV	NONE	N/A	NEY	CGE
OUTARDES IV 4	158.00	OPR	1969	HY	WAT	CONV	NONE	N/A	NEY	CGE
PARENT IC1	0.40	UNK	1970	IC	OIL	DIESEL	NONE	N/A	CAT	TAMPER
PARENT IC2	0.80	UNK	1977	IC	OIL	DIESEL	NONE	N/A	CAT	BBC
PARENT IC3	0.80	UNK	1980	IC	OIL	DIESEL	NONE	N/A	CAT	BBC
PARENT IC4	0.40	UNK	1983	IC	OIL	DIESEL	NONE	N/A	CAT	KATO
PAUGAN 01	24.23	STN	1928	HY	WAT	CONV	NONE	N/A	DEW	WHC

DIRECTORY OF POWER PLANTS IN CANADA
UDI/MCGRAW-HILL

PLANT/UNIT NAME	MW	STA	YEAR	TYPE	FUEL	FUEL TYPE	ALT FUEL	SSS MFR	TURB MFR	GEN MFR
OPERATOR: HYDRO-QUEBEC										
PAUGAN 02	24.23	STN	1928	HY	WAT	CONV	NONE	N/A	DEW	WHC
PAUGAN 03	24.23	STN	1928	HY	WAT	CONV	NONE	N/A	DEW	WHC
PAUGAN 04	24.23	OPR	1931	HY	WAT	CONV	NONE	N/A	DEW	WHC
PAUGAN 05	32.40	OPR	1956	HY	WAT	CONV	NONE	N/A	DEW	CGE
PAUGAN 06	31.10	OPR	1983	HY	WAT	CONV	NONE	N/A	WHC	CGE
PAUGAN 07	31.10	OPR	1984	HY	WAT	CONV	NONE	N/A	WHC	CGE
PAUGAN 08	31.10	OPR	1985	HY	WAT	CONV	NONE	N/A	WHC	CGE
PAUGAN 09	31.10	OPR	1986	HY	WAT	CONV	NONE	N/A	WHC	CGE
PAUGAN 10	31.10	OPR	1987	HY	WAT	CONV	NONE	N/A	WHC	CGE
PAUGAN 11	31.10	OPR	1988	HY	WAT	CONV	NONE	N/A	WHC	CGE
PONT ARNAUD 1	1.70	OPR	1912	HY	WAT	CONV	NONE	N/A	SMS	WHC
PONT ARNAUD 2	1.88	OPR	1917	HY	WAT	CONV	NONE	N/A	SMS	WHC
PONT ARNAUD 3	1.88	OPR	1917	HY	WAT	CONV	NONE	N/A	SMS	WHC
PORT MENIER IC1	0.80	UNK	1983	IC	OIL	DIESEL	NONE	N/A	CAT	KATO
PORT MENIER IC2	0.50	UNK	1983	IC	OIL	DIESEL	NONE	N/A	CAT	KATO
PORT MENIER IC3	0.80	OPR	1984	IC	OIL	DIESEL	NONE	N/A	CAT	BBC
PORT MENIER IC4	0.80	OPR	1983	IC	OIL	DIESEL	NONE	N/A	CAT	BBC
PORT MENIER IC5	0.40	OPR	1987	IC	OIL	DIESEL	NONE	N/A	CAT	BBC
POSTE-DE-LA-BALEINE IC1	0.80	OPR	1973	IC	OIL	DIESEL	NONE	N/A	CAT	TAMPER
POSTE-DE-LA-BALEINE IC2	0.80	OPR	1974	IC	OIL	DIESEL	NONE	N/A	CAT	TAMPER
POSTE-DE-LA-BALEINE IC3	0.80	OPR	1978	IC	OIL	DIESEL	NONE	N/A	CAT	BBC
POVUNGITUK IC1	0.60	OPR	1981	IC	OIL	DIESEL	NONE	N/A	CAT	BBC
POVUNGITUK IC2	0.60	OPR	1985	IC	OIL	DIESEL	NONE	N/A	CAT	LAW
POVUNGITUK IC3	0.60	OPR	1985	IC	OIL	DIESEL	NONE	N/A	CAT	LAW
PREMIERE CHUTE 1	31.05	OPR	1968	HY	WAT	CONV	NONE	N/A	DEW	WHC
PREMIERE CHUTE 2	31.05	OPR	1969	HY	WAT	CONV	NONE	N/A	DEW	WHC
PREMIERE CHUTE 3	31.05	OPR	1969	HY	WAT	CONV	NONE	N/A	DEW	WHC
PREMIERE CHUTE 4	31.05	OPR	1975	HY	WAT	CONV	NONE	N/A	DEW	WHC
QUATAQ IC1	0.14	UNK	1981	IC	OIL	DIESEL	NONE	N/A	CAT	BBC
QUATAQ IC2	0.25	OPR	1981	IC	OIL	DIESEL	NONE	N/A	CAT	BBC
QUATAQ IC3	0.25	OPR	1982	IC	OIL	DIESEL	NONE	N/A	CAT	LIMA

DIRECTORY OF POWER PLANTS IN CANADA
UDI/MCGRAW-HILL

PLANT/UNIT NAME	MW	STA	YEAR	TYPE	FUEL	FUEL TYPE	ALT FUEL	SSS MFR	TURB MFR	GEN MFR
OPERATOR: HYDRO-QUEBEC										
QUATAQ IC4	0.40	OPR	1987	IC	OIL	DIESEL	NONE	N/A		
RAPIDE BLANC 1	30.60	STN	1934	HY	WAT	CONV	NONE	N/A	IPM	WHC
RAPIDE BLANC 2	30.60	STN	1934	HY	WAT	CONV	NONE	N/A	IPM	WHC
RAPIDE BLANC 3	30.60	OPR	1934	HY	WAT	CONV	NONE	N/A	IPM	WHC
RAPIDE BLANC 4	30.60	OPR	1943	HY	WAT	CONV	NONE	N/A	IPM	WHC
RAPIDE BLANC 5	30.60	OPR	1955	HY	WAT	CONV	NONE	N/A	DEW	ASEA
RAPIDE BLANC 6	30.60	OPR	1985	HY	WAT	CONV	NONE	N/A	DEW	DEW
RAPIDE BLANC 7	30.60	OPR	1987	HY	WAT	CONV	NONE	N/A	DEW	DEW
RAPIDE BLANC 8	30.60	OPR	1988	HY	WAT	CONV	NONE	N/A	DEW	DEW
RAPIDE DES ILES 1	36.63	OPR	1966	HY	WAT	CONV	NONE	N/A	DEW	WHC
RAPIDE DES ILES 2	36.63	OPR	1967	HY	WAT	CONV	NONE	N/A	DEW	WHC
RAPIDE DES ILES 3	36.63	OPR	1967	HY	WAT	CONV	NONE	N/A	DEW	WHC
RAPIDE DES ILES 4	36.63	OPR	1973	HY	WAT	CONV	NONE	N/A	DEW	WHC
RAPIDE DES QUINZE 1	8.00	OPR	1923	HY	WAT	CONV	NONE	N/A	DEW	ASEA
RAPIDE DES QUINZE 2	8.00	OPR	1923	HY	WAT	CONV	NONE	N/A	DEW	ASEA
RAPIDE DES QUINZE 3	26.00	OPR	1951	HY	WAT	CONV	NONE	N/A	ACC	CGE
RAPIDE DES QUINZE 4	26.00	OPR	1955	HY	WAT	CONV	NONE	N/A	ACC	CGE
RAPIDE DES QUINZE 5	11.00	OPR	1984	HY	WAT	CONV	NONE	N/A	DEW	ASEA
RAPIDE DES QUINZE 6	11.00	OPR	1985	HY	WAT	CONV	NONE	N/A	DEW	ASEA
RAPIDE FARMERS 1	19.13	OPR	1927	HY	WAT	CONV	NONE	N/A	DEW	CGE
RAPIDE FARMERS 2	20.00	OPR	1927	HY	WAT	CONV	NONE	N/A	DEW	CGE
RAPIDE FARMERS 3	20.00	OPR	1927	HY	WAT	CONV	NONE	N/A	DEW	CGE
RAPIDE FARMERS 4	20.00	OPR	1929	HY	WAT	CONV	NONE	N/A	DEW	CGE
RAPIDE FARMERS 5	19.13	OPR	1947	HY	WAT	CONV	NONE	N/A	DEW	CGE
RAPIDE II 1	12.00	OPR	1954	HY	WAT	CONV	NONE	N/A	DEW	WHC
RAPIDE II 2	12.00	OPR	1954	HY	WAT	CONV	NONE	N/A	DEW	WHC
RAPIDE II 3	12.00	OPR	1956	HY	WAT	CONV	NONE	N/A	DEW	CGE
RAPIDE II 4	12.00	OPR	1956	HY	WAT	CONV	NONE	N/A	DEW	CGE
RAPIDE VII 1	14.25	OPR	1941	HY	WAT	CONV	NONE	N/A	DEW	WHC
RAPIDE VII 2	14.25	OPR	1941	HY	WAT	CONV	NONE	N/A	DEW	WHC
RAPIDE VII 3	14.25	OPR	1941	HY	WAT	CONV	NONE	N/A	DEW	WHC

DIRECTORY OF POWER PLANTS IN CANADA
UDI/MCGRAW-HILL

PLANT/UNIT NAME	MW	STA	YEAR	TYPE	FUEL	FUEL TYPE	ALT FUEL	SSS MFR	TURB MFR	GEN MFR
OPERATOR: HYDRO-QUEBEC										
RAPIDE VII 4	14.25	OPR	1949	HY	WAT	CONV	NONE	N/A	DEW	WHC
RAWDON 1	1.72	OPR	1928	HY	WAT	CONV	NONE	N/A	DEW	ASEA
RIVIERE DES PRAIRIES 1	7.50	STN	1929	HY	WAT	CONV	NONE	N/A	DEW	CGE
RIVIERE DES PRAIRIES 2	7.50	STN	1929	HY	WAT	CONV	NONE	N/A	DEW	CGE
RIVIERE DES PRAIRIES 3	7.50	OPR	1929	HY	WAT	CONV	NONE	N/A	ACC	CGE
RIVIERE DES PRAIRIES 4	7.50	OPR	1930	HY	WAT	CONV	NONE	N/A	DEW	CGE
RIVIERE DES PRAIRIES 5	7.50	OPR	1930	HY	WAT	CONV	NONE	N/A	ACC	CGE
RIVIERE DES PRAIRIES 6	8.60	OPR	1985	HY	WAT	CONV	NONE	N/A	DEW	CGE
RIVIERE DES PRAIRIES 7	8.60	OPR	1986	HY	WAT	CONV	NONE	N/A	DEW	CGE
RIVIERE DES PRAIRIES 8	8.60	OPR	1987	HY	WAT	CONV	NONE	N/A	DEW	CGE
SAINT-AUGUSTIN IC1	0.40	OPR	1970	IC	OIL	DIESEL	NONE	N/A	CAT	COEL
SAINT-AUGUSTIN IC2	0.40	OPR	1972	IC	OIL	DIESEL	NONE	N/A	CAT	TAMPER
SAINT-AUGUSTIN IC3	0.60	OPR	1974	IC	OIL	DIESEL	NONE	N/A	CAT	TAMPER
SAINT-AUGUSTIN IC4	0.80	OPR	1980	IC	OIL	DIESEL	NONE	N/A	CAT	BBC
SAINT-AUGUSTIN IC5	0.80	OPR	1980	IC	OIL	DIESEL	NONE	N/A	CAT	BBC
SALLUIT IC1	0.40	OPR	1982	IC	OIL	DIESEL	NONE	N/A	CAT	BBC
SALLUIT IC2	0.40	OPR	1983	IC	OIL	DIESEL	NONE	N/A	CAT	LAW
SALLUIT IC3	0.40	OPR	1984	IC	OIL	DIESEL	NONE	N/A	CAT	LIMA
SALLUIT IC4	0.46	OPR	1989	IC	OIL	DIESEL	NONE	N/A		
SEPT CHUTES 1	4.68	OPR	1916	HY	WAT	CONV	NONE	N/A	AC	CGE
SEPT CHUTES 2	4.68	OPR	1916	HY	WAT	CONV	NONE	N/A	AC	CGE
SEPT CHUTES 3	4.68	OPR	1916	HY	WAT	CONV	NONE	N/A	AC	CGE
SEPT CHUTES 4	4.68	OPR	1916	HY	WAT	CONV	NONE	N/A	AC	CGE
SHAWINIGAN II 01	14.00	OPR	1911	HY	WAT	CONV	NONE	N/A	IPM	WHC
SHAWINIGAN II 02	14.00	OPR	1911	HY	WAT	CONV	NONE	N/A	IPM	WHC
SHAWINIGAN II 03	15.00	STN	1913	HY	WAT	CONV	NONE	N/A	IPM	WHC
SHAWINIGAN II 04	15.00	STN	1914	HY	WAT	CONV	NONE	N/A	IPM	WHC
SHAWINIGAN II 05	15.00	STN	1914	HY	WAT	CONV	NONE	N/A	IPM	WHC
SHAWINIGAN II 06	30.00	STN	1922	HY	WAT	CONV	NONE	N/A	IPM	CGE
SHAWINIGAN II 07	30.00	STN	1928	HY	WAT	CONV	NONE	N/A	IPM	CGE
SHAWINIGAN II 08	30.00	OPR	1929	HY	WAT	CONV	NONE	N/A	IPM	CGE

DIRECTORY OF POWER PLANTS IN CANADA
UDI/MCGRAW-HILL

PLANT/UNIT NAME	MW	STA	YEAR	TYPE	FUEL	FUEL TYPE	ALT FUEL	SSS MFR	TURB MFR	GEN MFR
OPERATOR: HYDRO-QUEBEC										
SHAWINIGAN II 09	15.30	OPR	1986	HY	WAT	CONV	NONE	N/A		
SHAWINIGAN II 10	39.80	OPR	1986	HY	WAT	CONV	NONE	N/A		
SHAWINIGAN II 11	15.30	OPR	1987	HY	WAT	CONV	NONE	N/A		
SHAWINIGAN II 12	39.80	OPR	1987	HY	WAT	CONV	NONE	N/A		
SHAWINIGAN II 13	15.30	OPR	1988	HY	WAT	CONV	NONE	N/A		
SHAWINIGAN III 1	57.30	OPR	1983	HY	WAT	CONV	NONE	N/A	DEW	CGE
SHAWINIGAN III 2	57.30	OPR	1984	HY	WAT	CONV	NONE	N/A	DEW	CGE
SHAWINIGAN III 3	57.30	OPR	1984	HY	WAT	CONV	NONE	N/A	DEW	CGE
ST ALBAN 1	3.00	OPR	1927	HY	WAT	CONV	NONE	N/A	MV	CGE
ST NARCISSE 1	7.50	OPR	1926	HY	WAT	CONV	NONE	N/A	DEW	WHC
ST NARCISSE 2	7.50	OPR	1926	HY	WAT	CONV	NONE	N/A	DEW	WHC
ST RAPHAEL 1	0.85	OPR	1921	HY	WAT	CONV	NONE	N/A	BV	WHC
ST RAPHAEL 2	0.85	OPR	1921	HY	WAT	CONV	NONE	N/A	BV	WHC
ST RAPHAEL 3	0.85	OPR	1921	HY	WAT	CONV	NONE	N/A	BV	WHC
STE-MARGUERITE III 1	398.00	PLN	2006	HY	WAT	CONV	NONE	N/A		
STE-MARGUERITE III 2	398.00	PLN	2006	HY	WAT	CONV	NONE	N/A		
TASIUJAQ IC1	0.09	OPR	1981	IC	OIL	DIESEL	NONE	N/A	CAT	CAT
TASIUJAQ IC2	0.18	OPR	1981	IC	OIL	DIESEL	NONE	N/A	CAT	BBC
TASIUJAQ IC3	0.18	OPR	1981	IC	OIL	DIESEL	NONE	N/A	CAT	BBC
TRACY 1	150.00	OPR	1964	ST	OIL	HFO	NONE	CE	PAR	PAR
TRACY 2	150.00	OPR	1965	ST	OIL	HFO	NONE	CE	PAR	PAR
TRACY 3	150.00	OPR	1967	ST	OIL	HFO	NONE	CE	PAR	PAR
TRACY 4	150.00	OPR	1968	ST	OIL	HFO	NONE	CE	PAR	PAR
TRENCHE 1	47.70	OPR	1950	HY	WAT	CONV	NONE	N/A	DEW	CGE
TRENCHE 2	47.70	OPR	1951	HY	WAT	CONV	NONE	N/A	DEW	CGE
TRENCHE 3	50.40	OPR	1982	HY	WAT	CONV	NONE	N/A	DEW	CGE
TRENCHE 4	50.40	OPR	1983	HY	WAT	CONV	NONE	N/A	DEW	CGE
TRENCHE 5	50.40	OPR	1984	HY	WAT	CONV	NONE	N/A	DEW	CGE
TRENCHE 6	50.40	OPR	1985	HY	WAT	CONV	NONE	N/A	DEW	CGE
UMIUJAG IC1	0.25	OPR	1988	IC	OIL	DIESEL	NONE	N/A		
UMIUJAG IC2	0.25	OPR	1988	IC	OIL	DIESEL	NONE	N/A		

DIRECTORY OF POWER PLANTS IN CANADA
UDI/MCGRAW-HILL

PLANT/UNIT NAME	MW	STA	YEAR	TYPE	FUEL	FUEL TYPE	ALT FUEL	SSS MFR	TURB MFR	GEN MFR
OPERATOR: HYDRO-SHERBROOKE										
ABENQUAIS 1	0.80	OPR	1910	HY	WAT	CONV	NONE	N/A		
ABENQUAIS 2	0.80	OPR	1910	HY	WAT	CONV	NONE	N/A		
ABENQUAIS 3	0.80	OPR	1910	HY	WAT	CONV	NONE	N/A		
DRUMMOND 1	0.58	OPR	1928	HY	WAT	CONV	NONE	N/A	DEW	CGE
DRUMMOND 2	0.30	OPR	1928	HY	WAT	CONV	NONE	N/A	SMS	CGE
EUSTIS 2	0.70	OPR	1987	HY	WAT	CONV	NONE	N/A	OSS	
FRONTENAC 1	0.80	OPR	1917	HY	WAT	CONV	NONE	N/A	BV	CGE
FRONTENAC 2	0.80	OPR	1917	HY	WAT	CONV	NONE	N/A	BV	CGE
PATON 1	0.72	OPR	1926	HY	WAT	CONV	NONE	N/A	DEW	CGE
PATON 2	0.72	OPR	1926	HY	WAT	CONV	NONE	N/A	DEW	CGE
ROCK FOREST 1	0.94	OPR	1911	HY	WAT	CONV	NONE	N/A	SMS	WHC
ROCK FOREST 2	0.94	OPR	1911	HY	WAT	CONV	NONE	N/A	SMS	WHC
WEEDON 1	1.04	OPR	1920	HY	WAT	CONV	NONE	N/A	BV	WHC
WEEDON 2	1.04	OPR	1920	HY	WAT	CONV	NONE	N/A	BV	WHC
WEEDON 3	1.04	OPR	1926	HY	WAT	CONV	NONE	N/A	BV	CGE
WESTBURY 1	2.00	OPR	1928	HY	WAT	CONV	NONE	N/A	DEW	CGE
WESTBURY 2	2.00	OPR	1928	HY	WAT	CONV	NONE	N/A	DEW	CGE
OPERATOR: HYDROMEGA DEVELOPMENTS INC										
COTE ST CATHERINE 1	0.55	OPR	1989	HY	WAT	CONV	NONE	N/A		
COTE ST CATHERINE 2	0.55	OPR	1989	HY	WAT	CONV	NONE	N/A		
COTE ST CATHERINE 3	0.55	OPR	1989	HY	WAT	CONV	NONE	N/A		
COTE ST CATHERINE 4	0.55	OPR	1989	HY	WAT	CONV	NONE	N/A		
MONT LAURIER 1	0.56	OPR	1937	HY	WAT	CONV	NONE	N/A	LEIT	CGE
MONT LAURIER 2	0.90	OPR	1951	HY	WAT	CONV	NONE	N/A	DB	CGE
MONT LAURIER 3	0.90	OPR	1951	HY	WAT	CONV	NONE	N/A	DB	CGE
OPERATOR: INCO METALS CO										
BIG EDDY 1	7.20	OPR	1929	HY	WAT	CONV	NONE	N/A	IPM	WHC
BIG EDDY 2	7.20	OPR	1929	HY	WAT	CONV	NONE	N/A	IPM	WHC
BIG EDDY 3	6.70	OPR	1985	HY	WAT	CONV	NONE	N/A	IPM	CGE

DIRECTORY OF POWER PLANTS IN CANADA
UDI/MCGRAW-HILL

PLANT/UNIT NAME	MW	STA	YEAR	TYPE	FUEL	FUEL TYPE	ALT FUEL	SSS MFR	TURB MFR	GEN MFR
OPERATOR: INCO METALS CO										
HIGH FALLS (IMC) 1	5.55	OPR	1918	HY	WAT	CONV	NONE	N/A	IPM	WHC
HIGH FALLS (IMC) 2	3.00	OPR	1966	HY	WAT	CONV	NONE	N/A	DEW	CGE
HIGH FALLS (IMC) 3	3.00	OPR	1966	HY	WAT	CONV	NONE	N/A	DEW	CGE
HIGH FALLS (IMC) 4	3.00	OPR	1966	HY	WAT	CONV	NONE	N/A	DEW	CGE
HIGH FALLS (IMC) 5	3.00	OPR	1966	HY	WAT	CONV	NONE	N/A	DEW	CGE
NAIRN 1	1.50	OPR	1919	HY	WAT	CONV	NONE	N/A	AC	AC
NAIRN 2	1.50	OPR	1919	HY	WAT	CONV	NONE	N/A	AC	AC
NAIRN 3	1.50	OPR	1919	HY	WAT	CONV	NONE	N/A	AC	CGE
ONTARIO IRON 1	9.38	OPR	1963	ST	WSTH		NONE	DB	GE	GE
ONTARIO IRON 2	9.38	OPR	1963	ST	WSTH		NONE	DB	GE	GE
WABAGESHIK 1	1.60	OPR	1912	HY	WAT	CONV	NONE	N/A	AC	AC
WABAGESHIK 2	2.14	OPR	1935	HY	WAT	CONV	NONE	N/A	LF	CGE
OPERATOR: INDECK ENERGY SERVICES/GAZPLUS										
MONTREAL EST (HULL)	216.00	PLN	1995	CC	GAS		NONE			
OPERATOR: INDEPENDENT PRODUCER POWER INC										
DICKSON (RED DEER) 1	14.90	OPR	1992	HY	WAT	CONV	NONE	N/A	BARBER	IDEAL
OPERATOR: INNERGEX INC										
MONT ROLLAND 1	2.00	CON	1992	HY	WAT	CONV	NONE	N/A	OSS	OSS
OPERATOR: IRON ORE COMPANY OF CANADA										
LABRADOR CITY IC1	1.00	OPR	1962	IC	OIL	DIESEL	NONE	N/A	GM	GM
MENIHEK 1	4.25	OPR	1954	HY	WAT	CONV	NONE	N/A	ACC	WHC
MENIHEK 2	4.25	OPR	1954	HY	WAT	CONV	NONE	N/A	ACC	WHC
MENIHEK 3	10.20	OPR	1960	HY	WAT	CONV	NONE	N/A	KMW	WHC
MOBILE (IOC) IC10	1.00	OPR	1956	IC	OIL	DIESEL	NONE	N/A	GM	GM
MOBILE (IOC) IC11	1.00	OPR	1956	IC	OIL	DIESEL	NONE	N/A	GM	GM
MOBILE (IOC) IC12	1.00	OPR	1956	IC	OIL	DIESEL	NONE	N/A	GM	GM
MOBILE (IOC) IC13	1.00	OPR	1962	IC	OIL	DIESEL	NONE	N/A	GM	GM

DIRECTORY OF POWER PLANTS IN CANADA
UDI/MCGRAW-HILL

PLANT/UNIT NAME	MW	STA	YEAR	TYPE	FUEL	FUEL TYPE	ALT FUEL	SSS MFR	TURB MFR	GEN MFR
OPERATOR: IRON ORE COMPANY OF CANADA										
STE MARGUERITE 1	8.80	OPR	1954	HY	WAT	CONV	NONE	N/A	ACC	CGE
STE MARGUERITE 2	8.80	OPR	1954	HY	WAT	CONV	NONE	N/A	ACC	CGE
OPERATOR: IRVING PULP & PAPER LTD										
SAINT JOHN (IPP) 1	10.00	OPR	1956	ST/S	LIQ		NONE	CE/BW	GE	GE
SAINT JOHN (IPP) 2	12.50	OPR	1960	ST/S	LIQ		NONE	CE/BW	GE	GE
OPERATOR: JAMES RIVER MARATHON LTD										
MARATHON 1	7.50	OPR	1946	ST	LIQ		NONE	CE/BW	WH	WH
MARATHON 2	4.00	OPR	1948	ST	LIQ		NONE	CE/BW	GE	GE
MARATHON 3	4.00	OPR	1948	ST	LIQ		NONE	CE/BW	GE	GE
OPERATOR: JONQUIERE ELECTRIC SERVICE										
JONQUIERE 1	1.28	OPR	1924	HY	WAT	CONV	NONE	N/A	WH	CGE
JONQUIERE 2	2.81	OPR	1948	HY	WAT	CONV	NONE	N/A	SMS	CGE
OPERATOR: JORDAN HYDRO INC										
CLYDE RIVER	0.80	PLN	1993	HY	WAT	CONV	NONE	N/A		
OPERATOR: KAGAWONG POWER CORP										
CHARLTON DAM 1	0.50	OPR	1991	HY	WAT	CONV	NONE	N/A		
OPERATOR: KAISER COAL										
KAISER COAL 1	2.50	OPR	1972	HY	WAT	CONV	NONE	N/A		ABB
KAISER COAL 2	2.50	OPR	1972	HY	WAT	CONV	NONE	N/A		ABB
OPERATOR: KALIUM CHEMICALS										
BELLE PLAINE 1	7.50	OPR	1964	ST/S	GAS		NONE	BW	GE	GE
BELLE PLAINE 2	7.50	OPR	1964	ST/S	GAS		NONE	BW	GE	GE
BELLE PLAINE 3	20.00	OPR	1981	ST/S	GAS		NONE	BW	GE	GE

DIRECTORY OF POWER PLANTS IN CANADA
UDI/MCGRAW-HILL

PLANT/UNIT NAME	MW	STA	YEAR	TYPE	FUEL	FUEL TYPE	ALT FUEL	SSS MFR	TURB MFR	GEN MFR
OPERATOR: KIRKLAND LAKE POWER CORP										
KIRKLAND LAKE GT1	22.60	OPR	1991	GT/C	GAS		NONE	N/A	GE	BRUSH
KIRKLAND LAKE GT2	22.60	OPR	1991	GT/C	GAS		NONE	N/A	GE	BRUSH
KIRKLAND LAKE GT3	22.60	OPR	1991	GT/C	GAS		NONE	N/A	GE	BRUSH
OPERATOR: LA CIE GASPESIA LTEE										
CHANDLER 1	6.00	OPR	1954	ST/S	OIL	HFO	NONE		BBC	BBC
OPERATOR: LA CIE HYDROELEC MANICOUAGAN										
MCCORMICK DAM 1	35.63	OPR	1951	HY	WAT	CONV	NONE	N/A	SMS	CGE
MCCORMICK DAM 2	35.63	OPR	1952	HY	WAT	CONV	NONE	N/A	SMS	CGE
MCCORMICK DAM 3	40.00	OPR	1957	HY	WAT	CONV	NONE	N/A	AC	CGE
MCCORMICK DAM 4	40.00	OPR	1958	HY	WAT	CONV	NONE	N/A	AC	CGE
MCCORMICK DAM 5	40.00	OPR	1958	HY	WAT	CONV	NONE	N/A	AC	CGE
MCCORMICK DAM 6	56.25	OPR	1965	HY	WAT	CONV	NONE	N/A	AC	CGE
MCCORMICK DAM 7	56.25	OPR	1965	HY	WAT	CONV	NONE	N/A	AC	CGE
OPERATOR: LA CIE PRICE LTEE										
ADAM CUNNINGHAM 1	6.38	OPR	1953	HY	WAT	CONV	NONE	N/A	ACC	CGE
CHICOUTIMI 1	9.90	OPR	1923	HY	WAT	CONV	NONE	N/A	DEW	WHC
CHUTE AUX GALETS 1	6.80	OPR	1921	HY	WAT	CONV	NONE	N/A	SMS	CGE
CHUTE AUX GALETS 2	6.80	OPR	1921	HY	WAT	CONV	NONE	N/A	SMS	CGE
JIM GRAY 1	25.50	OPR	1982	HY	WAT	CONV	NONE	N/A	DEW	WHC
JIM GRAY 2	25.50	OPR	1982	HY	WAT	CONV	NONE	N/A	DEW	WHC
JONQUIERE MILL 1	1.20	OPR	1916	HY	WAT	CONV	NONE	N/A	SMS	CGE
JONQUIERE MILL 2	1.20	OPR	1916	HY	WAT	CONV	NONE	N/A	SMS	EE
KENOGAMI 3	14.75	OPR	1968	ST/S	OIL	HFO	NONE	FW/CE	STAL	STAL
MURDOCK WILSON 1	51.00	OPR	1957	HY	WAT	CONV	NONE	N/A	KMW	WHC
OPERATOR: LAIDLAW WASTE SYSTEMS										
SWARU (LWS) 1	4.23	OPR	1987	ST	REF		NONE			
SWARU (LWS) 2	8.25	OPR	1989	ST	REF		NONE			

DIRECTORY OF POWER PLANTS IN CANADA
UDI/MCGRAW-HILL

PLANT/UNIT NAME	MW	STA	YEAR	TYPE	FUEL	FUEL TYPE	ALT FUEL	SSS MFR	TURB MFR	GEN MFR
OPERATOR: LAKE SUPERIOR POWER LP										
ST MARYS PAPER GT1	42.00	PLN	1994	GT/C	GAS		NONE	N/A	GE	GE
ST MARYS PAPER GT2	42.00	PLN	1994	GT/C	GAS		NONE	N/A	GE	GE
ST MARYS PAPER SC1	25.00	PLN	1994	ST/C	WSTH		NONE		GE	GE
OPERATOR: LES CHAUDIERES FOSTER WHEELER										
MONTREAL WTE 1	35.00	PLN	1993	ST	REF		NONE	FW		
OPERATOR: LONG SLIDE POWER										
LONG SLIDE RAPIDS 1	0.66	OPR	1991	HY	WAT	CONV	NONE	N/A		
OPERATOR: MACLAREN FOREST PRODUCTS INC										
GASPE IC1	1.00	OPR	1953	IC	OIL	DIESEL	NONE	N/A	FM	WHC
GASPE IC2	1.00	OPR	1954	IC	OIL	DIESEL	NONE	N/A	FM	GE
GASPE IC3	0.90	OPR	1981	IC	OIL	DIESEL	NONE	N/A	CAT	BBC
OPERATOR: MACLAREN-QUEBEC POWER CO										
DUFFERIN FALLS 1	19.13	OPR	1958	HY	WAT	CONV	NONE	N/A	EE	WHC
DUFFERIN FALLS 2	19.13	OPR	1959	HY	WAT	CONV	NONE	N/A	EE	WHC
HIGH FALLS (MQP) 1	21.25	OPR	1929	HY	WAT	CONV	NONE	N/A	SMS	WHC
HIGH FALLS (MQP) 2	21.25	OPR	1929	HY	WAT	CONV	NONE	N/A	SMS	WHC
HIGH FALLS (MQP) 3	21.25	OPR	1929	HY	WAT	CONV	NONE	N/A	SMS	WHC
HIGH FALLS (MQP) 4	21.25	OPR	1933	HY	WAT	CONV	NONE	N/A	ACC	WHC
MASSON 1	23.80	OPR	1933	HY	WAT	CONV	NONE	N/A	ACC	WHC
MASSON 2	23.80	OPR	1933	HY	WAT	CONV	NONE	N/A	ACC	WHC
MASSON 3	23.80	OPR	1933	HY	WAT	CONV	NONE	N/A	ACC	WHC
MASSON 4	23.80	OPR	1933	HY	WAT	CONV	NONE	N/A	ACC	WHC
OPERATOR: MACMILLAN BLOEDEL LTD										
HARMAC 1	1.25	OPR	1953	ST	LIQ		NONE			
HARMAC 2	4.00	OPR	1963	ST	LIQ		NONE			
HARMAC 3	31.50	OPR	1963	ST	LIQ		NONE			

DIRECTORY OF POWER PLANTS IN CANADA
UDI/MCGRAW-HILL

PLANT/UNIT NAME	MW	STA	YEAR	TYPE	FUEL	FUEL TYPE	ALT FUEL	SSS MFR	TURB MFR	GEN MFR
OPERATOR: MACMILLAN BLOEDEL LTD										
PORT ALBERNI 1	26.00	OPR	1963	ST	WOOD		NONE			
POWELL RIVER 1	3.00	OPR	1911	HY	WAT	CONV	NONE	N/A	PI	CGE
POWELL RIVER 2	2.24	OPR	1911	HY	WAT	CONV	NONE	N/A	AC	CGE
POWELL RIVER 3	2.24	OPR	1911	HY	WAT	CONV	NONE	N/A	AC	CGE
POWELL RIVER 4	11.52	OPR	1926	HY	WAT	CONV	NONE	N/A	DEW	CGE
POWELL RIVER 5	25.50	OPR	1976	HY	WAT	CONV	NONE	N/A	AC	CGE
POWELL RIVER S1	10.50	OPR	1951	ST	LIQ		NONE	BW/FW	BBC	BBC
POWELL RIVER S2	36.00	OPR	1967	ST	LIQ		NONE	BW/FW	GE	GE
STILLWATER 1	16.00	OPR	1930	HY	WAT	CONV	NONE	N/A	DEW	CGE
STILLWATER 2	14.40	OPR	1948	HY	WAT	CONV	NONE	N/A	DEW	CGE
STURGEON FALLS (MBI) 1	1.80	OPR	1951	HY	WAT	CONV	NONE	N/A	WK	WHC
STURGEON FALLS (MBI) 2	1.42	OPR	1932	HY	WAT	CONV	NONE	N/A	HOK	CGE
STURGEON FALLS (MBI) 3	1.69	OPR	1942	HY	WAT	CONV	NONE	N/A	SMS	WHC
STURGEON FALLS (MBI) 4	1.69	OPR	1942	HY	WAT	CONV	NONE	N/A	HOK	WHC
STURGEON FALLS (MBI) 5	1.35	OPR	1942	HY	WAT	CONV	NONE	N/A	HOK	WHC
STURGEON FALLS (MBI) 6	1.42	OPR	1964	HY	WAT	CONV	NONE	N/A	SMS	WHC
OPERATOR: MAGOG ELECTRIC DEPT										
MAGOG 1	0.90	OPR	1911	HY	WAT	CONV	NONE	N/A	GE	CGE
MAGOG 2	0.90	OPR	1911	HY	WAT	CONV	NONE	N/A	GE	CGE
OPERATOR: MAGPIE RIVER DEVELOPERS										
MAGPIE RIVER 1	13.00	OPR	1989	HY	WAT	CONV	NONE	N/A		ABB
MAGPIE RIVER 2	13.00	OPR	1989	HY	WAT	CONV	NONE	N/A		ABB
MAGPIE RIVER 3	13.00	OPR	1989	HY	WAT	CONV	NONE	N/A		ABB
OPERATOR: MAINE-NEW BRUNSWICK ELEC POWER										
TINKER 1	1.50	OPR	1922	HY	WAT	CONV	NONE	N/A	DEW	WHC
TINKER 2	1.50	OPR	1923	HY	WAT	CONV	NONE	N/A	DEW	WHC
TINKER 3	3.52	OPR	1926	HY	WAT	CONV	NONE	N/A	DEW	WHC
TINKER 4	3.52	OPR	1952	HY	WAT	CONV	NONE	N/A	SMS	WHC

DIRECTORY OF POWER PLANTS IN CANADA
UDI/MCGRAW-HILL

PLANT/UNIT NAME	MW	STA	YEAR	TYPE	FUEL	FUEL TYPE	ALT FUEL	SSS MFR	TURB MFR	GEN MFR
OPERATOR: MAINE-NEW BRUNSWICK ELEC POWER										
TINKER 5	20.80	OPR	1965	HY	WAT	CONV	NONE	N/A	ACC	WHC
TINKER IC1	1.00	OPR	1949	IC	OIL	DIESEL	NONE	N/A	SUP	GE
OPERATOR: MALETTE KRAFT PULP & PAPER										
SMOOTH ROCK FALLS S1	15.00	OPR	1976	ST	LIQ		NONE	BW	WH	EMC
SMOOTH ROCK FALLS S2	10.90	OPR	1991	ST	LIQ		NONE			
OPERATOR: MANITOBA FORESTRY RESOURCES										
THE PAS (MFR) 1	9.80	OPR	1970	ST/S	WOOD		NONE	FW/CE	WH	EE
THE PAS (MFR) 2	13.00	OPR	1970	ST/S	WOOD		NONE	FW/CE	WH	EE
OPERATOR: MANITOBA HYDRO										
BRANDON 1	33.00	OPR	1957	ST	COAL	LIG	NONE	CE	MV	MV
BRANDON 2	33.00	OPR	1958	ST	COAL	LIG	NONE	CE	MV	MV
BRANDON 3	33.00	OPR	1958	ST	COAL	LIG	NONE	CE	MV	MV
BRANDON 4	33.00	OPR	1958	ST	COAL	LIG	NONE	CE	MV	MV
BRANDON 5	105.00	OPR	1970	ST	COAL	LIG	NONE	BW	BBC	BBC
BROCHET IC1	0.18	OPR	1973	IC	OIL	DIESEL	NONE	N/A	CAT	TAMPER
BROCHET IC2	0.18	UNK	1974	IC	OIL	DIESEL	NONE	N/A	CAT	TAMPER
BROCHET IC3	0.30	UNK	1976	IC	OIL	DIESEL	NONE	N/A	CAT	CAT
BROCHET IC4	0.33	OPR	1988	IC	OIL	DIESEL	NONE	N/A		
BROCHET IC5	0.33	OPR	1988	IC	OIL	DIESEL	NONE	N/A		
CONAWAPA 01	139.00	PLN	2001	HY	WAT	CONV	NONE	N/A		
CONAWAPA 02	139.00	PLN	2001	HY	WAT	CONV	NONE	N/A		
CONAWAPA 03	139.00	PLN	2001	HY	WAT	CONV	NONE	N/A		
CONAWAPA 04	139.00	PLN	2001	HY	WAT	CONV	NONE	N/A		
CONAWAPA 05	139.00	PLN	2002	HY	WAT	CONV	NONE	N/A		
CONAWAPA 06	139.00	PLN	2002	HY	WAT	CONV	NONE	N/A		
CONAWAPA 07	139.00	PLN	2002	HY	WAT	CONV	NONE	N/A		
CONAWAPA 08	139.00	PLN	2002	HY	WAT	CONV	NONE	N/A		
CONAWAPA 09	139.00	PLN	2004	HY	WAT	CONV	NONE	N/A		

DIRECTORY OF POWER PLANTS IN CANADA
UDI/MCGRAW-HILL

PLANT/UNIT NAME	MW	STA	YEAR	TYPE	FUEL	FUEL TYPE	ALT FUEL	SSS MFR	TURB MFR	GEN MFR
OPERATOR: MANITOBA HYDRO										
CONAWAPA 10	139.00	PLN	2004	HY	WAT	CONV	NONE	N/A		
FORT CHURCHILL IC1	1.14	UNK	1953	IC	OIL	DIESEL	NONE	N/A	FM	FM
FORT CHURCHILL IC2	1.14	UNK	1959	IC	OIL	DIESEL	NONE	N/A	FM	FM
FORT CHURCHILL IC3	1.00	UNK	1961	IC	OIL	DIESEL	NONE	N/A	GM	GE
FORT CHURCHILL IC4	1.00	UNK	1962	IC	OIL	DIESEL	NONE	N/A	GM	GM
FORT CHURCHILL IC5	1.14	UNK	1963	IC	OIL	DIESEL	NONE	N/A	FM	FM
FORT CHURCHILL IC6	2.50	UNK	1971	IC	OIL	DIESEL	NONE	N/A	GM	GM
FORT CHURCHILL IC7	1.00	UNK	1971	IC	OIL	DIESEL	NONE	N/A	GM	GE
FORT CHURCHILL IC8	2.34	UNK	1974	IC	OIL	DIESEL	NONE	N/A	MIRR	BRUSH
GARDEN HILL IC1	0.30	OPR	1970	IC	OIL	DIESEL	NONE	N/A	CAT	TAMPER
GARDEN HILL IC2	0.30	OPR	1974	IC	OIL	DIESEL	NONE	N/A	CAT	KATO
GARDEN HILL IC3	0.50	OPR	1979	IC	OIL	DIESEL	NONE	N/A	CAT	TAMPER
GARDEN HILL IC4	0.50	UNK	1979	IC	OIL	DIESEL	NONE	N/A	CAT	KATO
GARDEN HILL IC5	0.86	OPR	1988	IC	OIL	DIESEL	NONE	N/A		
GARDEN HILL IC6	0.86	OPR	1988	IC	OIL	DIESEL	NONE	N/A		
GARDEN HILL IC7	0.86	OPR	1988	IC	OIL	DIESEL	NONE	N/A		
GOD'S LAKE NARROWS IC1	0.30	OPR	1972	IC	OIL	DIESEL	NONE	N/A	CAT	TAMPER
GOD'S LAKE NARROWS IC2	0.30	UNK	1972	IC	OIL	DIESEL	NONE	N/A	CAT	TAMPER
GOD'S LAKE NARROWS IC3	0.30	OPR	1980	IC	OIL	DIESEL	NONE	N/A	CAT	TAMPER
GOD'S LAKE NARROWS IC4	0.35	UNK	1982	IC	OIL	DIESEL	NONE	N/A	DD	TAMPER
GOD'S LAKE NARROWS IC5	0.35	UNK	1984	IC	OIL	DIESEL	NONE	N/A	DD	TAMPER
GOD'S RIVER IC1	0.18	OPR	1979	IC	OIL	DIESEL	NONE	N/A	CAT	TAMPER
GOD'S RIVER IC2	0.18	OPR	1979	IC	OIL	DIESEL	NONE	N/A	CAT	TAMPER
GRAND RAPIDS 1	109.25	OPR	1965	HY	WAT	CONV	NONE	N/A	JI	CGE
GRAND RAPIDS 2	109.25	OPR	1965	HY	WAT	CONV	NONE	N/A	JI	CGE
GRAND RAPIDS 3	109.25	OPR	1965	HY	WAT	CONV	NONE	N/A	JI	CGE
GRAND RAPIDS 4	109.25	OPR	1968	HY	WAT	CONV	NONE	N/A	ACC	CGE
GREAT FALLS (MH) 1	22.00	OPR	1923	HY	WAT	CONV	NONE	N/A	DEW	CGE
GREAT FALLS (MH) 2	22.00	OPR	1923	HY	WAT	CONV	NONE	N/A	DEW	CGE
GREAT FALLS (MH) 3	22.00	OPR	1926	HY	WAT	CONV	NONE	N/A	DEW	CGE
GREAT FALLS (MH) 4	22.00	OPR	1927	HY	WAT	CONV	NONE	N/A	SMS	CGE

DIRECTORY OF POWER PLANTS IN CANADA
UDI/MCGRAW-HILL

PLANT/UNIT NAME	MW	STA	YEAR	TYPE	FUEL	FUEL TYPE	ALT FUEL	SSS MFR	TURB MFR	GEN MFR
OPERATOR: MANITOBA HYDRO										
GREAT FALLS (MH) 5	22.00	OPR	1928	HY	WAT	CONV	NONE	N/A	DEW	CGE
GREAT FALLS (MH) 6	22.00	OPR	1928	HY	WAT	CONV	NONE	N/A	DEW	CGE
JENPEG 1	31.00	OPR	1977	HY	WAT	CONV	NONE	N/A	LMZ	LMZ
JENPEG 2	31.00	OPR	1978	HY	WAT	CONV	NONE	N/A	LMZ	LMZ
JENPEG 3	31.00	OPR	1978	HY	WAT	CONV	NONE	N/A	LMZ	LMZ
JENPEG 4	31.00	OPR	1978	HY	WAT	CONV	NONE	N/A	LMZ	LMZ
JENPEG 5	31.00	OPR	1979	HY	WAT	CONV	NONE	N/A	LMZ	LMZ
JENPEG 6	31.00	OPR	1979	HY	WAT	CONV	NONE	N/A	LMZ	LMZ
KELSEY 1	33.75	OPR	1960	HY	WAT	CONV	NONE	N/A	DEW	CGE
KELSEY 2	33.75	OPR	1960	HY	WAT	CONV	NONE	N/A	DEW	CGE
KELSEY 3	33.75	OPR	1960	HY	WAT	CONV	NONE	N/A	DEW	CGE
KELSEY 4	33.75	OPR	1960	HY	WAT	CONV	NONE	N/A	DEW	CGE
KELSEY 5	33.75	OPR	1961	HY	WAT	CONV	NONE	N/A	DEW	CGE
KELSEY 6	33.75	OPR	1969	HY	WAT	CONV	NONE	N/A	DEW	CGE
KELSEY 7	33.75	OPR	1972	HY	WAT	CONV	NONE	N/A	DEW	CGE
KETTLE RAPIDS 01	102.00	OPR	1970	HY	WAT	CONV	NONE	N/A	DEW	MTS
KETTLE RAPIDS 02	102.00	OPR	1971	HY	WAT	CONV	NONE	N/A	DEW	MTS
KETTLE RAPIDS 03	102.00	OPR	1971	HY	WAT	CONV	NONE	N/A	DEW	MTS
KETTLE RAPIDS 04	102.00	OPR	1971	HY	WAT	CONV	NONE	N/A	DEW	MTS
KETTLE RAPIDS 05	102.00	OPR	1972	HY	WAT	CONV	NONE	N/A	DEW	MTS
KETTLE RAPIDS 06	102.00	OPR	1972	HY	WAT	CONV	NONE	N/A	DEW	MTS
KETTLE RAPIDS 07	102.00	OPR	1973	HY	WAT	CONV	NONE	N/A	DEW	MTS
KETTLE RAPIDS 08	102.00	OPR	1973	HY	WAT	CONV	NONE	N/A	DEW	MTS
KETTLE RAPIDS 09	102.00	OPR	1973	HY	WAT	CONV	NONE	N/A	DEW	MTS
KETTLE RAPIDS 10	102.00	OPR	1974	HY	WAT	CONV	NONE	N/A	DEW	MTS
KETTLE RAPIDS 11	102.00	OPR	1974	HY	WAT	CONV	NONE	N/A	DEW	MTS
KETTLE RAPIDS 12	102.00	OPR	1974	HY	WAT	CONV	NONE	N/A	DEW	MTS
LAC BROCHET IC1	0.18	OPR	1981	IC	OIL	DIESEL	NONE	N/A	CAT	TAMPER
LAC BROCHET IC2	0.18	OPR	1981	IC	OIL	DIESEL	NONE	N/A	CAT	TAMPER
LAC BROCHET IC3	0.18	OPR	1981	IC	OIL	DIESEL	NONE	N/A	CAT	BBC
LAURIE RIVER I 1	2.48	OPR	1952	HY	WAT	CONV	NONE	N/A	AC	CGE

DIRECTORY OF POWER PLANTS IN CANADA
UDI/MCGRAW-HILL

PLANT/UNIT NAME	MW	STA	YEAR	TYPE	FUEL	FUEL TYPE	ALT FUEL	SSS MFR	TURB MFR	GEN MFR
OPERATOR: MANITOBA HYDRO										
LAURIE RIVER I 2	2.48	OPR	1952	HY	WAT	CONV	NONE	N/A	AC	CGE
LAURIE RIVER II 1	5.40	OPR	1958	HY	WAT	CONV	NONE	N/A	JI	CGE
LIMESTONE 01	133.00	OPR	1990	HY	WAT	CONV	NONE	N/A		CGE
LIMESTONE 02	133.00	OPR	1990	HY	WAT	CONV	NONE	N/A		CGE
LIMESTONE 03	133.00	OPR	1990	HY	WAT	CONV	NONE	N/A		CGE
LIMESTONE 04	133.00	OPR	1991	HY	WAT	CONV	NONE	N/A		CGE
LIMESTONE 05	133.00	OPR	1991	HY	WAT	CONV	NONE	N/A		CGE
LIMESTONE 06	133.00	OPR	1991	HY	WAT	CONV	NONE	N/A		CGE
LIMESTONE 07	133.00	OPR	1991	HY	WAT	CONV	NONE	N/A		CGE
LIMESTONE 08	133.00	OPR	1992	HY	WAT	CONV	NONE	N/A		CGE
LIMESTONE 09	133.00	OPR	1992	HY	WAT	CONV	NONE	N/A		CGE
LIMESTONE 10	133.00	OPR	1992	HY	WAT	CONV	NONE	N/A		CGE
LITTLE GRAND RAPIDS IC1	0.18	UNK	1976	IC	OIL	DIESEL	NONE	N/A	CAT	TAMPER
LITTLE GRAND RAPIDS IC2	0.18	UNK	1976	IC	OIL	DIESEL	NONE	N/A	CAT	TAMPER
LITTLE GRAND RAPIDS IC3	0.18	UNK	1984	IC	OIL	DIESEL	NONE	N/A	DD	EMC
LONG SPRUCE 01	98.00	OPR	1977	HY	WAT	CONV	NONE	N/A	DEW	CGE
LONG SPRUCE 02	98.00	OPR	1977	HY	WAT	CONV	NONE	N/A	DEW	CGE
LONG SPRUCE 03	98.00	OPR	1978	HY	WAT	CONV	NONE	N/A	DEW	CGE
LONG SPRUCE 04	98.00	OPR	1978	HY	WAT	CONV	NONE	N/A	DEW	CGE
LONG SPRUCE 05	98.00	OPR	1978	HY	WAT	CONV	NONE	N/A	DEW	CGE
LONG SPRUCE 06	98.00	OPR	1978	HY	WAT	CONV	NONE	N/A	DEW	CGE
LONG SPRUCE 07	98.00	OPR	1979	HY	WAT	CONV	NONE	N/A	DEW	CGE
LONG SPRUCE 08	98.00	OPR	1979	HY	WAT	CONV	NONE	N/A	DEW	CGE
LONG SPRUCE 09	98.00	OPR	1979	HY	WAT	CONV	NONE	N/A	DEW	CGE
LONG SPRUCE 10	98.00	OPR	1979	HY	WAT	CONV	NONE	N/A	DEW	CGE
MCARTHUR 1	7.65	OPR	1954	HY	WAT	CONV	NONE	N/A	DEW	CGE
MCARTHUR 2	7.65	OPR	1954	HY	WAT	CONV	NONE	N/A	DEW	CGE
MCARTHUR 3	7.65	OPR	1954	HY	WAT	CONV	NONE	N/A	DEW	CGE
MCARTHUR 4	7.65	OPR	1954	HY	WAT	CONV	NONE	N/A	DEW	CGE
MCARTHUR 5	7.65	OPR	1955	HY	WAT	CONV	NONE	N/A	DEW	CGE
MCARTHUR 6	7.65	OPR	1955	HY	WAT	CONV	NONE	N/A	DEW	CGE

DIRECTORY OF POWER PLANTS IN CANADA
UDI/MCGRAW-HILL

PLANT/UNIT NAME	MW	STA	YEAR	TYPE	FUEL	FUEL TYPE	ALT FUEL	SSS MFR	TURB MFR	GEN MFR
OPERATOR: MANITOBA HYDRO										
MCARTHUR 7	7.65	OPR	1955	HY	WAT	CONV	NONE	N/A	DEW	CGE
MCARTHUR 8	7.65	OPR	1955	HY	WAT	CONV	NONE	N/A	DEW	CGE
OXFORD HOUSE IC1	0.30	UNK	1974	IC	OIL	DIESEL	NONE	N/A	CAT	KATO
OXFORD HOUSE IC2	0.30	UNK	1974	IC	OIL	DIESEL	NONE	N/A	CAT	KATO
OXFORD HOUSE IC3	0.30	UNK	1976	IC	OIL	DIESEL	NONE	N/A	CAT	TAMPER
OXFORD HOUSE IC4	0.50	UNK	1980	IC	OIL	DIESEL	NONE	N/A	CAT	GE
OXFORD HOUSE IC5	0.43	OPR	1989	IC	OIL	DIESEL	NONE	N/A		
OXFORD HOUSE IC6	0.43	OPR	1989	IC	OIL	DIESEL	NONE	N/A		
OXFORD HOUSE IC7	0.43	OPR	1989	IC	OIL	DIESEL	NONE	N/A		
OXFORD HOUSE IC8	0.43	OPR	1989	IC	OIL	DIESEL	NONE	N/A		
PAUINGASSI IC1	0.08	UNK	1976	IC	OIL	DIESEL	NONE	N/A	CAT	TAMPER
PAUINGASSI IC2	0.08	UNK	1976	IC	OIL	DIESEL	NONE	N/A	CAT	TAMPER
PIKWITONEI IC1	0.18	OPR	1976	IC	OIL	DIESEL	NONE	N/A	CAT	TAMPER
PIKWITONEI IC2	0.18	OPR	1976	IC	OIL	DIESEL	NONE	N/A	CAT	TAMPER
PIKWITONEI IC3	0.08	OPR	1979	IC	OIL	DIESEL	NONE	N/A	CAT	TAMPER
PIKWITONEI IC4	0.08	OPR	1979	IC	OIL	DIESEL	NONE	N/A	CAT	TAMPER
PINE FALLS 1	13.95	OPR	1951	HY	WAT	CONV	NONE	N/A	DEW	CGE
PINE FALLS 2	13.95	OPR	1951	HY	WAT	CONV	NONE	N/A	DEW	CGE
PINE FALLS 3	13.95	OPR	1952	HY	WAT	CONV	NONE	N/A	DEW	CGE
PINE FALLS 4	13.95	OPR	1952	HY	WAT	CONV	NONE	N/A	DEW	CGE
PINE FALLS 5	13.95	OPR	1952	HY	WAT	CONV	NONE	N/A	DEW	CGE
PINE FALLS 6	13.95	OPR	1952	HY	WAT	CONV	NONE	N/A	DEW	CGE
RED SUCKER LAKE IC1	0.30	OPR	1975	IC	OIL	DIESEL	NONE	N/A	CAT	TAMPER
RED SUCKER LAKE IC2	0.18	OPR	1976	IC	OIL	DIESEL	NONE	N/A	CAT	TAMPER
RED SUCKER LAKE IC3	0.18	OPR	1976	IC	OIL	DIESEL	NONE	N/A	CAT	TAMPER
RED SUCKER LAKE IC4	0.30	OPR	1976	IC	OIL	DIESEL	NONE	N/A	CAT	TAMPER
SELKIRK 1	66.00	OPR	1960	ST	COAL	LIG	NONE	BW	PAR	PAR
SELKIRK 2	66.00	OPR	1960	ST	COAL	LIG	NONE	BW	PAR	PAR
SELKIRK GT1	11.90	OPR	1967	GT	JET		NONE	N/A	PW	BBC
SELKIRK GT2	11.90	OPR	1968	GT	JET		NONE	N/A	PW	BBC
SEVEN SISTERS 1	25.00	OPR	1931	HY	WAT	CONV	NONE	N/A	AC	CGE

DIRECTORY OF POWER PLANTS IN CANADA
UDI/MCGRAW-HILL

PLANT/UNIT NAME	MW	STA	YEAR	TYPE	FUEL	FUEL TYPE	ALT FUEL	SSS MFR	TURB MFR	GEN MFR
OPERATOR: MANITOBA HYDRO										
SEVEN SISTERS 2	25.00	OPR	1931	HY	WAT	CONV	NONE	N/A	DEW	CGE
SEVEN SISTERS 3	25.00	OPR	1931	HY	WAT	CONV	NONE	N/A	SMS	CGE
SEVEN SISTERS 4	25.00	OPR	1949	HY	WAT	CONV	NONE	N/A	DEW	CGE
SEVEN SISTERS 5	25.00	OPR	1950	HY	WAT	CONV	NONE	N/A	DEW	CGE
SEVEN SISTERS 6	25.00	OPR	1952	HY	WAT	CONV	NONE	N/A	DEW	CGE
SHAMATTAWA IC1	0.18	OPR	1973	IC	OIL	DIESEL	NONE	N/A	CAT	TAMPER
ST THERESA IC1	0.18	OPR	1971	IC	OIL	DIESEL	NONE	N/A	CAT	TAMPER
ST THERESA IC2	0.30	OPR	1975	IC	OIL	DIESEL	NONE	N/A	CAT	TAMPER
ST THERESA IC3	0.50	OPR	1982	IC	OIL	DIESEL	NONE	N/A	CAT	GE
ST THERESA IC4	0.50	OPR	1985	IC	OIL	DIESEL	NONE	N/A	CAT	GE
TADOULE LAKE IC1	0.18	OPR	1982	IC	OIL	DIESEL	NONE	N/A	CAT	GE
TADOULE LAKE IC2	0.18	OPR	1982	IC	OIL	DIESEL	NONE	N/A	CAT	GE
TADOULE LAKE IC3	0.13	UNK	1982	IC	OIL	DIESEL	NONE	N/A	CAT	GE
TADOULE LAKE IC4	0.13	UNK	1982	IC	OIL	DIESEL	NONE	N/A	CAT	GE
TADOULE LAKE IC5	0.18	UNK	1973	IC	OIL	DIESEL	NONE	N/A	CAT	TAMPER
TADOULE LAKE IC6	0.18	UNK	1983	IC	OIL	DIESEL	NONE	N/A	CAT	TAMPER
THICKET PORTAGE IC1	0.08	UNK	1972	IC	OIL	DIESEL	NONE	N/A	DD	EMC
THICKET PORTAGE IC2	0.08	UNK	1972	IC	OIL	DIESEL	NONE	N/A	DD	EMC
THICKET PORTAGE IC3	0.08	OPR	1976	IC	OIL	DIESEL	NONE	N/A	DD	EMC
THICKET PORTAGE IC4	0.08	OPR	1976	IC	OIL	DIESEL	NONE	N/A	DD	EMC
THICKET PORTAGE IC5	0.18	OPR	1972	IC	OIL	DIESEL	NONE	N/A	CAT	TAMPER
THICKET PORTAGE IC6	0.18	UNK	1977	IC	OIL	DIESEL	NONE	N/A	CAT	TAMPER
WAASAGOMACH IC1	0.30	OPR	1975	IC	OIL	DIESEL	NONE	N/A	CAT	TAMPER
WAASAGOMACH IC2	0.30	OPR	1979	IC	OIL	DIESEL	NONE	N/A	CAT	TAMPER
WAASAGOMACH IC3	0.50	UNK	1979	IC	OIL	DIESEL	NONE	N/A	CAT	KATO
OPERATOR: MARITIME ELECTRIC COMPANY LTD										
BORDEN GT1	14.85	OPR	1971	GT	OIL	DIESEL	NONE	N/A	RR	EE
BORDEN GT2	26.00	OPR	1973	GT	OIL	DIESEL	NONE	N/A	JBE	BRUSH
CHARLOTTETOWN 1	1.50	OPR	1946	ST	OIL	HFO	NONE	BW	AC	AC
CHARLOTTETOWN 2	4.00	OPR	1947	ST	OIL	HFO	NONE	DB	PAR	PAR

DIRECTORY OF POWER PLANTS IN CANADA
UDI/MCGRAW-HILL

PLANT/UNIT NAME	MW	STA	YEAR	TYPE	FUEL	FUEL TYPE	ALT FUEL	SSS MFR	TURB MFR	GEN MFR
OPERATOR: MARITIME ELECTRIC COMPANY LTD										
CHARLOTTETOWN 3	7.50	OPR	1951	ST	OIL	HFO	NONE	BW	PAR	PAR
CHARLOTTETOWN 4	7.50	OPR	1955	ST	OIL	HFO	NONE	FW	BBC	BBC
CHARLOTTETOWN 5	10.00	OPR	1960	ST	OIL	HFO	NONE	BW	PAR	PAR
CHARLOTTETOWN 6	20.00	OPR	1963	ST	OIL	HFO	NONE	BW	MV	MV
CHARLOTTETOWN 7	20.00	OPR	1968	ST	OIL	HFO	NONE	FW	MV	MV
CHARLOTTETOWN GT1	24.00	PLN	1996	GT	OIL		NONE			
OPERATOR: MEDICINE HAT ELECTRIC UTILITY										
MEDICINE HAT 3	15.00	OPR	1974	ST	GAS		NONE	TIW	PAR	PAR
MEDICINE HAT 4	3.00	OPR	1929	ST	GAS		NONE	FW	PAR	PAR
MEDICINE HAT 6	5.00	OPR	1949	ST	GAS		NONE	FW	PAR	PAR
MEDICINE HAT 7	30.00	OPR	1953	ST	GAS		NONE	FW	PAR	PAR
MEDICINE HAT 8	30.00	PLN	1996	ST	GAS		NONE			
MEDICINE HAT GT1	19.50	OPR	1975	GT	GAS		NONE	N/A	WH	WH
MEDICINE HAT GT2	35.00	OPR	1979	GT	GAS		NONE	N/A	WH	WH
MEDICINE HAT GT3	35.00	OPR	1979	GT	GAS		NONE	N/A	WH	WH
MEDICINE HAT GT4	34.00	CON	1993	GT	GAS		OIL	N/A	GE	GE
MEDICINE HAT GT5	34.00	CON	1993	GT	GAS		OIL	N/A	GE	GE
OPERATOR: MILLAR WESTERN POWER LTD										
WHITECOURT 1	21.50	PLN	1994	ST	WOOD		NONE	OUTO	GE	GE
OPERATOR: MINAS BASIN PULP & PAPER CO										
SALMON HOLE 1	2.00	OPR	1974	HY	WAT	CONV	NONE	N/A	BARBER	CGE
ST CROIX 1	3.00	OPR	1974	HY	WAT	CONV	NONE	N/A	DEW	CGE
OPERATOR: MIRAMICHI PULP & PAPER LTD										
NEWCASTLE 1	17.60	OPR	1966	ST/S	LIQ		NONE	CE	GE	GE
OPERATOR: NBIP FOREST PRODUCTS LTD										
DALHOUSIE (NBIP) 1	6.00	OPR	1929	ST/S	OIL	HFO	NONE	CE/BW	GE	GE

DIRECTORY OF POWER PLANTS IN CANADA
UDI/MCGRAW-HILL

PLANT/UNIT NAME	MW	STA	YEAR	TYPE	FUEL	FUEL TYPE	ALT FUEL	SSS MFR	TURB MFR	GEN MFR
OPERATOR: NBIP FOREST PRODUCTS LTD										
DALHOUSIE (NBIP) 2	0.75	OPR	1930	ST/S	OIL	HFO	NONE	CE/BW	WHA	WHA
DALHOUSIE (NBIP) 3	0.75	OPR	1930	ST/S	OIL	HFO	NONE	CE/BW	WHA	WHA
OPERATOR: NELSON LIGHT DEPARTMENT										
BONNINGTON 1	0.75	STN		HY	WAT	CONV	NONE	N/A		
BONNINGTON 2	1.20	OPR	1908	HY	WAT	CONV	NONE	N/A		
BONNINGTON 3	2.40	OPR	1929	HY	WAT	CONV	NONE	N/A	ACC	CGE
BONNINGTON 4	6.00	OPR	1948	HY	WAT	CONV	NONE	N/A	ACC	CGE
OPERATOR: NERCO CONSOLIDATED MINE LTD										
YELLOWKNIFE 1	3.36	OPR	1941	HY	WAT	CONV	NONE	N/A	AC	WH
OPERATOR: NEW BRUNSWICK POWER CORP										
BEECHWOOD 1	36.00	OPR	1957	HY	WAT	CONV	NONE	N/A	DEW	CGE
BEECHWOOD 2	36.00	OPR	1958	HY	WAT	CONV	NONE	N/A	DEW	CGE
BEECHWOOD 3	40.50	OPR	1962	HY	WAT	CONV	NONE	N/A	ACC	WH
BELLEDUNE 1	440.00	CON	1993	ST	COAL	BIT	NONE	CE	TS	TS
BELLEDUNE 2	440.00	PLN	2004	ST	COAL	BIT	NONE	CE		
CHATHAM 1	12.50	OPR	1948	ST	COAL	BIT	NONE	FW	PAR	PAR
CHATHAM 2	20.00	OPR	1956	ST	COAL	BIT	NONE	CE	BBC	BBC
COLESON COVE 1	350.00	OPR	1976	ST	OIL	HFO	NONE	BW	HT	HT
COLESON COVE 2	350.00	OPR	1976	ST	OIL	HFO	NONE	BW	HT	HT
COLESON COVE 3	350.00	OPR	1977	ST	OIL	HFO	NONE	BW	HT	HT
COURTENAY BAY 1	50.00	OPR	1961	ST	OIL	HFO	NONE	CE	EE	EE
COURTENAY BAY 2	13.37	OPR	1965	ST	OIL	HFO	NONE	BW	BBC	BBC
COURTENAY BAY 3	100.00	OPR	1966	ST	OIL	HFO	NONE	BW	BBC	BBC
COURTENAY BAY 4	100.00	OPR	1967	ST	OIL	HFO	NONE	BW	BBC	BBC
DALHOUSIE 1	100.00	OPR	1969	ST	COAL	BIT	NONE	CE	BBC	BBC
DALHOUSIE 2	200.00	OPR	1980	ST	COAL	BIT	NONE	CE	BBC	BBC
GRAND FALLS 1	15.75	OPR	1928	HY	WAT	CONV	NONE	N/A	ACC	CGE
GRAND FALLS 2	15.75	OPR	1929	HY	WAT	CONV	NONE	N/A	ACC	CGE

DIRECTORY OF POWER PLANTS IN CANADA
UDI/MCGRAW-HILL

PLANT/UNIT NAME	MW	STA	YEAR	TYPE	FUEL	FUEL TYPE	ALT FUEL	SSS MFR	TURB MFR	GEN MFR
OPERATOR: NEW BRUNSWICK POWER CORP										
GRAND FALLS 3	15.75	OPR	1930	HY	WAT	CONV	NONE	N/A	ACC	CGE
GRAND FALLS 4	15.75	OPR	1931	HY	WAT	CONV	NONE	N/A	ACC	CGE
GRAND LAKE II 1	5.00	OPR	1951	ST	COAL	BIT	NONE	CE	PAR	PAR
GRAND LAKE II 2	5.00	OPR	1952	ST	COAL	BIT	NONE	CE	PAR	PAR
GRAND LAKE II 3	15.00	OPR	1953	ST	COAL	BIT	NONE	FW	PAR	PAR
GRAND LAKE II 4	60.00	OPR	1964	ST	COAL	BIT	NONE	BWGM	PAR	PAR
GRAND MANAN GT1	25.00	OPR	1989	GT	OIL	DIESEL	NONE	N/A		
GRAND MANAN IC1	0.70	OPR	1963	IC	OIL	DIESEL	NONE	N/A	MIRR	HAWK
GRAND MANAN IC2	0.53	OPR	1965	IC	OIL	DIESEL	NONE	N/A	MIRR	HAWK
GRAND MANAN IC3	0.71	OPR	1967	IC	OIL	DIESEL	NONE	N/A	MIRR	HAWK
GRAND MANAN IC4	0.90	OPR	1969	IC	OIL	DIESEL	NONE	N/A	HAWK	HAWK
GRAND MANAN IC5	1.00	OPR	1974	IC	OIL	DIESEL	NONE	N/A	DD	KATO
MACTAQUAC 1	102.60	OPR	1968	HY	WAT	CONV	NONE	N/A	DEW	WH
MACTAQUAC 2	102.60	OPR	1968	HY	WAT	CONV	NONE	N/A	DEW	WH
MACTAQUAC 3	102.60	OPR	1968	HY	WAT	CONV	NONE	N/A	DEW	WH
MACTAQUAC 4	110.00	OPR	1972	HY	WAT	CONV	NONE	N/A	DEW	WH
MACTAQUAC 5	110.00	OPR	1979	HY	WAT	CONV	NONE	N/A	LMZ	CGE
MACTAQUAC 6	110.00	OPR	1979	HY	WAT	CONV	NONE	N/A	LMZ	CGE
MILLBANK GT1	100.00	OPR	1991	GT	GAS		NONE	N/A	ABB	ABB
MILLBANK GT2	100.00	OPR	1991	GT	GAS		NONE	N/A	ABB	ABB
MILLBANK GT3	100.00	OPR	1991	GT	GAS		NONE	N/A	ABB	ABB
MILLBANK GT4	100.00	OPR	1991	GT	GAS		NONE	N/A	ABB	ABB
MILLTOWN 1	0.70	OPR	1920	HY	WAT	CONV	NONE	N/A	WH	CGE
MILLTOWN 2	0.70	OPR	1920	HY	WAT	CONV	NONE	N/A	WH	CGE
MILLTOWN 3	0.70	OPR	1920	HY	WAT	CONV	NONE	N/A	WH	CGE
MILLTOWN 4	0.25	OPR	1947	HY	WAT	CONV	NONE	N/A	SMS	CGE
MILLTOWN 5	0.30	OPR	1962	HY	WAT	CONV	NONE	N/A	AC	CGE
MILLTOWN 6	0.40	OPR	1968	HY	WAT	CONV	NONE	N/A	AC	ASEA
MILLTOWN 7	0.60	OPR	1969	HY	WAT	CONV	NONE	N/A	DEW	WH
MONCTON GT1	23.38	OPR	1971	GT	OIL	DIESEL	NONE	N/A	PW	BRUSH
MUSQUASH 1	2.32	OPR	1920	HY	WAT	CONV	NONE	N/A	SMS	CGE

DIRECTORY OF POWER PLANTS IN CANADA
UDI/MCGRAW-HILL

PLANT/UNIT NAME	MW	STA	YEAR	TYPE	FUEL	FUEL TYPE	ALT FUEL	SSS MFR	TURB MFR	GEN MFR
OPERATOR: *NEW BRUNSWICK POWER CORP*										
MUSQUASH 2	2.32	OPR	1920	HY	WAT	CONV	NONE	N/A	SMS	CGE
NBP GT1	100.00	PLN	1999	GT	OIL		NONE	N/A		
NBP GT2	100.00	PLN	2000	GT	OIL		NONE	N/A		
NBP GT3	100.00	PLN	2001	GT	OIL		NONE	N/A		
POINT LEPREAU 1	680.00	OPR	1983	ST	UR	CANDU	NONE	AECL	PAR	PAR
POINT LEPREAU IC1	4.80	OPR	1977	IC	OIL	DIESEL	NONE	N/A		
POINT LEPREAU IC2	4.80	OPR	1977	IC	OIL	DIESEL	NONE	N/A		
POINT LEPREAU IC3	0.95	OPR	1977	IC	OIL	DIESEL	NONE	N/A		
POINT LEPREAU IC4	0.95	OPR	1977	IC	OIL	DIESEL	NONE	N/A		
SISSON 1	10.00	OPR	1965	HY	WAT	CONV	NONE	N/A	ACC	WHC
STE-ROSE GT1	100.00	OPR	1991	GT	GAS		NONE	N/A	ABB	ABB
TOBIQUE 1	10.00	OPR	1943	HY	WAT	CONV	NONE	N/A	SMS	CGE
TOBIQUE 2	10.00	OPR	1943	HY	WAT	CONV	NONE	N/A	SMS	CGE
OPERATOR: *NEWFOUNDLAND & LABRADOR HYDRO*										
BAY D'ESPOIR 1	76.50	OPR	1967	HY	WAT	CONV	NONE	N/A	ACC	CGE
BAY D'ESPOIR 2	76.50	OPR	1967	HY	WAT	CONV	NONE	N/A	ACC	CGE
BAY D'ESPOIR 3	76.50	OPR	1967	HY	WAT	CONV	NONE	N/A	ACC	CGE
BAY D'ESPOIR 4	76.50	OPR	1968	HY	WAT	CONV	NONE	N/A	ACC	CGE
BAY D'ESPOIR 5	76.50	OPR	1970	HY	WAT	CONV	NONE	N/A	ACC	CGE
BAY D'ESPOIR 6	76.50	OPR	1970	HY	WAT	CONV	NONE	N/A	ACC	CGE
BAY D'ESPOIR 7	154.00	OPR	1977	HY	WAT	CONV	NONE	N/A	DEW	CGE
BLACK TICKLE IC1	0.25	OPR	1978	IC	OIL	DIESEL	NONE	N/A	CAT	BBC
BLACK TICKLE IC2	0.25	OPR	1978	IC	OIL	DIESEL	NONE	N/A	CAT	BBC
BLACK TICKLE IC3	0.30	OPR	1978	IC	OIL	DIESEL	NONE	N/A	CAT	BBC
CARTWRIGHT IC1	0.30	OPR	1975	IC	OIL	DIESEL	NONE	N/A	CAT	BBC
CARTWRIGHT IC2	0.30	OPR	1978	IC	OIL	DIESEL	NONE	N/A	CAT	KATO
CARTWRIGHT IC4	0.45	OPR	1987	IC	OIL	DIESEL	NONE	N/A		
CARTWRIGHT IC5	0.45	OPR	1987	IC	OIL	DIESEL	NONE	N/A		
CAT ARM 1	71.72	OPR	1985	HY	WAT	CONV	NONE	N/A	DB	CGE
CAT ARM 2	71.72	OPR	1985	HY	WAT	CONV	NONE	N/A	DB	CGE

DIRECTORY OF POWER PLANTS IN CANADA
UDI/MCGRAW-HILL

PLANT/UNIT NAME	MW	STA	YEAR	TYPE	FUEL	FUEL TYPE	ALT FUEL	SSS MFR	TURB MFR	GEN MFR
OPERATOR: NEWFOUNDLAND & LABRADOR HYDRO										
CHARLOTTETOWN (NLH) IC1	0.06	UNK	1971	IC	OIL	DIESEL	NONE	N/A	DEUTZ	TAMPER
CHARLOTTETOWN (NLH) IC2	0.14	UNK	1975	IC	OIL	DIESEL	NONE	N/A	CAT	TAMPER
CHARLOTTETOWN (NLH) IC3	0.14	OPR	1978	IC	OIL	DIESEL	NONE	N/A	CAT	BBC
CHARLOTTETOWN (NLH) IC4	0.30	OPR	1975	IC	OIL	DIESEL	NONE	N/A		
CHARLOTTETOWN (NLH) IC5	0.25	OPR	1986	IC	OIL	DIESEL	NONE	N/A		
DAVIS INLET IC1	0.06	OPR	1970	IC	OIL	DIESEL	NONE	N/A	CAT	TAMPER
DAVIS INLET IC2	0.14	OPR	1975	IC	OIL	DIESEL	NONE	N/A	CAT	TAMPER
DAVIS INLET IC3	0.14	OPR	1975	IC	OIL	DIESEL	NONE	N/A	CAT	TAMPER
DAVIS INLET IC4	0.10	OPR	1989	IC	OIL	DIESEL	NONE	N/A	CAT	TAMPER
FLOWERS COVE IC1	0.60	OPR	1970	IC	OIL	DIESEL	NONE	N/A	CAT	TAMPER
FLOWERS COVE IC2	0.60	OPR	1972	IC	OIL	DIESEL	NONE	N/A	CAT	TAMPER
FLOWERS COVE IC3	0.70	OPR	1973	IC	OIL	DIESEL	NONE	N/A	CAT	TAMPER
FLOWERS COVE IC4	0.80	OPR	1975	IC	OIL	DIESEL	NONE	N/A	CAT	TAMPER
FLOWERS COVE IC5	0.80	OPR	1987	IC	OIL	DIESEL	NONE	N/A		
FOX HARBOUR IC1	0.22	OPR	1974	IC	OIL	DIESEL	NONE	N/A	CAT	TAMPER
FOX HARBOUR IC2	0.14	OPR	1978	IC	OIL	DIESEL	NONE	N/A	CAT	BBC
FOX HARBOUR IC3	0.14	OPR	1978	IC	OIL	DIESEL	NONE	N/A	CAT	BBC
FOX HARBOUR IC4	0.08	OPR	1980	IC	OIL	DIESEL	NONE	N/A	CAT	TAMPER
FRANCOIS IC1	0.10	OPR	1971	IC	OIL	DIESEL	NONE	N/A	CUM	ON
FRANCOIS IC2	0.18	UNK	1980	IC	OIL	DIESEL	NONE	N/A	CAT	BBC
FRANCOIS IC3	0.20	OPR	1980	IC	OIL	DIESEL	NONE	N/A	CUM	ON
GOOSE BAY NORTH IC1	0.75	OPR	1952	IC	OIL	DIESEL	NONE	N/A	MIRR	GE
GOOSE BAY NORTH IC2	0.75	OPR	1952	IC	OIL	DIESEL	NONE	N/A	MIRR	GE
GOOSE BAY NORTH IC3	0.75	OPR	1952	IC	OIL	DIESEL	NONE	N/A	MIRR	GE
GOOSE BAY NORTH IC4	0.75	OPR	1952	IC	OIL	DIESEL	NONE	N/A	MIRR	GE
GOOSE BAY NORTH IC5	1.00	OPR	1958	IC	OIL	DIESEL	NONE	N/A	GM	GM
GOOSE BAY NORTH IC6	2.50	OPR	1968	IC	OIL	DIESEL	NONE	N/A	GM	GM
GOOSE BAY NORTH IC7	2.60	OPR	1969	IC	OIL	DIESEL	NONE	N/A	GM	GM
GOOSE BAY NORTH IC8	2.60	OPR	1974	IC	OIL	DIESEL	NONE	N/A	GM	GM
GREY RIVER IC1	0.06	UNK	1970	IC	OIL	DIESEL	NONE	N/A	CAT	TAMPER
GREY RIVER IC2	0.14	OPR	1975	IC	OIL	DIESEL	NONE	N/A	CAT	TAMPER

DIRECTORY OF POWER PLANTS IN CANADA
UDI/MCGRAW-HILL

PLANT/UNIT NAME	MW	STA	YEAR	TYPE	FUEL	FUEL TYPE	ALT FUEL	SSS MFR	TURB MFR	GEN MFR
OPERATOR: NEWFOUNDLAND & LABRADOR HYDRO										
GREY RIVER IC3	0.14	OPR	1975	IC	OIL	DIESEL	NONE	N/A	CAT	TAMPER
GREY RIVER IC4	0.25	OPR	1989	IC	OIL	DIESEL	NONE	N/A		
HAPPY VALLEY GT1	25.00	OPR	1992	GT	OIL	DIST	NONE	N/A	P&W	
HARBOUR DEEP IC1	0.25	OPR	1974	IC	OIL	DIESEL	NONE	N/A	CAT	TAMPER
HARBOUR DEEP IC2	0.14	OPR	1975	IC	OIL	DIESEL	NONE	N/A	CAT	TAMPER
HARBOUR DEEP IC3	0.14	OPR	1979	IC	OIL	DIESEL	NONE	N/A	CAT	BBC
HARBOUR DEEP IC4	0.14	OPR	1980	IC	OIL	DIESEL	NONE	N/A	CAT	LS
HARDWOODS GT1	54.00	OPR	1977	GT	OIL	DIST	NONE	N/A	RR	BRUSH
HAWKES BAY IC1	2.50	OPR	1971	IC	OIL	DIESEL	NONE	N/A	GM	GM
HAWKES BAY IC2	2.50	OPR	1971	IC	OIL	DIESEL	NONE	N/A	GM	GM
HINDS LAKE 1	75.00	OPR	1980	HY	WAT	CONV	NONE	N/A	NOHAB	HT
HOLYROOD 1	175.00	OPR	1970	ST	OIL	HFO	NONE	CE	GE	GE
HOLYROOD 2	175.00	OPR	1971	ST	OIL	HFO	NONE	CE	GE	GE
HOLYROOD 3	150.00	OPR	1979	ST	OIL	HFO	NONE	BW	HT	HT
HOLYROOD GT1	14.15	OPR	1966	GT	OIL	DIST	NONE	N/A	RR	AEI
HOPEDALE IC1	0.18	OPR	1975	IC	OIL	DIESEL	NONE	N/A	CAT	BBC
HOPEDALE IC2	0.20	OPR	1980	IC	OIL	DIESEL	NONE	N/A	CAT	CAT
HOPEDALE IC3	0.25	OPR	1984	IC	OIL	DIESEL	NONE	N/A	CAT	LS
HOPEDALE IC4	0.30	OPR	1974	IC	OIL	DIESEL	NONE	N/A		
L'ANSE AU LOUP IC1	0.60	OPR	1974	IC	OIL	DIESEL	NONE	N/A	CAT	TAMPER
L'ANSE AU LOUP IC2	0.60	OPR	1974	IC	OIL	DIESEL	NONE	N/A	CAT	TAMPER
L'ANSE AU LOUP IC3	0.80	OPR	1976	IC	OIL	DIESEL	NONE	N/A	CAT	TAMPER
L'ANSE AU LOUP IC4	1.10	OPR	1984	IC	OIL	DIESEL	NONE	N/A	CAT	KATO
L'ANSE AU LOUP IC5	0.80	OPR	1981	IC	OIL	DIESEL	NONE	N/A		
LA POILE IC1	0.04	UNK	1975	IC	OIL	DIESEL	NONE	N/A	DEUTZ	TAMPER
LA POILE IC2	0.06	OPR	1975	IC	OIL	DIESEL	NONE	N/A	DEUTZ	TAMPER
LA POILE IC3	0.10	OPR	1975	IC	OIL	DIESEL	NONE	N/A	DEUTZ	TAMPER
LA POILE IC4	0.14	OPR	1986	IC	OIL	DIESEL	NONE	N/A		
LITTLE BAY ISLANDS IC1	0.10	UNK	1970	IC	OIL	DIESEL	NONE	N/A	BUDA	AC
LITTLE BAY ISLANDS IC2	0.10	UNK	1975	IC	OIL	DIESEL	NONE	N/A	CUM	MR
LITTLE BAY ISLANDS IC3	0.30	OPR	1979	IC	OIL	DIESEL	NONE	N/A	CAT	BBC

DIRECTORY OF POWER PLANTS IN CANADA
UDI/MCGRAW-HILL

PLANT/UNIT NAME	MW	STA	YEAR	TYPE	FUEL	FUEL TYPE	ALT FUEL	SSS MFR	TURB MFR	GEN MFR
OPERATOR: _NEWFOUNDLAND & LABRADOR HYDRO_										
LITTLE BAY ISLANDS IC4	0.30	OPR	1980	IC	OIL	DIESEL	NONE	N/A	CAT	LS
LITTLE BAY ISLANDS IC5	0.45	OPR	1987	IC	OIL	DIESEL	NONE	N/A		
MAIN BROOK IC1	0.25	OPR	1970	IC	OIL	DIESEL	NONE	N/A	DEUTZ	TAMPER
MAIN BROOK IC2	0.25	OPR	1974	IC	OIL	DIESEL	NONE	N/A	CAT	TAMPER
MAIN BROOK IC3	0.25	OPR	1980	IC	OIL	DIESEL	NONE	N/A	CAT	LS
MAIN BROOK IC4	0.45	UNK	1984	IC	OIL	DIESEL	NONE	N/A	DEUTZ	STM
MAKKOVIK IC1	0.25	OPR	1974	IC	OIL	DIESEL	NONE	N/A	CAT	CAT
MAKKOVIK IC2	0.25	OPR	1978	IC	OIL	DIESEL	NONE	N/A	CAT	TAMPER
MAKKOVIK IC3	0.45	OPR	1980	IC	OIL	DIESEL	NONE	N/A	CAT	CAT
MARYS HARBOUR IC1	0.30	OPR	1974	IC	OIL	DIESEL	NONE	N/A	CAT	GE
MARYS HARBOUR IC2	0.25	OPR	1975	IC	OIL	DIESEL	NONE	N/A	CAT	TAMPER
MARYS HARBOUR IC3	0.25	OPR	1975	IC	OIL	DIESEL	NONE	N/A	CAT	TAMPER
MARYS HARBOUR IC4	0.18	OPR	1980	IC	OIL	DIESEL	NONE	N/A		
MCCALLUM IC1	0.14	OPR	1975	IC	OIL	DIESEL	NONE	N/A	CAT	TAMPER
MCCALLUM IC2	0.14	OPR	1975	IC	OIL	DIESEL	NONE	N/A	CAT	TAMPER
MCCALLUM IC3	0.06	UNK	1975	IC	OIL	DIESEL	NONE	N/A	CAT	TAMPER
MCCALLUM IC4	0.25	UNK	1975	IC	OIL	DIESEL	NONE	N/A		
MUD LAKE IC1	0.06	OPR	1975	IC	OIL	DIESEL	NONE	N/A	DEUTZ	TAMPER
MUD LAKE IC2	0.05	OPR	1980	IC	OIL	DIESEL	NONE	N/A	CAT	CAT
MUD LAKE IC3	0.05	OPR	1980	IC	OIL	DIESEL	NONE	N/A	CAT	CAT
NAIN IC1	0.30	OPR	1974	IC	OIL	DIESEL	NONE	N/A	CAT	TAMPER
NAIN IC2	0.35	OPR	1975	IC	OIL	DIESEL	NONE	N/A	DD	KOHLER
NAIN IC3	0.30	OPR	1975	IC	OIL	DIESEL	NONE	N/A	CAT	TAMPER
NAIN IC4	0.30	OPR	1978	IC	OIL	DIESEL	NONE	N/A		
NORMAN BAY IC1	0.03	OPR	1987	IC	OIL	DIESEL	NONE	N/A		
NORMAN BAY IC2	0.03	OPR	1987	IC	OIL	DIESEL	NONE	N/A		
NORMAN BAY IC3	0.03	OPR	1987	IC	OIL	DIESEL	NONE	N/A		
PARADISE RIVER H1	8.01	OPR	1987	HY	WAT	CONV	NONE	N/A		
PARADISE RIVER IC1	0.04	OPR	1971	IC	OIL	DIESEL	NONE	N/A	DEUTZ	TAMPER
PARADISE RIVER IC2	0.04	OPR	1971	IC	OIL	DIESEL	NONE	N/A	DEUTZ	TAMPER
PARADISE RIVER IC3	0.06	OPR	1971	IC	OIL	DIESEL	NONE	N/A	DEUTZ	TAMPER

DIRECTORY OF POWER PLANTS IN CANADA
UDI/MCGRAW-HILL

PLANT/UNIT NAME	MW	STA	YEAR	TYPE	FUEL	FUEL TYPE	ALT FUEL	SSS MFR	TURB MFR	GEN MFR
OPERATOR: **NEWFOUNDLAND & LABRADOR HYDRO**										
PETIT FORTE IC1	0.06	OPR	1971	IC	OIL	DIESEL	NONE	N/A	DEUTZ	TAMPER
PETIT FORTE IC2	0.06	UNK	1971	IC	OIL	DIESEL	NONE	N/A	DEUTZ	TAMPER
PETIT FORTE IC3	0.14	OPR	1980	IC	OIL	DIESEL	NONE	N/A	CAT	BBC
PETIT FORTE IC4	0.14	OPR	1980	IC	OIL	DIESEL	NONE	N/A		
PETITES IC1	0.10	OPR	1974	IC	OIL	DIESEL	NONE	N/A	DEUTZ	TAMPER
PETITES IC2	0.10	OPR	1974	IC	OIL	DIESEL	NONE	N/A	DEUTZ	TAMPER
PETITES IC3	0.06	OPR	1975	IC	OIL	DIESEL	NONE	N/A	CUM	TAMPER
POND COVE IC1	0.92	OPR	1978	IC	OIL	DIESEL	NONE	N/A	DD	EP
POND COVE IC2	0.85	UNK	1980	IC	OIL	DIESEL	NONE	N/A	CAT	KATO
POND COVE IC3	0.80	UNK	1981	IC	OIL	DIESEL	NONE	N/A	CAT	BBC
POND COVE IC4	0.92	UNK	1978	IC	OIL	DIESEL	NONE	N/A		
POND COVE IC5	0.70	UNK	1985	IC	OIL	DIESEL	NONE	N/A		
PORT HOPE SIMPSON IC1	0.25	OPR	1974	IC	OIL	DIESEL	NONE	N/A	CAT	TAMPER
PORT HOPE SIMPSON IC2	0.25	OPR	1974	IC	OIL	DIESEL	NONE	N/A	CAT	TAMPER
PORT HOPE SIMPSON IC3	0.14	OPR	1975	IC	OIL	DIESEL	NONE	N/A	CAT	TAMPER
PORT HOPE SIMPSON IC4	0.35	OPR	1971	IC	OIL	DIESEL	NONE	N/A		
POSTVILLE IC1	0.08	OPR	1973	IC	OIL	DIESEL	NONE	N/A	CAT	TAMPER
POSTVILLE IC2	0.08	OPR	1973	IC	OIL	DIESEL	NONE	N/A	CAT	TAMPER
POSTVILLE IC3	0.08	OPR	1976	IC	OIL	DIESEL	NONE	N/A	CAT	TAMPER
POSTVILLE IC4	0.05	UNK	1980	IC	OIL	DIESEL	NONE	N/A	CAT	TAMPER
POSTVILLE IC5	0.23	OPR	1974	IC	OIL	DIESEL	NONE	N/A		
POSTVILLE IC6	0.16	OPR	1987	IC	OIL	DIESEL	NONE	N/A		
RAMEA IC1	0.30	OPR	1970	IC	OIL	DIESEL	NONE	N/A	LB	TAMPER
RAMEA IC2	0.30	OPR	1970	IC	OIL	DIESEL	NONE	N/A	LB	TAMPER
RAMEA IC3	0.44	OPR	1972	IC	OIL	DIESEL	NONE	N/A	LB	TAMPER
RAMEA IC4	0.43	OPR	1974	IC	OIL	DIESEL	NONE	N/A	LIST	TAMPER
RAMEA IC5	0.57	OPR	1977	IC	OIL	DIESEL	NONE	N/A	LB	TAMPER
RAMEA IC6	1.00	OPR	1980	IC	OIL	DIESEL	NONE	N/A	PAX	GE
RENCONTRE EAST IC1	0.06	UNK	1974	IC	OIL	DIESEL	NONE	N/A	CAT	TAMPER
RENCONTRE EAST IC2	0.14	OPR	1980	IC	OIL	DIESEL	NONE	N/A	CAT	BBC
RENCONTRE EAST IC3	0.14	OPR	1980	IC	OIL	DIESEL	NONE	N/A	CAT	BBC

DIRECTORY OF POWER PLANTS IN CANADA
UDI/MCGRAW-HILL

PLANT/UNIT NAME	MW	STA	YEAR	TYPE	FUEL	FUEL TYPE	ALT FUEL	SSS MFR	TURB MFR	GEN MFR
OPERATOR: NEWFOUNDLAND & LABRADOR HYDRO										
RENCONTRE EAST IC4	0.30	OPR	1980	IC	OIL	DIESEL	NONE	N/A		
RENCONTRE EAST IC5	0.25	OPR	1986	IC	OIL	DIESEL	NONE	N/A		
RIGOLET IC1	0.25	UNK	1974	IC	OIL	DIESEL	NONE	N/A	CAT	TAMPER
RIGOLET IC2	0.06	UNK	1974	IC	OIL	DIESEL	NONE	N/A	CAT	TAMPER
RIGOLET IC3	0.13	OPR	1980	IC	OIL	DIESEL	NONE	N/A	CAT	CAT
RIGOLET IC4	0.25	UNK	1980	IC	OIL	DIESEL	NONE	N/A	CAT	GE
RIGOLET IC5	0.10	OPR	1982	IC	OIL	DIESEL	NONE	N/A		
RIGOLET IC6	0.25	OPR	1988	IC	OIL	DIESEL	NONE	N/A		
RODDICKTON 1	5.00	OPR	1989	ST	WOOD		NONE			
RODDICKTON IC1	1.00	OPR	1975	IC	OIL	DIESEL	NONE	N/A	RHL	TAMPER
RODDICKTON IC2	0.56	OPR	1975	IC	OIL	DIESEL	NONE	N/A	LIST	TAMPER
RODDICKTON IC3	0.45	OPR	1975	IC	OIL	DIESEL	NONE	N/A	DD	KOHLER
RODDICKTON IC4	1.00	OPR	1977	IC	OIL	DIESEL	NONE	N/A	RHL	TAMPER
RODDICKTON IC5	0.45	OPR	1981	IC	OIL	DIESEL	NONE	N/A	CAT	CAT
SNOOKS ARM 1	0.56	OPR	1957	HY	WAT	CONV	NONE	N/A	GGG	LANCS
SOUTH EAST BIGHT IC1	0.06	OPR	1974	IC	OIL	DIESEL	NONE	N/A	DEUTZ	TAMPER
SOUTH EAST BIGHT IC2	0.06	UNK	1974	IC	OIL	DIESEL	NONE	N/A	DEUTZ	TAMPER
SOUTH EAST BIGHT IC3	0.06	UNK	1974	IC	OIL	DIESEL	NONE	N/A	DEUTZ	TAMPER
SOUTH EAST BIGHT IC4	0.14	OPR	1985	IC	OIL	DIESEL	NONE	N/A		
SOUTH EAST BIGHT IC5	0.14	OPR	1987	IC	OIL	DIESEL	NONE	N/A		
ST ANTHONY IC1	1.00	OPR	1973	IC	OIL	DIESEL	NONE	N/A	PAX	TAMPER
ST ANTHONY IC2	1.00	OPR	1973	IC	OIL	DIESEL	NONE	N/A	PAX	TAMPER
ST ANTHONY IC3	1.00	OPR	1973	IC	OIL	DIESEL	NONE	N/A	PAX	TAMPER
ST ANTHONY IC4	1.00	OPR	1975	IC	OIL	DIESEL	NONE	N/A	PAX	TAMPER
ST ANTHONY IC5	2.00	OPR	1980	IC	OIL	DIESEL	NONE	N/A	PAX	GEC
ST ANTHONY IC6	2.00	OPR	1982	IC	OIL	DIESEL	NONE	N/A	PAX	GEC
ST BRENDANS IC1	0.06	UNK	1970	IC	OIL	DIESEL	NONE	N/A	CAT	TAMPER
ST BRENDANS IC2	0.25	OPR	1974	IC	OIL	DIESEL	NONE	N/A	CAT	TAMPER
ST BRENDANS IC3	0.25	UNK	1974	IC	OIL	DIESEL	NONE	N/A	CAT	TAMPER
ST BRENDANS IC4	0.30	UNK	1974	IC	OIL	DIESEL	NONE	N/A		
ST BRENDANS IC5	0.30	UNK	1974	IC	OIL	DIESEL	NONE	N/A		

UDI-033-92 *December 1992*

DIRECTORY OF POWER PLANTS IN CANADA
UDI/MCGRAW-HILL

PLANT/UNIT NAME	MW	STA	YEAR	TYPE	FUEL	FUEL TYPE	ALT FUEL	SSS MFR	TURB MFR	GEN MFR
OPERATOR: NEWFOUNDLAND & LABRADOR HYDRO										
ST LEWIS IC1	0.14	OPR	1976	IC	OIL	DIESEL	NONE	N/A		
ST LEWIS IC2	0.14	OPR	1978	IC	OIL	DIESEL	NONE	N/A		
ST LEWIS IC3	0.22	OPR	1974	IC	OIL	DIESEL	NONE	N/A		
ST LEWIS IC4	0.25	OPR	1987	IC	OIL	DIESEL	NONE	N/A		
STEPHENVILLE GT1	54.00	OPR	1976	GT	OIL	DIST	NONE	N/A	RR	BRUSH
UPPER SALMON 1	84.00	OPR	1982	HY	WAT	CONV	NONE	N/A		
VENAMS BIGHT 1	0.36	OPR	1957	HY	WAT	CONV	NONE	N/A	GGG	LANCS
WESTPORT IC1	0.06	UNK	1970	IC	OIL	DIESEL	NONE	N/A	CAT	TAMPER
WESTPORT IC2	0.25	OPR	1974	IC	OIL	DIESEL	NONE	N/A	CAT	TAMPER
WESTPORT IC3	0.25	OPR	1980	IC	OIL	DIESEL	NONE	N/A	CAT	TAMPER
WESTPORT IC4	0.25	OPR	1980	IC	OIL	DIESEL	NONE	N/A	CAT	TAMPER
WILLIAMS HARBOR IC1	0.14	OPR	1975	IC	OIL	DIESEL	NONE	N/A	CAT	TAMPER
WILLIAMS HARBOR IC2	0.14	OPR	1975	IC	OIL	DIESEL	NONE	N/A	CAT	TAMPER
WILLIAMS HARBOR IC3	0.06	OPR	1980	IC	OIL	DIESEL	NONE	N/A	DEUTZ	TAMPER
OPERATOR: NEWFOUNDLAND LIGHT & POWER CO										
AGUATHUNA IC1	1.20	OPR	1962	IC	OIL	DIESEL	NONE	N/A	HOW	HOW
BELL ISLAND 1	0.34	OPR		WTG	WIND		NONE	N/A		
CAPE BROYLE 1	6.00	OPR	1952	HY	WAT	CONV	NONE	N/A	CVIC	WHC
FALL POND 1	0.40	OPR	1939	HY	WAT	CONV	NONE	N/A	VOI	WH
GREENHILL GT1	26.80	OPR	1975	GT	OIL	DIESEL	NONE	N/A	RR	BRUSH
HEARTS CONTENT 1	2.40	OPR	1960	HY	WAT	CONV	NONE	N/A	EE	PEEB
HORSE CHOPS 1	7.65	OPR	1953	HY	WAT	CONV	NONE	N/A	DEW	CGE
LAWN 1	0.71	OPR	1983	HY	WAT	CONV	NONE	N/A	BARBER	IDEAL
LOCKSTON 1	1.50	OPR	1955	HY	WAT	CONV	NONE	N/A	GGG	CGE
LOCKSTON 2	1.50	OPR	1961	HY	WAT	CONV	NONE	N/A	GGG	CGE
LOOKOUT BROOK 1	2.40	OPR	1958	HY	WAT	CONV	NONE	N/A	GGG	CGE
LOOKOUT BROOK 2	2.67	OPR	1983	HY	WAT	CONV	NONE	N/A	BARBER	IDEAL
MOBILE (NLP) GT1	7.29	OPR	1974	GT	OIL	DIESEL	NONE	N/A	OREND	EMC
MOBILE (NLP) IC1	0.70	OPR	1973	IC	OIL	DIESEL	NONE	N/A	CAT	CANR
MOBILE (NLP) IC2	0.67	OPR	1976	IC	OIL	DIESEL	NONE	N/A	CAT	BBC

DIRECTORY OF POWER PLANTS IN CANADA
UDI/MCGRAW-HILL

PLANT/UNIT NAME	MW	STA	YEAR	TYPE	FUEL	FUEL TYPE	ALT FUEL	SSS MFR	TURB MFR	GEN MFR
OPERATOR: *NEWFOUNDLAND LIGHT & POWER CO*										
MOBILE RIVER 1	9.35	OPR	1951	HY	WAT	CONV	NONE	N/A	DEW	WH
MORRIS 1	1.09	OPR	1983	HY	WAT	CONV	NONE	N/A	BARBER	IDEAL
NEW CHELSEA 1	4.00	OPR	1957	HY	WAT	CONV	NONE	N/A	DEW	WH
PALMQUIST IC1	1.00	OPR	1948	IC	OIL	DIESEL	NONE	N/A	NOHAB	GE
PALMQUIST IC2	1.00	OPR	1953	IC	OIL	DIESEL	NONE	N/A	NOHAB	GE
PALMQUIST IC3	1.00	OPR	1957	IC	OIL	DIESEL	NONE	N/A	NOHAB	GE
PETTY HARBOUR 1	1.60	OPR	1908	HY	WAT	CONV	NONE	N/A	VOI	WH
PETTY HARBOUR 2	1.60	OPR	1911	HY	WAT	CONV	NONE	N/A	VOI	WH
PETTY HARBOUR 3	1.51	OPR	1926	HY	WAT	CONV	NONE	N/A	ARG	CGE
PIERRES BROOK 1	3.20	OPR	1931	HY	WAT	CONV	NONE	N/A	VOI	GEC
PITMANS POND 1	0.80	OPR	1959	HY	WAT	CONV	NONE	N/A	GGG	WH
PORT AUX BASQUES GT1	25.00	CON	1993	GT	OIL		NONE	N/A		
PORT AUX BASQUES IC1	0.25	OPR	1949	IC	OIL	DIESEL	NONE	N/A	CAT	GE
PORT AUX BASQUES IC2	0.35	OPR	1954	IC	OIL	DIESEL	NONE	N/A	CAT	GE
PORT AUX BASQUES IC3	0.35	OPR	1957	IC	OIL	DIESEL	NONE	N/A	CAT	GE
PORT AUX BASQUES IC4	0.21	OPR	1957	IC	OIL	DIESEL	NONE	N/A	CAT	GE
PORT AUX BASQUES IC5	0.25	OPR	1964	IC	OIL	DIESEL	NONE	N/A	CAT	GE
PORT AUX BASQUES IC6	0.25	OPR	1964	IC	OIL	DIESEL	NONE	N/A	CAT	GE
PORT AUX BASQUES IC7	2.50	OPR	1969	IC	OIL	DIESEL	NONE	N/A	GM	GM
PORT UNION 1	0.28	OPR	1918	HY	WAT	CONV	NONE	N/A	PWW	CGE
PORT UNION 2	0.28	OPR	1918	HY	WAT	CONV	NONE	N/A	PWW	CGE
PORT UNION IC2	0.50	OPR	1961	IC	OIL	DIESEL	NONE	N/A	CAT	CAT
RATTLING BROOK 1	6.38	OPR	1958	HY	WAT	CONV	NONE	N/A	ACC	CGE
RATTLING BROOK 2	6.38	OPR	1958	HY	WAT	CONV	NONE	N/A	ACC	CGE
ROCKY POND 1	3.20	OPR	1943	HY	WAT	CONV	NONE	N/A	DEW	WH
ROSE BLANCHE	6.00	CON	1993	HY	WAT	CONV	NONE	N/A		
SALT POND GT1	14.15	OPR	1968	GT	OIL	DIESEL	NONE	N/A	RR	AEI
SALT POND IC1	0.50	OPR	1963	IC	OIL	DIESEL	NONE	N/A	WORT	EMC
SALT POND IC2	0.50	OPR	1963	IC	OIL	DIESEL	NONE	N/A	WORT	EMC
SALT POND IC3	0.50	OPR	1963	IC	OIL	DIESEL	NONE	N/A	WORT	EMC
SANDY BROOK 1	5.95	OPR	1963	HY	WAT	CONV	NONE	N/A	DEW	WH

DIRECTORY OF POWER PLANTS IN CANADA
UDI/MCGRAW-HILL

PLANT/UNIT NAME	MW	STA	YEAR	TYPE	FUEL	FUEL TYPE	ALT FUEL	SSS MFR	TURB MFR	GEN MFR
OPERATOR: NEWFOUNDLAND LIGHT & POWER CO										
SEAL COVE 1	1.20	OPR	1922	HY	WAT	CONV	NONE	N/A	AC	AC
SEAL COVE 2	2.54	OPR	1927	HY	WAT	CONV	NONE	N/A	VOI	WH
ST JOHNS 1	10.00	OPR	1957	ST	OIL	HFO	NONE	BWGM	AEI	AEI
ST JOHNS 2	20.00	OPR	1959	ST	OIL	HFO	NONE	BWGM	AEI	AEI
ST JOHNS IC1	2.50	OPR	1956	IC	OIL	DIESEL	NONE	N/A	NBG	GE
TOPSAIL 1	2.28	OPR	1983	HY	WAT	CONV	NONE	N/A	BARBER	IDEAL
TORS COVE 1	2.00	OPR	1942	HY	WAT	CONV	NONE	N/A	EE	EE
TORS COVE 2	2.00	OPR	1942	HY	WAT	CONV	NONE	N/A	EE	EE
TORS COVE 3	2.50	OPR	1951	HY	WAT	CONV	NONE	N/A	EE	EE
VICTORIA 1	0.45	OPR	1914	HY	WAT	CONV	NONE	N/A	VOI	WH
WEST BROOK 1	0.70	OPR	1942	HY	WAT	CONV	NONE	N/A	LF	WH
OPERATOR: NORANDA MINES LTD										
MURDOCHVILLE 1	5.40	OPR	1955	ST	WSTH		NONE	CE	BBC	BBC
NORANDA SMELTER 1	2.60	OPR	1934	ST/S	WGAS		NONE	JI	PAR	PAR
NORANDA SMELTER 2	1.50	OPR	1982	ST/S	WGAS		NONE	JI	WAU	LA
OPERATOR: NORDIC POWER INC										
WINDSOR	102.00	PLN	1995	CC	GAS		NONE			
OPERATOR: NORTH PACIFIC POWER CORP										
MURPHY CREEK	275.00	PLN	9999	HY	WAT	CONV	NONE	N/A		
TRAIL 1	55.00	PLN	9999	HY	WAT	CONV	NONE	N/A		
TRAIL 2	55.00	PLN	9999	HY	WAT	CONV	NONE	N/A		
TRAIL 3	55.00	PLN	9999	HY	WAT	CONV	NONE	N/A		
TRAIL 4	55.00	PLN	9999	HY	WAT	CONV	NONE	N/A		
TRAIL 5	55.00	PLN	9999	HY	WAT	CONV	NONE	N/A		
OPERATOR: NORTHLAND POWER										
COCHRANE 1	15.00	OPR	1990	ST	WOOD		NONE		WH	WH
COCHRANE GT1	30.00	OPR	1990	GT/C	GAS		NONE	N/A	GE	BRUSH

DIRECTORY OF POWER PLANTS IN CANADA
UDI/MCGRAW-HILL

PLANT/UNIT NAME	MW	STA	YEAR	TYPE	FUEL	FUEL TYPE	ALT FUEL	SSS MFR	TURB MFR	GEN MFR
OPERATOR: NORTHLAND POWER										
KIRKLAND LAKE 1	15.00	CON	1991	ST	WOOD		NONE	DELT		
OPERATOR: NORTHLAND UTILITIES (NWT) LTD										
DORY POINT IC1	0.10	OPR	1970	IC	OIL	DIESEL	NONE	N/A	CAT	TAMPER
DORY POINT IC2	0.04	OPR	1974	IC	OIL	DIESEL	NONE	N/A	CAT	CAT
DORY POINT IC3	0.08	OPR	1983	IC	OIL	DIESEL	NONE	N/A	CAT	BBC
FORT PROVIDENCE IC1	0.23	OPR	1968	IC	OIL	DIESEL	NONE	N/A	CAT	GE
FORT PROVIDENCE IC2	0.50	OPR	1973	IC	OIL	DIESEL	NONE	N/A	CAT	TAMPER
FORT PROVIDENCE IC3	0.25	OPR	1984	IC	OIL	DIESEL	NONE	N/A	VOLV	MR
HAY RIVER IC1	0.50	OPR	1959	IC	OIL	DIESEL	NONE	N/A	COOP	EE
HAY RIVER IC2	0.65	OPR	1962	IC	OIL	DIESEL	NONE	N/A	COOP	EE
HAY RIVER IC3	0.60	OPR	1969	IC	OIL	DIESEL	NONE	N/A	CAT	TAMPER
HAY RIVER IC4	0.50	OPR	1970	IC	OIL	DIESEL	NONE	N/A	CAT	TAMPER
HAY RIVER IC5	1.20	OPR	1972	IC	OIL	DIESEL	NONE	N/A	WAU	KATO
HAY RIVER IC6	0.80	OPR	1974	IC	OIL	DIESEL	NONE	N/A	CAT	TAMPER
HAY RIVER IC7	0.88	OPR	1974	IC	OIL	DIESEL	NONE	N/A	CAT	TAMPER
HAY RIVER IC8	2.75	OPR	1975	IC	OIL	DIESEL	NONE	N/A	GM	GM
HAY RIVER IC9	1.20	OPR	1978	IC	OIL	DIESEL	NONE	N/A	WAU	KATO
OPERATOR: NORTHWEST TERRITORIES POWER										
AKLAVIK IC1	0.60	OPR	1975	IC	OIL	DIESEL	NONE	N/A	CAT	TAMPER
AKLAVIK IC2	0.30	OPR	1976	IC	OIL	DIESEL	NONE	N/A	CAT	TAMPER
AKLAVIK IC3	0.60	OPR	1981	IC	OIL	DIESEL	NONE	N/A	CAT	KATO
ARTIC BAY IC1	0.23	OPR	1975	IC	OIL	DIESEL	NONE	N/A	CAT	GE
ARTIC BAY IC2	0.40	OPR	1980	IC	OIL	DIESEL	NONE	N/A	CAT	TAMPER
ARTIC BAY IC3	0.40	OPR	1983	IC	OIL	DIESEL	NONE	N/A	CAT	KATO
ARTIC RED RIVER IC1	0.05	OPR	1974	IC	OIL	DIESEL	NONE	N/A	CUM	TAMPER
ARTIC RED RIVER IC2	0.15	OPR	1974	IC	OIL	DIESEL	NONE	N/A	CUM	TAMPER
ARTIC RED RIVER IC3	0.08	OPR	1980	IC	OIL	DIESEL	NONE	N/A	GM	TAMPER
BAKER LAKE IC1	0.55	OPR	1973	IC	OIL	DIESEL	NONE	N/A	CAT	KATO
BAKER LAKE IC2	0.72	OPR	1975	IC	OIL	DIESEL	NONE	N/A	CAT	KATO

DIRECTORY OF POWER PLANTS IN CANADA
UDI/MCGRAW-HILL

PLANT/UNIT NAME	MW	STA	YEAR	TYPE	FUEL	FUEL TYPE	ALT FUEL	SSS MFR	TURB MFR	GEN MFR
OPERATOR: NORTHWEST TERRITORIES POWER										
BAKER LAKE IC3	0.72	OPR	1985	IC	OIL	DIESEL	NONE	N/A	CAT	GE
BROUGHTON ISLAND IC1	0.15	OPR	1972	IC	OIL	DIESEL	NONE	N/A	CAT	KATO
BROUGHTON ISLAND IC2	0.15	OPR	1973	IC	OIL	DIESEL	NONE	N/A	CAT	KATO
BROUGHTON ISLAND IC3	0.30	OPR	1978	IC	OIL	DIESEL	NONE	N/A	CAT	BBC
CAMBRIDGE BAY IC1	0.38	OPR	1972	IC	OIL	DIESEL	NONE	N/A	LB	TAMPER
CAMBRIDGE BAY IC2	0.56	OPR	1972	IC	OIL	DIESEL	NONE	N/A	LB	BRUSH
CAMBRIDGE BAY IC3	0.72	OPR	1973	IC	OIL	DIESEL	NONE	N/A	CAT	GE
CAMBRIDGE BAY IC4	0.72	OPR	1973	IC	OIL	DIESEL	NONE	N/A	CAT	GE
CAPE DORSET IC1	0.30	OPR	1973	IC	OIL	DIESEL	NONE	N/A	CAT	KATO
CAPE DORSET IC2	0.54	OPR	1975	IC	OIL	DIESEL	NONE	N/A	CAT	TAMPER
CAPE DORSET IC3	0.54	OPR	1980	IC	OIL	DIESEL	NONE	N/A	CAT	BBC
CHESTERFIELD INLET IC1	0.20	OPR	1968	IC	OIL	DIESEL	NONE	N/A	CAT	GE
CHESTERFIELD INLET IC2	0.30	OPR	1972	IC	OIL	DIESEL	NONE	N/A	CAT	KATO
CHESTERFIELD INLET IC3	0.40	OPR	1985	IC	OIL	DIESEL	NONE	N/A	CAT	KATO
CLYDE IC1	0.30	OPR	1973	IC	OIL	DIESEL	NONE	N/A	CAT	GE
CLYDE IC2	0.30	OPR	1976	IC	OIL	DIESEL	NONE	N/A	CAT	BBC
CLYDE IC3	0.50	OPR	1976	IC	OIL	DIESEL	NONE	N/A	CAT	GE
COPPERMINE IC1	0.20	OPR	1967	IC	OIL	DIESEL	NONE	N/A	LIST	GE
COPPERMINE IC2	0.20	OPR	1967	IC	OIL	DIESEL	NONE	N/A	LIST	GE
COPPERMINE IC3	0.20	OPR	1967	IC	OIL	DIESEL	NONE	N/A	LIST	GE
COPPERMINE IC4	0.38	OPR	1972	IC	OIL	DIESEL	NONE	N/A	LB	TAMPER
COPPERMINE IC5	0.54	OPR	1976	IC	OIL	DIESEL	NONE	N/A	CAT	TAMPER
CORAL HARBOR IC1	0.25	OPR	1957	IC	OIL	DIESEL	NONE	N/A	CAT	KATO
CORAL HARBOR IC2	0.25	OPR	1957	IC	OIL	DIESEL	NONE	N/A	CAT	KATO
CORAL HARBOR IC3	0.25	OPR	1957	IC	OIL	DIESEL	NONE	N/A	CAT	KATO
CORAL HARBOR IC4	0.30	OPR	1974	IC	OIL	DIESEL	NONE	N/A	CAT	KATO
CORAL HARBOR IC5	0.30	OPR	1975	IC	OIL	DIESEL	NONE	N/A	CAT	GE
DAWSON CITY IC1	0.50	OPR	1971	IC	OIL	DIESEL	NONE	N/A	CAT	KATO
DAWSON CITY IC2	0.50	OPR	1971	IC	OIL	DIESEL	NONE	N/A	CAT	KATO
DAWSON CITY IC3	0.72	OPR	1975	IC	OIL	DIESEL	NONE	N/A	CAT	TAMPER
DAWSON CITY IC4	0.30	OPR	1981	IC	OIL	DIESEL	NONE	N/A	CAT	TAMPER

DIRECTORY OF POWER PLANTS IN CANADA
UDI/MCGRAW-HILL

PLANT/UNIT NAME	MW	STA	YEAR	TYPE	FUEL	FUEL TYPE	ALT FUEL	SSS MFR	TURB MFR	GEN MFR
OPERATOR: _NORTHWEST TERRITORIES POWER_										
DAWSON CITY IC5	0.50	OPR	1981	IC	OIL	DIESEL	NONE	N/A	CAT	TAMPER
ESKIMO POINT IC1	0.30	OPR	1972	IC	OIL	DIESEL	NONE	N/A	CAT	KATO
ESKIMO POINT IC2	0.30	OPR	1973	IC	OIL	DIESEL	NONE	N/A	CAT	KATO
ESKIMO POINT IC3	0.60	OPR	1975	IC	OIL	DIESEL	NONE	N/A	CAT	TAMPER
ESKIMO POINT IC4	0.60	OPR	1980	IC	OIL	DIESEL	NONE	N/A	CAT	BBC
FARO IC1	5.15	OPR	1970	IC	OIL	DIESEL	NONE	N/A	MIRR	BRUSH
FORT FRANKLIN IC1	0.20	OPR	1971	IC	OIL	DIESEL	NONE	N/A	CUM	TAMPER
FORT FRANKLIN IC2	0.30	OPR	1972	IC	OIL	DIESEL	NONE	N/A	CAT	KATO
FORT FRANKLIN IC3	0.30	OPR	1979	IC	OIL	DIESEL	NONE	N/A	CAT	KATO
FORT GOOD HOPE IC1	0.30	OPR	1971	IC	OIL	DIESEL	NONE	N/A	CAT	KATO
FORT GOOD HOPE IC2	0.30	OPR	1974	IC	OIL	DIESEL	NONE	N/A	CAT	GE
FORT GOOD HOPE IC3	0.30	OPR	1983	IC	OIL	DIESEL	NONE	N/A	CAT	GE
FORT LIARD IC1	0.15	OPR	1975	IC	OIL	DIESEL	NONE	N/A	CUM	TAMPER
FORT LIARD IC2	0.20	OPR	1975	IC	OIL	DIESEL	NONE	N/A	CUM	ON
FORT LIARD IC3	0.19	OPR	1982	IC	OIL	DIESEL	NONE	N/A	CAT	TAMPER
FORT LIARD IC4	0.20	OPR	1982	IC	OIL	DIESEL	NONE	N/A	CUM	TAMPER
FORT MCPHERSON IC1	0.38	OPR	1974	IC	OIL	DIESEL	NONE	N/A	LB	TAMPER
FORT MCPHERSON IC2	0.38	OPR	1974	IC	OIL	DIESEL	NONE	N/A	LB	TAMPER
FORT MCPHERSON IC3	0.60	OPR	1974	IC	OIL	DIESEL	NONE	N/A	CAT	KATO
FORT NORMAN IC1	0.30	OPR	1977	IC	OIL	DIESEL	NONE	N/A	GM	TAMPER
FORT NORMAN IC2	0.30	OPR	1979	IC	OIL	DIESEL	NONE	N/A	GM	TAMPER
FORT NORMAN IC3	0.60	OPR	1983	IC	OIL	DIESEL	NONE	N/A	CUM	TAMPER
FORT RESOLUTION IC1	0.15	OPR	1960	IC	OIL	DIESEL	NONE	N/A	MIRR	EE
FORT RESOLUTION IC2	0.20	OPR	1979	IC	OIL	DIESEL	NONE	N/A	LB	GE
FORT RESOLUTION IC3	0.40	OPR	1983	IC	OIL	DIESEL	NONE	N/A	CUM	TAMPER
FORT SIMPSON IC1	0.60	OPR	1962	IC	OIL	DIESEL	NONE	N/A	RHN	GE
FORT SIMPSON IC2	1.00	OPR	1973	IC	OIL	DIESEL	NONE	N/A	RHN	BRUSH
FORT SIMPSON IC3	2.00	OPR	1975	IC	OIL	DIESEL	NONE	N/A	MLW	TAMPER
FORT SMITH IC1	2.00	OPR	1975	IC	OIL	DIESEL	NONE	N/A	MLW	TAMPER
FORT SMITH IC2	1.50	OPR	1975	IC	OIL	DIESEL	NONE	N/A	MLW	BBC
FORT SMITH IC3	2.50	OPR	1983	IC	OIL	DIESEL	NONE	N/A	MLW	BBC

DIRECTORY OF POWER PLANTS IN CANADA
UDI/MCGRAW-HILL

PLANT/UNIT NAME	MW	STA	YEAR	TYPE	FUEL	FUEL TYPE	ALT FUEL	SSS MFR	TURB MFR	GEN MFR
OPERATOR: NORTHWEST TERRITORIES POWER										
FROBISHER BAY IC1	1.00	OPR	1964	IC	OIL	DIESEL	NONE	N/A	MIRR	GE
FROBISHER BAY IC2	2.59	OPR	1969	IC	OIL	DIESEL	NONE	N/A	MIRR	BRUSH
FROBISHER BAY IC3	3.92	OPR	1970	IC	OIL	DIESEL	NONE	N/A	MIRR	BRUSH
FROBISHER BAY IC4	2.50	OPR	1976	IC	OIL	DIESEL	NONE	N/A	GM	EMC
GJOA HAVEN IC1	0.30	OPR	1976	IC	OIL	DIESEL	NONE	N/A	CAT	TAMPER
GJOA HAVEN IC2	0.30	OPR	1979	IC	OIL	DIESEL	NONE	N/A	CAT	TAMPER
GJOA HAVEN IC3	0.60	OPR	1984	IC	OIL	DIESEL	NONE	N/A	CAT	TAMPER
GRISE FIORD IC1	0.18	OPR	1975	IC	OIL	DIESEL	NONE	N/A	CUM	TAMPER
GRISE FIORD IC2	0.15	OPR	1981	IC	OIL	DIESEL	NONE	N/A	CAT	ON
GRISE FIORD IC3	0.08	OPR	1982	IC	OIL	DIESEL	NONE	N/A	DD	DELCO
HALL BEACH IC1	0.30	OPR	1977	IC	OIL	DIESEL	NONE	N/A	CAT	BBC
HALL BEACH IC2	0.30	OPR	1982	IC	OIL	DIESEL	NONE	N/A	CAT	BBC
HALL BEACH IC3	0.20	OPR	1982	IC	OIL	DIESEL	NONE	N/A	CAT	BBC
HOLMAN ISLAND IC1	0.15	OPR	1972	IC	OIL	DIESEL	NONE	N/A	CAT	KATO
HOLMAN ISLAND IC2	0.30	OPR	1979	IC	OIL	DIESEL	NONE	N/A	CAT	KATO
HOLMAN ISLAND IC3	0.40	OPR	1984	IC	OIL	DIESEL	NONE	N/A	CAT	KATO
IGLOOLIK IC1	0.30	OPR	1975	IC	OIL	DIESEL	NONE	N/A	CAT	TAMPER
IGLOOLIK IC2	0.54	OPR	1976	IC	OIL	DIESEL	NONE	N/A	CAT	KATO
IGLOOLIK IC3	0.50	OPR	1985	IC	OIL	DIESEL	NONE	N/A	CAT	KATO
INUVIK IC1	5.18	OPR	1970	IC	OIL	DIESEL	NONE	N/A	MIRR	BRUSH
INUVIK IC2	2.50	OPR	1975	IC	OIL	DIESEL	NONE	N/A	GM	EMC
INUVIK IC3	2.50	OPR	1975	IC	OIL	DIESEL	NONE	N/A	GM	EMC
INUVIK IC4	2.08	OPR	1975	IC	OIL	DIESEL	NONE	N/A	MIRR	BRUSH
INUVIK IC5	2.87	OPR	1984	IC	OIL	DIESEL	NONE	N/A	GM	EMC
INUVIK IC6	0.30	OPR	1984	IC	OIL	DIESEL	NONE	N/A	CAT	GE
JEAN MARIE RIVER IC1	0.04	OPR	1973	IC	OIL	DIESEL	NONE	N/A	GM	DELCO
JEAN MARIE RIVER IC2	0.02	OPR	1979	IC	OIL	DIESEL	NONE	N/A	GM	DELCO
JOHNSONS CROSSINGS IC1	0.02	OPR	1975	IC	OIL	DIESEL	NONE	N/A	DELCO	TAMPER
JOHNSONS CROSSINGS IC2	0.03	OPR	1975	IC	OIL	DIESEL	NONE	N/A	DELCO	TAMPER
JOHNSONS CROSSINGS IC3	0.03	OPR	1980	IC	OIL	DIESEL	NONE	N/A	CAT	CAT
LAC LA MARTE IC1	0.08	OPR	1979	IC	OIL	DIESEL	NONE	N/A	CAT	TAMPER

DIRECTORY OF POWER PLANTS IN CANADA
UDI/MCGRAW-HILL

PLANT/UNIT NAME	MW	STA	YEAR	TYPE	FUEL	FUEL TYPE	ALT FUEL	SSS MFR	TURB MFR	GEN MFR
OPERATOR: NORTHWEST TERRITORIES POWER										
LAC LA MARTE IC2	0.15	OPR	1981	IC	OIL	DIESEL	NONE	N/A	CAT	KATO
LAC LA MARTE IC3	0.19	OPR	1983	IC	OIL	DIESEL	NONE	N/A	CAT	KATO
LAKE HARBOUR IC1	0.15	OPR	1973	IC	OIL	DIESEL	NONE	N/A	CAT	GE
LAKE HARBOUR IC2	0.30	OPR	1976	IC	OIL	DIESEL	NONE	N/A	CAT	BARB
LAKE HARBOUR IC3	0.30	OPR	1983	IC	OIL	DIESEL	NONE	N/A	CAT	BBC
MAYO IC1	0.80	OPR	1975	IC	OIL	DIESEL	NONE	N/A	CAT	TAMPER
MAYO IC2	0.35	OPR	1979	IC	OIL	DIESEL	NONE	N/A	CUM	BBC
NAHANNI BUTTE IC1	0.02	OPR	1973	IC	OIL	DIESEL	NONE	N/A	GM	DELCO
NAHANNI BUTTE IC2	0.05	OPR	1975	IC	OIL	DIESEL	NONE	N/A	GM	DELCO
NAHANNI BUTTE IC3	0.08	OPR	1981	IC	OIL	DIESEL	NONE	N/A	GM	DELCO
NORMAN WELLS (NCPC) IC1	0.50	OPR	1970	IC	OIL	DIESEL	NONE	N/A	CAT	KATO
NORMAN WELLS (NCPC) IC2	0.70	OPR	1972	IC	OIL	DIESEL	NONE	N/A	CAT	GE
NORMAN WELLS (NCPC) IC3	0.55	OPR	1980	IC	OIL	DIESEL	NONE	N/A	CUM	BBC
PANGNIRTUNG IC1	0.30	OPR	1972	IC	OIL	DIESEL	NONE	N/A	CAT	ACC
PANGNIRTUNG IC2	0.30	OPR	1973	IC	OIL	DIESEL	NONE	N/A	CAT	TAMPER
PANGNIRTUNG IC3	0.54	OPR	1976	IC	OIL	DIESEL	NONE	N/A	CAT	TAMPER
PANGNIRTUNG IC4	0.54	OPR	1981	IC	OIL	DIESEL	NONE	N/A	CAT	BBC
PAULATUK IC1	0.08	OPR	1970	IC	OIL	DIESEL	NONE	N/A	GM	DELCO
PAULATUK IC2	0.15	OPR	1979	IC	OIL	DIESEL	NONE	N/A	CAT	KATO
PAULATUK IC3	0.15	OPR	1980	IC	OIL	DIESEL	NONE	N/A	CAT	DELCO
PELLY BAY IC1	0.20	OPR	1979	IC	OIL	DIESEL	NONE	N/A	GM	GE
PELLY BAY IC2	0.30	OPR	1980	IC	OIL	DIESEL	NONE	N/A	CAT	GE
PELLY BAY IC3	0.30	OPR	1981	IC	OIL	DIESEL	NONE	N/A	CAT	BBC
PINE POINT IC1	5.18	OPR	1970	IC	OIL	DIESEL	NONE	N/A	MIRR	BRUSH
PINE POINT IC2	2.50	OPR	1978	IC	OIL	DIESEL	NONE	N/A	RHN	GEC
PINE POINT IC3	2.50	OPR	1978	IC	OIL	DIESEL	NONE	N/A	RHN	GEC
PINE POINT IC4	2.50	OPR	1978	IC	OIL	DIESEL	NONE	N/A	RHN	GEC
POND INLET IC1	0.30	OPR	1975	IC	OIL	DIESEL	NONE	N/A	CAT	TAMPER
POND INLET IC2	0.54	OPR	1979	IC	OIL	DIESEL	NONE	N/A	CAT	TAMPER
POND INLET IC3	0.72	OPR	1983	IC	OIL	DIESEL	NONE	N/A	CAT	BBC
RAE LAKES IC1	0.08	OPR	1975	IC	OIL	DIESEL	NONE	N/A	GM	DELCO

DIRECTORY OF POWER PLANTS IN CANADA
UDI/MCGRAW-HILL

PLANT/UNIT NAME	MW	STA	YEAR	TYPE	FUEL	FUEL TYPE	ALT FUEL	SSS MFR	TURB MFR	GEN MFR
OPERATOR: NORTHWEST TERRITORIES POWER										
RAE LAKES IC2	0.08	OPR	1981	IC	OIL	DIESEL	NONE	N/A	GM	BBC
RAE LAKES IC3	0.10	OPR	1984	IC	OIL	DIESEL	NONE	N/A	CAT	DELCO
RANKIN INLET IC1	0.70	OPR	1973	IC	OIL	DIESEL	NONE	N/A	CAT	GE
RANKIN INLET IC2	0.70	OPR	1973	IC	OIL	DIESEL	NONE	N/A	CAT	GE
RANKIN INLET IC3	0.72	OPR	1975	IC	OIL	DIESEL	NONE	N/A	CAT	ACC
RANKIN INLET IC4	0.54	OPR	1976	IC	OIL	DIESEL	NONE	N/A	CAT	KATO
REPULSE BAY IC1	0.15	OPR	1973	IC	OIL	DIESEL	NONE	N/A	CAT	KATO
REPULSE BAY IC2	0.30	OPR	1976	IC	OIL	DIESEL	NONE	N/A	CAT	BBC
REPULSE BAY IC3	0.30	OPR	1982	IC	OIL	DIESEL	NONE	N/A	CAT	CANR
RESOLUTE BAY IC1	0.35	OPR	1976	IC	OIL	DIESEL	NONE	N/A	WAU	KATO
RESOLUTE BAY IC2	0.90	OPR	1976	IC	OIL	DIESEL	NONE	N/A	WAU	TAMPER
RESOLUTE BAY IC3	0.90	OPR	1976	IC	OIL	DIESEL	NONE	N/A	WAU	TAMPER
RESOLUTE BAY IC4	0.90	OPR	1976	IC	OIL	DIESEL	NONE	N/A	WAU	BBC
RESOLUTE BAY IC5	0.90	OPR	1976	IC	OIL	DIESEL	NONE	N/A	WAU	BBC
SACHS HARBOUR IC1	0.30	OPR	1975	IC	OIL	DIESEL	NONE	N/A	CAT	TAMPER
SACHS HARBOUR IC2	0.30	OPR	1976	IC	OIL	DIESEL	NONE	N/A	CAT	TAMPER
SACHS HARBOUR IC3	0.20	OPR	1984	IC	OIL	DIESEL	NONE	N/A	CAT	TAMPER
SNOWDRIFT IC1	0.08	OPR	1970	IC	OIL	DIESEL	NONE	N/A	GM	TAMPER
SNOWDRIFT IC2	0.20	OPR	1976	IC	OIL	DIESEL	NONE	N/A	GM	DELCO
SNOWDRIFT IC3	0.15	OPR	1980	IC	OIL	DIESEL	NONE	N/A	CAT	KATO
SPENCE BAY IC1	0.15	OPR	1971	IC	OIL	DIESEL	NONE	N/A	CAT	KATO
SPENCE BAY IC2	0.30	OPR	1973	IC	OIL	DIESEL	NONE	N/A	CAT	GE
SPENCE BAY IC3	0.15	OPR	1975	IC	OIL	DIESEL	NONE	N/A	CAT	KATO
SPENCE BAY IC4	0.30	OPR	1976	IC	OIL	DIESEL	NONE	N/A	CAT	KATO
TUKTOYAKTUK IC1	0.72	OPR	1974	IC	OIL	DIESEL	NONE	N/A	CAT	GE
TUKTOYAKTUK IC2	0.54	OPR	1980	IC	OIL	DIESEL	NONE	N/A	CAT	GE
TUKTOYAKTUK IC3	0.72	OPR	1983	IC	OIL	DIESEL	NONE	N/A	CAT	GE
WHALE COVE IC1	0.15	OPR	1972	IC	OIL	DIESEL	NONE	N/A	CAT	CAT
WHALE COVE IC2	0.20	OPR	1976	IC	OIL	DIESEL	NONE	N/A	CUM	VS
WHALE COVE IC3	0.30	OPR	1981	IC	OIL	DIESEL	NONE	N/A	CAT	TAMPER
WHITEHORSE IC1	3.92	OPR	1968	IC	OIL	DIESEL	NONE	N/A	MIRR	BRUSH

DIRECTORY OF POWER PLANTS IN CANADA
UDI/MCGRAW-HILL

PLANT/UNIT NAME	MW	STA	YEAR	TYPE	FUEL	FUEL TYPE	ALT FUEL	SSS MFR	TURB MFR	GEN MFR
OPERATOR: NORTHWEST TERRITORIES POWER										
WHITEHORSE IC2	5.15	OPR	1968	IC	OIL	DIESEL	NONE	N/A	MIRR	BRUSH
WHITEHORSE IC3	5.15	OPR	1970	IC	OIL	DIESEL	NONE	N/A	MIRR	BRUSH
WHITEHORSE IC4	2.50	OPR	1975	IC	OIL	DIESEL	NONE	N/A	GM	EMC
WHITEHORSE IC5	2.50	OPR	1975	IC	OIL	DIESEL	NONE	N/A	GM	EMC
WHITEHORSE IC6	2.50	OPR	1977	IC	OIL	DIESEL	NONE	N/A	GM	EMC
WRIGLEY IC1	0.15	OPR	1975	IC	OIL	DIESEL	NONE	N/A	GM	TAMPER
WRIGLEY IC2	0.20	OPR	1975	IC	OIL	DIESEL	NONE	N/A	GM	TAMPER
WRIGLEY IC3	0.13	OPR	1983	IC	OIL	DIESEL	NONE	N/A	CAT	STM
YELLOWKNIFE IC1	5.15	OPR	1969	IC	OIL	DIESEL	NONE	N/A	MIRR	BRUSH
YELLOWKNIFE IC2	0.68	OPR	1973	IC	OIL	DIESEL	NONE	N/A	CAT	TAMPER
YELLOWKNIFE IC3	0.68	OPR	1973	IC	OIL	DIESEL	NONE	N/A	CAT	TAMPER
YELLOWKNIFE IC4	2.50	OPR	1974	IC	OIL	DIESEL	NONE	N/A	GM	EMC
YELLOWKNIFE IC5	2.50	OPR	1974	IC	OIL	DIESEL	NONE	N/A	GM	EMC
OPERATOR: NORTHWOOD PULP & PAPER LTD										
FRASER FLATS 1	28.80	OPR	1973	ST	GAS		NONE	FW/CE	STAL	STAL
FRASER FLATS 2	28.00	OPR	1981	ST	GAS		NONE	FW/CE	STAL	STAL
OPERATOR: NOVA SCOTIA FOREST INDUST LTD										
PORT HAWKESBURY 1	10.00	OPR	1961	ST/S	OIL	HFO	NONE	FW	WH	WH
PORT HAWKESBURY 2	17.56	OPR	1971	ST/S	OIL	HFO	NONE	GOT	STAL	STAL
OPERATOR: NOVA SCOTIA POWER CORP										
AVON I 1	3.75	OPR	1958	HY	WAT	CONV	NONE	N/A	MV	BBC
AVON II 1	3.00	OPR	1929	HY	WAT	CONV	NONE	N/A	DEW	CGE
BIG FALLS 1	4.50	OPR	1929	HY	WAT	CONV	NONE	N/A	SMS	ASEA
BIG FALLS 2	4.50	OPR	1929	HY	WAT	CONV	NONE	N/A	SMS	ASEA
BURNSIDE GT1	30.00	OPR	1976	GT	OIL	DIESEL	NONE	N/A	PW	BRUSH
BURNSIDE GT2	30.00	OPR	1976	GT	OIL	DIESEL	NONE	N/A	PW	BRUSH
BURNSIDE GT3	30.00	OPR	1976	GT	OIL	DIESEL	NONE	N/A	PW	BRUSH
BURNSIDE GT4	30.00	OPR	1976	GT	OIL	DIESEL	NONE	N/A	PW	BRUSH

DIRECTORY OF POWER PLANTS IN CANADA
UDI/MCGRAW-HILL

PLANT/UNIT NAME	MW	STA	YEAR	TYPE	FUEL	FUEL TYPE	ALT FUEL	SSS MFR	TURB MFR	GEN MFR
OPERATOR: NOVA SCOTIA POWER CORP										
COWIE FALLS 1	3.60	OPR	1938	HY	WAT	CONV	NONE	N/A	SMS	OERL
COWIE FALLS 2	3.60	OPR	1938	HY	WAT	CONV	NONE	N/A	SMS	OERL
DEEP BROOK 1	4.50	OPR	1950	HY	WAT	CONV	NONE	N/A	SMS	WHC
DEEP BROOK 2	4.50	OPR	1950	HY	WAT	CONV	NONE	N/A	SMS	WHC
DICKIE BROOK 1	1.20	OPR	1948	HY	WAT	CONV	NONE	N/A	ACC	WHC
DICKIE BROOK 2	2.60	OPR	1948	HY	WAT	CONV	NONE	N/A	ACC	ASEA
FALL RIVER 1	0.50	OPR	1985	HY	WAT	CONV	NONE	N/A	DALE	WHC
FOURTH LAKE 1	3.00	OPR	1983	HY	WAT	CONV	NONE	N/A	DBSZ	CGE
GISBORNE 1	3.50	OPR	1982	HY	WAT	CONV	NONE	N/A	AC	SHINK
GLACE BAY 1	15.00	OPR	1951	ST	COAL	BIT	NONE	FW	PAR	PAR
GLACE BAY 2	15.00	OPR	1954	ST	COAL	BIT	NONE	FW	PAR	PAR
GLACE BAY 3	15.00	OPR	1955	ST	COAL	BIT	NONE	FW	PAR	PAR
GLACE BAY 4	15.00	OPR	1959	ST	COAL	BIT	NONE	FW	PAR	PAR
GLACE BAY 5	36.00	OPR	1967	ST	COAL	BIT	NONE	BWGM	SS	SS
GULCH 1	6.00	OPR	1952	HY	WAT	CONV	NONE	N/A	CVIC	WHC
HARMONY 1	0.60	OPR	1943	HY	WAT	CONV	NONE	N/A	RHM	WH
HELLS GATE 1	3.36	OPR	1930	HY	WAT	CONV	NONE	N/A	DEW	CGE
HELLS GATE 2	3.57	OPR	1949	HY	WAT	CONV	NONE	N/A	DEW	WHC
HOLLOW BRIDGE 1	5.31	OPR	1940	HY	WAT	CONV	NONE	N/A	DEW	CGE
LEQUILLE 1	11.18	OPR	1968	HY	WAT	CONV	NONE	N/A	DEW	BBC
LINGAN 1	158.20	OPR	1979	ST	COAL	BIT	NONE	CE	TS	TS
LINGAN 2	158.20	OPR	1980	ST	COAL	BIT	NONE	CE	TS	TS
LINGAN 3	158.20	OPR	1983	ST	COAL	BIT	NONE	CE	TS	TS
LINGAN 4	158.20	OPR	1984	ST	COAL	BIT	NONE	CE	TS	TS
LOWER GREAT BROOK 1	2.25	OPR	1955	HY	WAT	CONV	NONE	N/A	SMS	WHC
LOWER GREAT BROOK 2	2.25	OPR	1955	HY	WAT	CONV	NONE	N/A	SMS	WHC
LOWER LAKE FALLS 1	3.69	OPR	1929	HY	WAT	CONV	NONE	N/A	SMS	CGE
LOWER LAKE FALLS 2	3.69	OPR	1929	HY	WAT	CONV	NONE	N/A	SMS	CGE
LOWER WATER STREET 1	10.00	OPR	1944	ST	OIL	HFO	NONE	BWGM	PAR	PAR
LOWER WATER STREET 2	20.00	OPR	1951	ST	OIL	HFO	NONE	BWGM	PAR	PAR
LOWER WATER STREET 3	20.00	OPR	1953	ST	OIL	HFO	NONE	BWGM	MV	MV

DIRECTORY OF POWER PLANTS IN CANADA
UDI/MCGRAW-HILL

PLANT/UNIT NAME	MW	STA	YEAR	TYPE	FUEL	FUEL TYPE	ALT FUEL	SSS MFR	TURB MFR	GEN MFR
OPERATOR: NOVA SCOTIA POWER CORP										
LOWER WATER STREET 4	25.00	OPR	1955	ST	OIL	HFO	NONE	BWGM	MV	MV
LOWER WATER STREET 5	45.00	OPR	1957	ST	OIL	HFO	NONE	BWGM	EE	EE
LOWER WATER STREET 6	45.00	OPR	1959	ST	OIL	HFO	NONE	BWGM	EE	EE
LUMSDEN 1	2.80	OPR	1942	HY	WAT	CONV	NONE	N/A	DEW	WHC
MACCAN 1	15.00	OPR	1949	ST	COAL	BIT	NONE	BWGM	PAR	PAR
MALAY FALLS 1	1.20	OPR	1924	HY	WAT	CONV	NONE	N/A	WSM	WHC
MALAY FALLS 2	1.20	OPR	1924	HY	WAT	CONV	NONE	N/A	WSM	WHC
MALAY FALLS 3	1.20	OPR	1924	HY	WAT	CONV	NONE	N/A	WSM	WHC
METHALS 1	3.40	OPR	1949	HY	WAT	CONV	NONE	N/A	DEW	WHC
MILL LAKE 1	1.28	OPR	1922	HY	WAT	CONV	NONE	N/A	SMS	CGE
MILL LAKE 2	1.28	OPR	1922	HY	WAT	CONV	NONE	N/A	SMS	CGE
NICTAUX 1	6.80	OPR	1954	HY	WAT	CONV	NONE	N/A	DEW	WHC
PARADISE 1	3.60	OPR	1950	HY	WAT	CONV	NONE	N/A	CVIC	WHC
POINT ACONI 1	165.00	CON	1993	ST	COAL	BIT	NONE	PYR		
POINT ACONI 2	165.00	PLN	2001	ST	COAL	BIT	NONE			
POINT TUPPER 1	78.51	OPR	1969	ST	COAL	BIT	NONE	BW	GE/STA	GE
POINT TUPPER 2	150.00	OPR	1973	ST	COAL	BIT	NONE	CE	HP	PAR
RIDGE 1	4.00	OPR	1957	HY	WAT	CONV	NONE	N/A	SMS	CGE
ROSEWAY 1	0.60	OPR	1959	HY	WAT	CONV	NONE	N/A	LF	CGE
ROSEWAY 2	0.32	OPR	1931	HY	WAT	CONV	NONE	N/A	CVIC	PEEB
RUTH FALLS 1	2.00	OPR	1925	HY	WAT	CONV	NONE	N/A	SMS	CGE
RUTH FALLS 2	2.00	OPR	1925	HY	WAT	CONV	NONE	N/A	SMS	CGE
RUTH FALLS 3	2.97	OPR	1936	HY	WAT	CONV	NONE	N/A	DEW	MTPL
SANDY LAKE 1	1.60	OPR	1928	HY	WAT	CONV	NONE	N/A	DEW	CGE
SANDY LAKE 2	1.60	OPR	1928	HY	WAT	CONV	NONE	N/A	DEW	CGE
SISSIBOO FALLS 1	6.00	OPR	1961	HY	WAT	CONV	NONE	N/A	KMW	WHC
TIDAL I 1	19.46	OPR	1983	HY	WAT	CONV	NONE	N/A	DBSZ	CGE
TIDEWATER 1	2.32	OPR	1922	HY	WAT	CONV	NONE	N/A	SMS	CGE
TIDEWATER 2	2.32	OPR	1922	HY	WAT	CONV	NONE	N/A	SMS	CGE
TRENTON 3	20.00	OPR	1955	ST	COAL	BIT	NONE	CE	PAR	PAR
TRENTON 4	20.00	OPR	1959	ST	COAL	BIT	NONE	BWGM	PAR	PAR

DIRECTORY OF POWER PLANTS IN CANADA
UDI/MCGRAW-HILL

PLANT/UNIT NAME	MW	STA	YEAR	TYPE	FUEL	FUEL TYPE	ALT FUEL	SSS MFR	TURB MFR	GEN MFR
OPERATOR: NOVA SCOTIA POWER CORP										
TRENTON 5	150.00	OPR	1969	ST	COAL	BIT	NONE	BW	HP	GE
TRENTON 6	150.00	OPR	1991	ST	COAL	BIT	NONE	BW		
TUFTS COVE 1	100.00	OPR	1965	ST	OIL	HFO	NONE	BWGM	AEI	AEI
TUFTS COVE 2	105.00	OPR	1972	ST	OIL	HFO	NONE	BW	HP	PAR
TUFTS COVE 3	150.00	OPR	1976	ST	OIL	HFO	NONE	BW	HP	PAR
TUSKET 1	0.72	OPR	1929	HY	WAT	CONV	NONE	N/A	SMS	WHC
TUSKET 2	0.72	OPR	1929	HY	WAT	CONV	NONE	N/A	SMS	WHC
TUSKET 3	0.72	OPR	1929	HY	WAT	CONV	NONE	N/A	SMS	WHC
TUSKET GT1	25.00	OPR	1971	GT	OIL	DIESEL	NONE	N/A	PW	BRUSH
UPPER LAKE FALLS 1	2.70	OPR	1929	HY	WAT	CONV	NONE	N/A	DEW	CGE
UPPER LAKE FALLS 2	2.70	OPR	1929	HY	WAT	CONV	NONE	N/A	DEW	CGE
VICTORIA JUNCTION GT1	30.00	OPR	1975	GT	OIL	DIESEL	NONE	N/A	PW	BRUSH
VICTORIA JUNCTION GT2	30.00	OPR	1975	GT	OIL	DIESEL	NONE	N/A	PW	BRUSH
WEYMOUTH FALLS 1	9.00	OPR	1961	HY	WAT	CONV	NONE	N/A	KMW	WHC
WEYMOUTH FALLS 2	9.00	OPR	1961	HY	WAT	CONV	NONE	N/A	KMW	WHC
WHITE ROCK 1	3.20	OPR	1952	HY	WAT	CONV	NONE	N/A	CVIC	WHC
WRECK COVE 1	100.00	OPR	1978	HY	WAT	CONV	NONE	N/A	MHI	CGE
WRECK COVE 2	100.00	OPR	1978	HY	WAT	CONV	NONE	N/A	MHI	CGE
OPERATOR: NW ENERGY (WILLIAMS LAKE) LP										
WILLIAMS LAKE 1	67.50	CON	1993	ST/S	WOOD		NONE			
OPERATOR: NWT POWER CORP										
SNARE FALLS 1	7.00	OPR	1960	HY	WAT	CONV	NONE	N/A	AC	CGE
SNARE FORKS 1	6.50	OPR	1976	HY	WAT	CONV	NONE	N/A	AC	CGE
SNARE FORKS 2	6.50	OPR	1976	HY	WAT	CONV	NONE	N/A	AC	CGE
SNARE RAPIDS 1	8.00	OPR	1948	HY	WAT	CONV	NONE	N/A	SMS	CGE
TALTSON 1	18.00	OPR	1965	HY	WAT	CONV	NONE	N/A	DEW	WHC
TALTSON 2	1.00	OPR	1976	HY	WAT	CONV	NONE	N/A	OSS	BBC
TALTSON 3	1.00	OPR	1976	HY	WAT	CONV	NONE	N/A	OSS	BBC
TALTSON 4	1.00	OPR	1976	HY	WAT	CONV	NONE	N/A	OSS	BBC

DIRECTORY OF POWER PLANTS IN CANADA
UDI/MCGRAW-HILL

PLANT/UNIT NAME	MW	STA	YEAR	TYPE	FUEL	FUEL TYPE	ALT FUEL	SSS MFR	TURB MFR	GEN MFR
OPERATOR: NWT POWER CORP										
TALTSON 5	1.00	OPR	1976	HY	WAT	CONV	NONE	N/A	OSS	BBC
OPERATOR: OGDEN MARTIN SYSTEMS LTD										
DARTMOUTH WTE 1	13.40	PLN	1995	ST	REF		NONE			
OPERATOR: ONSITE ENERGY INC										
CINOLA	11.00	PLN	9999	IC	OIL	DIESEL	NONE	N/A		
OPERATOR: ONTARIO HYDRO										
ABITIBI CANYON 1	48.50	OPR	1933	HY	WAT	CONV	NONE	N/A	ACC	CGE
ABITIBI CANYON 2	70.00	OPR	1977	HY	WAT	CONV	NONE	N/A	ACC	CGE
ABITIBI CANYON 3	70.00	OPR	1977	HY	WAT	CONV	NONE	N/A	ACC	CGE
ABITIBI CANYON 4	70.00	OPR	1977	HY	WAT	CONV	NONE	N/A	ACC	CGE
ABITIBI CANYON 5	70.00	OPR	1977	HY	WAT	CONV	NONE	N/A	ACC	CGE
AGUASABON 1	22.50	OPR	1948	HY	WAT	CONV	NONE	N/A	DEW	WHC
AGUASABON 2	22.50	OPR	1948	HY	WAT	CONV	NONE	N/A	DEW	WHC
ALEXANDER 1	18.00	OPR	1930	HY	WAT	CONV	NONE	N/A	SMS	CGE
ALEXANDER 2	18.00	OPR	1931	HY	WAT	CONV	NONE	N/A	SMS	CGE
ALEXANDER 3	18.00	OPR	1931	HY	WAT	CONV	NONE	N/A	SMS	CGE
ALEXANDER 4	18.00	OPR	1945	HY	WAT	CONV	NONE	N/A	SMS	CGE
ALEXANDER 5	18.00	OPR	1958	HY	WAT	CONV	NONE	N/A	DT	CGE
ARNPRIOR 1	39.00	OPR	1976	HY	WAT	CONV	NONE	N/A	DEW	CGE
ARNPRIOR 2	39.00	OPR	1976	HY	WAT	CONV	NONE	N/A	DEW	CGE
ATIKOKAN 1	230.00	OPR	1985	ST	COAL	LIG	NONE	BW	BBC	BBC
AUBREY FALLS 1	68.50	OPR	1969	HY	WAT	CONV	NONE	N/A	DEW	CGE
AUBREY FALLS 2	68.50	OPR	1969	HY	WAT	CONV	NONE	N/A	DEW	CGE
AUBURN 1	0.63	OPR	1911	HY	WAT	CONV	NONE	N/A	WH	CGE
AUBURN 2	0.63	OPR	1911	HY	WAT	CONV	NONE	N/A	WH	CGE
AUBURN 3	0.63	OPR	1912	HY	WAT	CONV	NONE	N/A	WH	CGE
BARRETT CHUTE 1	24.00	OPR	1942	HY	WAT	CONV	NONE	N/A	ACC	CGE
BARRETT CHUTE 2	24.00	OPR	1942	HY	WAT	CONV	NONE	N/A	ACC	CGE

DIRECTORY OF POWER PLANTS IN CANADA
UDI/MCGRAW-HILL

PLANT/UNIT NAME	MW	STA	YEAR	TYPE	FUEL	FUEL TYPE	ALT FUEL	SSS MFR	TURB MFR	GEN MFR
OPERATOR: ONTARIO HYDRO										
BARRETT CHUTE 3	62.00	OPR	1968	HY	WAT	CONV	NONE	N/A	ACC	CGE
BARRETT CHUTE 4	62.00	OPR	1968	HY	WAT	CONV	NONE	N/A	ACC	CGE
BIG CHUTE II 1	9.90	CON	1993	HY	WAT	CONV	NONE	N/A	DBSZ	DBSZ
BIG EDDY (OH) 1	4.50	OPR	1941	HY	WAT	CONV	NONE	N/A	SMS	WHC
BIG EDDY (OH) 2	4.50	OPR	1941	HY	WAT	CONV	NONE	N/A	SMS	WHC
BINGHAM CHUTE 1	0.45	OPR	1923	HY	WAT	CONV	NONE	N/A	WK	WHC
BINGHAM CHUTE 2	0.45	OPR	1924	HY	WAT	CONV	NONE	N/A	WK	WHC
BRUCE A GT1	12.10	OPR	1974	GT	OIL	DIST	NONE	N/A	RR	EE
BRUCE A GT2	12.10	OPR	1974	GT	OIL	DIST	NONE	N/A	RR	EE
BRUCE A GT3	12.10	OPR	1975	GT	OIL	DIST	NONE	N/A	RR	EE
BRUCE A GT4	12.10	OPR	1976	GT	OIL	DIST	NONE	N/A	RR	EE
BRUCE A1	791.00	OPR	1977	ST	UR	CANDU	NONE	AECL	PAR	PAR
BRUCE A2	791.00	OPR	1977	ST	UR	CANDU	NONE	AECL	PAR	PAR
BRUCE A3	791.00	OPR	1978	ST	UR	CANDU	NONE	AECL	PAR	PAR
BRUCE A4	791.00	OPR	1979	ST	UR	CANDU	NONE	AECL	PAR	PAR
BRUCE B GT1	12.10	OPR	1983	GT	OIL	DIST	NONE	N/A	RR	GE
BRUCE B GT2	12.10	OPR	1983	GT	OIL	DIST	NONE	N/A	RR	GE
BRUCE B GT3	12.10	OPR	1983	GT	OIL	DIST	NONE	N/A	RR	GE
BRUCE B GT4	12.10	OPR	1983	GT	OIL	DIST	NONE	N/A	RR	GE
BRUCE B GT5	4.05	OPR	1983	GT	OIL	DIST	NONE	N/A	SOLAR	BBC
BRUCE B GT6	4.05	OPR	1983	GT	OIL	DIST	NONE	N/A	SOLAR	BBC
BRUCE B1	807.00	OPR	1985	ST	UR	CANDU	NONE	AECL	PAR	PAR
BRUCE B2	807.00	OPR	1984	ST	UR	CANDU	NONE	AECL	PAR	PAR
BRUCE B3	807.00	OPR	1986	ST	UR	CANDU	NONE	AECL	PAR	PAR
BRUCE B4	807.00	OPR	1987	ST	UR	CANDU	NONE	AECL	PAR	PAR
BRUCE HWP GT1	12.10	OPR	1977	GT	OIL	DIST	NONE	N/A	RR	EE
BRUCE HWP GT2	12.10	OPR	1977	GT	OIL	DIST	NONE	N/A	RR	EE
BRUCE HWP GT3	12.10	OPR	1977	GT	OIL	DIST	NONE	N/A	RR	EE
CALABOGIE 1	2.50	OPR	1917	HY	WAT	CONV	NONE	N/A	AC	CGE
CALABOGIE 2	2.50	OPR	1917	HY	WAT	CONV	NONE	N/A	AC	CGE
CAMERON 1	10.60	OPR	1921	HY	WAT	CONV	NONE	N/A	IPM	WHC

DIRECTORY OF POWER PLANTS IN CANADA
UDI/MCGRAW-HILL

PLANT/UNIT NAME	MW	STA	YEAR	TYPE	FUEL	FUEL TYPE	ALT FUEL	SSS MFR	TURB MFR	GEN MFR
OPERATOR: ONTARIO HYDRO										
CAMERON 2	10.60	OPR	1921	HY	WAT	CONV	NONE	N/A	IPM	WHC
CAMERON 3	10.60	OPR	1924	HY	WAT	CONV	NONE	N/A	ACC	CGE
CAMERON 4	10.60	OPR	1924	HY	WAT	CONV	NONE	N/A	ACC	CGE
CAMERON 5	10.60	OPR	1925	HY	WAT	CONV	NONE	N/A	CVIC	CGE
CAMERON 6	10.60	OPR	1926	HY	WAT	CONV	NONE	N/A	CVIC	CGE
CAMERON 7	20.00	OPR	1958	HY	WAT	CONV	NONE	N/A	DEW	WHC
CARIBOU FALLS 1	28.50	OPR	1958	HY	WAT	CONV	NONE	N/A	DEW	CGE
CARIBOU FALLS 2	28.50	OPR	1958	HY	WAT	CONV	NONE	N/A	DEW	CGE
CARIBOU FALLS 3	28.50	OPR	1958	HY	WAT	CONV	NONE	N/A	DEW	CGE
CHATS FALLS 1	23.50	OPR	1931	HY	WAT	CONV	NONE	N/A	DEW	WHC
CHATS FALLS 2	23.50	OPR	1931	HY	WAT	CONV	NONE	N/A	DEW	WHC
CHATS FALLS 3	23.50	OPR	1931	HY	WAT	CONV	NONE	N/A	DEW	WHC
CHATS FALLS 4	23.50	OPR	1931	HY	WAT	CONV	NONE	N/A	DEW	WHC
CHENAUX 1	17.00	OPR	1950	HY	WAT	CONV	NONE	N/A	DEW	CGE
CHENAUX 2	17.00	OPR	1950	HY	WAT	CONV	NONE	N/A	DEW	CGE
CHENAUX 3	17.00	OPR	1951	HY	WAT	CONV	NONE	N/A	DEW	CGE
CHENAUX 4	17.00	OPR	1951	HY	WAT	CONV	NONE	N/A	DEW	CGE
CHENAUX 5	17.00	OPR	1951	HY	WAT	CONV	NONE	N/A	DEW	CGE
CHENAUX 6	17.00	OPR	1951	HY	WAT	CONV	NONE	N/A	DEW	CGE
CHENAUX 7	17.00	OPR	1951	HY	WAT	CONV	NONE	N/A	DEW	CGE
CHENAUX 8	17.00	OPR	1951	HY	WAT	CONV	NONE	N/A	DEW	CGE
CONISTON 1	1.00	OPR	1905	HY	WAT	CONV	NONE	N/A	JM	CGE
CONISTON 2	1.25	OPR	1907	HY	WAT	CONV	NONE	N/A	JM	CGE
CONISTON 3	2.50	OPR	1915	HY	WAT	CONV	NONE	N/A	AC	CGE
CRYSTAL FALLS 1	2.13	OPR	1921	HY	WAT	CONV	NONE	N/A	IPM	WH
CRYSTAL FALLS 2	2.13	OPR	1921	HY	WAT	CONV	NONE	N/A	IPM	WH
CRYSTAL FALLS 3	2.13	OPR	1921	HY	WAT	CONV	NONE	N/A	IPM	WH
CRYSTAL FALLS 4	2.13	OPR	1921	HY	WAT	CONV	NONE	N/A	IPM	WH
DARLINGTON 1	881.00	OPR	1991	ST	UR	CANDU	NONE	AECL	ABB	ABB
DARLINGTON 2	881.00	OPR	1990	ST	UR	CANDU	NONE	AECL	ABB	ABB
DARLINGTON 3	881.00	CON	1993	ST	UR	CANDU	NONE	AECL	ABB	ABB

DIRECTORY OF POWER PLANTS IN CANADA
UDI/MCGRAW-HILL

PLANT/UNIT NAME	MW	STA	YEAR	TYPE	FUEL	FUEL TYPE	ALT FUEL	SSS MFR	TURB MFR	GEN MFR
OPERATOR: ONTARIO HYDRO										
DARLINGTON 4	881.00	CON	1993	ST	UR	CANDU	NONE	AECL	ABB	ABB
DARLINGTON GT1	28.00	UNK	1985	GT	OIL		NONE	N/A	RR	BRUSH
DARLINGTON GT2	28.00	UNK	1985	GT	OIL		NONE	N/A	RR	BRUSH
DARLINGTON GT3	28.00	UNK	1985	GT	OIL		NONE	N/A	RR	BRUSH
DARLINGTON GT4	28.00	UNK	1985	GT	OIL		NONE	N/A	RR	BRUSH
DECEW FALLS I 1	5.00	OPR	1904	HY	WAT	CONV	NONE	N/A	VOI	WELEC
DECEW FALLS I 2	5.00	OPR	1904	HY	WAT	CONV	NONE	N/A	VOI	WELEC
DECEW FALLS I 3	6.40	OPR	1905	HY	WAT	CONV	NONE	N/A	VOI	WELEC
DECEW FALLS I 4	6.40	OPR	1905	HY	WAT	CONV	NONE	N/A	VOI	WELEC
DECEW FALLS I 5	6.40	OPR	1911	HY	WAT	CONV	NONE	N/A	VOI	WHC
DECEW FALLS I 6	6.40	OPR	1911	HY	WAT	CONV	NONE	N/A	VOI	WHC
DECEW FALLS II 1	64.00	OPR	1943	HY	WAT	CONV	NONE	N/A	ACC	CGE
DECEW FALLS II 2	64.00	OPR	1947	HY	WAT	CONV	NONE	N/A	ACC	CGE
DES JOACHIMS 1	50.00	OPR	1950	HY	WAT	CONV	NONE	N/A	DEW	WHC
DES JOACHIMS 2	50.00	OPR	1950	HY	WAT	CONV	NONE	N/A	DEW	WHC
DES JOACHIMS 3	50.00	OPR	1950	HY	WAT	CONV	NONE	N/A	DEW	WHC
DES JOACHIMS 4	50.00	OPR	1950	HY	WAT	CONV	NONE	N/A	DEW	WHC
DES JOACHIMS 5	50.00	OPR	1950	HY	WAT	CONV	NONE	N/A	DEW	WHC
DES JOACHIMS 6	50.00	OPR	1950	HY	WAT	CONV	NONE	N/A	DEW	WHC
DES JOACHIMS 7	50.00	OPR	1950	HY	WAT	CONV	NONE	N/A	DEW	WHC
DES JOACHIMS 8	50.00	OPR	1951	HY	WAT	CONV	NONE	N/A	DEW	WHC
EAR FALLS 1	5.00	OPR	1930	HY	WAT	CONV	NONE	N/A	DEW	WHC
EAR FALLS 2	4.50	OPR	1937	HY	WAT	CONV	NONE	N/A	SMS	OERL
EAR FALLS 3	6.00	OPR	1940	HY	WAT	CONV	NONE	N/A	SMS	WHC
EAR FALLS 4	6.00	OPR	1948	HY	WAT	CONV	NONE	N/A	SMS	WHC
ELLIOTT CHUTE 1	1.80	OPR	1929	HY	WAT	CONV	NONE	N/A	SMS	CGE
EUGENIA 1	1.20	OPR	1915	HY	WAT	CONV	NONE	N/A	EW	WHC
EUGENIA 2	2.40	OPR	1920	HY	WAT	CONV	NONE	N/A	AC	WHC
EUGENIA 3	1.40	OPR	1987	HY	WAT	CONV	NONE	N/A		
FRANKFORD 1	0.81	OPR	1913	HY	WAT	CONV	NONE	N/A	BV	CGE
FRANKFORD 2	0.81	OPR	1913	HY	WAT	CONV	NONE	N/A	BV	CGE

DIRECTORY OF POWER PLANTS IN CANADA
UDI/MCGRAW-HILL

PLANT/UNIT NAME	MW	STA	YEAR	TYPE	FUEL	FUEL TYPE	ALT FUEL	SSS MFR	TURB MFR	GEN MFR
OPERATOR: ONTARIO HYDRO										
FRANKFORD 3	0.81	OPR	1913	HY	WAT	CONV	NONE	N/A	BV	CGE
FRANKFORD 4	0.81	OPR	1913	HY	WAT	CONV	NONE	N/A	BV	CGE
GEORGE W RAYNER 1	23.50	OPR	1950	HY	WAT	CONV	NONE	N/A	ACC	WHC
GEORGE W RAYNER 2	23.50	OPR	1950	HY	WAT	CONV	NONE	N/A	ACC	WHC
HAGUES REACH 1	1.40	OPR	1925	HY	WAT	CONV	NONE	N/A	ACC	WHC
HAGUES REACH 2	1.40	OPR	1925	HY	WAT	CONV	NONE	N/A	ACC	WHC
HAGUES REACH 3	1.40	OPR	1925	HY	WAT	CONV	NONE	N/A	ACC	WHC
HANNA CHUTE 1	1.40	OPR	1926	HY	WAT	CONV	NONE	N/A	DEW	CGE
HARMON 1	68.00	OPR	1965	HY	WAT	CONV	NONE	N/A	JI	WHC
HARMON 2	68.00	OPR	1965	HY	WAT	CONV	NONE	N/A	JI	WHC
HARMON 3	68.00	PLN	9999	HY	WAT	CONV	NONE	N/A		
HEALEY FALLS 1	3.75	OPR	1919	HY	WAT	CONV	NONE	N/A	WSM	CGE
HEALEY FALLS 2	3.75	OPR	1913	HY	WAT	CONV	NONE	N/A	EW	CGE
HEALEY FALLS 3	3.75	OPR	1914	HY	WAT	CONV	NONE	N/A	EW	CGE
HIGH FALLS (OH) 1	0.88	OPR	1920	HY	WAT	CONV	NONE	N/A	LF	CGE
HIGH FALLS (OH) 2	0.70	OPR	1920	HY	WAT	CONV	NONE	N/A	LF	CGE
HIGH FALLS (OH) 3	0.70	OPR	1920	HY	WAT	CONV	NONE	N/A	LF	CGE
HOUND CHUTE 1	0.88	OPR	1910	HY	WAT	CONV	NONE	N/A	WK	CGE
HOUND CHUTE 2	0.88	OPR	1910	HY	WAT	CONV	NONE	N/A	WK	CGE
HOUND CHUTE 3	0.88	OPR	1910	HY	WAT	CONV	NONE	N/A	WK	CGE
HOUND CHUTE 4	0.88	OPR	1911	HY	WAT	CONV	NONE	N/A	WK	CGE
INDIAN CHUTE 1	1.80	OPR	1923	HY	WAT	CONV	NONE	N/A	BV	WHC
INDIAN CHUTE 2	1.80	OPR	1924	HY	WAT	CONV	NONE	N/A	WK	WHC
J CLARK KEITH 1	66.00	OPR	1952	ST	COAL	BIT	NONE	BWGM	EE	EE
J CLARK KEITH 2	66.00	OPR	1952	ST	COAL	BIT	NONE	BWGM	EE	EE
J CLARK KEITH 3	66.00	OPR	1953	ST	COAL	BIT	NONE	BWGM	EE	EE
J CLARK KEITH 4	66.00	OPR	1953	ST	COAL	BIT	NONE	BWGM	EE	EE
KAKABEKA FALLS 1	6.35	OPR	1906	HY	WAT	CONV	NONE	N/A	VOI	CGE
KAKABEKA FALLS 2	6.35	OPR	1906	HY	WAT	CONV	NONE	N/A	VOI	CGE
KAKABEKA FALLS 3	6.35	OPR	1911	HY	WAT	CONV	NONE	N/A	VOI	CGE
KAKABEKA FALLS 4	9.35	OPR	1914	HY	WAT	CONV	NONE	N/A	VOI	CGE

DIRECTORY OF POWER PLANTS IN CANADA
UDI/MCGRAW-HILL

PLANT/UNIT NAME	MW	STA	YEAR	TYPE	FUEL	FUEL TYPE	ALT FUEL	SSS MFR	TURB MFR	GEN MFR
OPERATOR: ONTARIO HYDRO										
KIPLING 1	66.00	OPR	1966	HY	WAT	CONV	NONE	N/A	DEW	WHC
KIPLING 2	66.00	OPR	1966	HY	WAT	CONV	NONE	N/A	DEW	WHC
KIPLING 3	66.00	PLN	9999	HY	WAT	CONV	NONE	N/A		
LAKEFIELD 1	2.50	OPR	1928	HY	WAT	CONV	NONE	N/A	ACC	CGE
LAKEVIEW 1	250.00	OPR	1962	ST	COAL	BIT	NONE	BWGM	PAR	PAR
LAKEVIEW 2	250.00	OPR	1963	ST	COAL	BIT	NONE	BWGM	PAR	PAR
LAKEVIEW 3	300.00	OPR	1965	ST	COAL	BIT	NONE	CE	AEI	AEI/GE
LAKEVIEW 4	300.00	OPR	1965	ST	COAL	BIT	NONE	CE	AEI	AEI/GE
LAKEVIEW 5	300.00	OPR	1967	ST	COAL	BIT	NONE	BW	AEI	AEI/GE
LAKEVIEW 6	300.00	OPR	1969	ST	COAL	BIT	NONE	BW	AEI	AEI/GE
LAKEVIEW 7	300.00	OPR	1969	ST	COAL	BIT	NONE	BW	PAR	PAR
LAKEVIEW 8	300.00	OPR	1969	ST	COAL	BIT	NONE	BW	PAR	PAR
LAKEVIEW GT1	6.40	OPR	1967	GT	OIL	DIST	NONE	N/A	OREND	BRUSH
LAKEVIEW GT2	6.40	OPR	1967	GT	OIL	DIST	NONE	N/A	OREND	BRUSH
LAKEVIEW GT3	6.40	OPR	1967	GT	OIL	DIST	NONE	N/A	OREND	BRUSH
LAMBTON 1	510.00	OPR	1969	ST	COAL	BIT	NONE	CE	GE	GE
LAMBTON 2	510.00	OPR	1970	ST	COAL	BIT	NONE	CE	GE	GE
LAMBTON 3	510.00	OPR	1970	ST	COAL	BIT	NONE	CE	GE	GE
LAMBTON 4	510.00	OPR	1970	ST	COAL	BIT	NONE	CE	GE	GE
LAMBTON GT1	6.40	OPR	1967	GT	OIL	DIST	NONE	N/A	OREND	BRUSH
LAMBTON GT2	6.40	OPR	1968	GT	OIL	DIST	NONE	N/A	OREND	BRUSH
LAMBTON GT3	6.40	OPR	1976	GT	OIL	DIST	NONE	N/A	OREND	BRUSH
LENNOX 1	573.75	OPR	1976	ST	OIL	HFO	NONE	CE	GE	GE
LENNOX 2	573.75	OPR	1976	ST	OIL	HFO	NONE	CE	GE	GE
LENNOX 3	573.75	OPR	1976	ST	OIL	HFO	NONE	CE	GE	GE
LENNOX 4	573.75	OPR	1976	ST	OIL	HFO	NONE	CE	GE	GE
LENNOX GT1	2.50	OPR	1968	GT	OIL	DIST	NONE	N/A	SOLAR	EMC
LENNOX GT2	2.50	OPR	1968	GT	OIL	DIST	NONE	N/A	SOLAR	EMC
LITTLE LONG 1	64.00	OPR	1963	HY	WAT	CONV	NONE	N/A	EE	WHC
LITTLE LONG 2	64.00	OPR	1963	HY	WAT	CONV	NONE	N/A	EE	WHC
LITTLE LONG 3	64.00	PLN	9999	HY	WAT	CONV	NONE	N/A		

UDI-033-92 December 1992

DIRECTORY OF POWER PLANTS IN CANADA
UDI/MCGRAW-HILL

PLANT/UNIT NAME	MW	STA	YEAR	TYPE	FUEL	FUEL TYPE	ALT FUEL	SSS MFR	TURB MFR	GEN MFR
OPERATOR: ONTARIO HYDRO										
LOWER NOTCH 1	120.00	OPR	1971	HY	WAT	CONV	NONE	N/A	DEW	CGE
LOWER NOTCH 2	120.00	OPR	1971	HY	WAT	CONV	NONE	N/A	DEW	CGE
LOWER STURGEON 1	4.00	OPR	1923	HY	WAT	CONV	NONE	N/A	DEW	CGE
LOWER STURGEON 2	4.00	OPR	1923	HY	WAT	CONV	NONE	N/A	DEW	CGE
MANITOU FALLS 1	16.00	OPR	1956	HY	WAT	CONV	NONE	N/A	DEW	CGE
MANITOU FALLS 2	16.00	OPR	1956	HY	WAT	CONV	NONE	N/A	DEW	CGE
MANITOU FALLS 3	16.00	OPR	1956	HY	WAT	CONV	NONE	N/A	DEW	CGE
MANITOU FALLS 4	16.00	OPR	1956	HY	WAT	CONV	NONE	N/A	DEW	CGE
MANITOU FALLS 5	16.00	OPR	1956	HY	WAT	CONV	NONE	N/A	DEW	CGE
MANITOU FALLS 6	16.00	OPR	1958	HY	WAT	CONV	NONE	N/A	DEW	CGE
MATABITCHUAN 1	1.88	OPR	1910	HY	WAT	CONV	NONE	N/A	IPM	CGE
MATABITCHUAN 2	1.88	OPR	1910	HY	WAT	CONV	NONE	N/A	IPM	CGE
MATABITCHUAN 3	1.88	OPR	1910	HY	WAT	CONV	NONE	N/A	IPM	CGE
MATABITCHUAN 4	1.88	OPR	1910	HY	WAT	CONV	NONE	N/A	IPM	CGE
MCVITTIE 1	1.25	OPR	1912	HY	WAT	CONV	NONE	N/A	WK	CGE
MCVITTIE 2	1.25	OPR	1912	HY	WAT	CONV	NONE	N/A	WK	CGE
MERRICKVILLE 1	0.55	OPR	1915	HY	WAT	CONV	NONE	N/A	WH	CGE
MERRICKVILLE 2	0.50	OPR	1919	HY	WAT	CONV	NONE	N/A	SMS	CGE
MEYERSBURG 1	2.00	OPR	1924	HY	WAT	CONV	NONE	N/A	ACC	CGE
MEYERSBURG 2	2.00	OPR	1924	HY	WAT	CONV	NONE	N/A	ACC	CGE
MEYERSBURG 3	2.00	OPR	1924	HY	WAT	CONV	NONE	N/A	ACC	CGE
MOUNTAIN CHUTE 1	75.00	OPR	1967	HY	WAT	CONV	NONE	N/A	EE	WHC
MOUNTAIN CHUTE 2	75.00	OPR	1967	HY	WAT	CONV	NONE	N/A	EE	WHC
NANTICOKE 1	512.00	OPR	1973	ST	COAL	BIT	NONE	BW	HP	PAR
NANTICOKE 2	512.00	OPR	1973	ST	COAL	BIT	NONE	BW	HP	PAR
NANTICOKE 3	512.00	OPR	1973	ST	COAL	BIT	NONE	BW	HP	PAR
NANTICOKE 4	512.00	OPR	1974	ST	COAL	BIT	NONE	BW	HP	PAR
NANTICOKE 5	512.00	OPR	1975	ST	COAL	BIT	NONE	BW	HP	PAR
NANTICOKE 6	512.00	OPR	1976	ST	COAL	BIT	NONE	BW	HP	PAR
NANTICOKE 7	512.00	OPR	1978	ST	COAL	BIT	NONE	BW	HP	PAR
NANTICOKE 8	512.00	OPR	1978	ST	COAL	BIT	NONE	BW	HP	PAR

DIRECTORY OF POWER PLANTS IN CANADA
UDI/MCGRAW-HILL

PLANT/UNIT NAME	MW	STA	YEAR	TYPE	FUEL	FUEL TYPE	ALT FUEL	SSS MFR	TURB MFR	GEN MFR
OPERATOR: ONTARIO HYDRO										
NANTICOKE GT1	6.40	OPR	1971	GT	OIL	DIST	NONE	N/A	OREND	BRUSH
NANTICOKE GT2	6.40	OPR	1971	GT	OIL	DIST	NONE	N/A	OREND	BRUSH
NANTICOKE GT3	6.40	OPR	1971	GT	OIL	DIST	NONE	N/A	OREND	BRUSH
NIPISSING 1	1.40	OPR	1921	HY	WAT	CONV	NONE	N/A	JM	WHC
NIPISSING 2	1.25	OPR	1924	HY	WAT	CONV	NONE	N/A	JM	CGE
ONTARIO POWER 01	7.50	OPR	1905	HY	WAT	CONV	NONE	N/A	VOI	WELEC
ONTARIO POWER 02	7.50	OPR	1905	HY	WAT	CONV	NONE	N/A	VOI	WELEC
ONTARIO POWER 03	7.50	OPR	1905	HY	WAT	CONV	NONE	N/A	VOI	WELEC
ONTARIO POWER 04	8.77	OPR	1906	HY	WAT	CONV	NONE	N/A	VOI	WELEC
ONTARIO POWER 05	8.77	OPR	1908	HY	WAT	CONV	NONE	N/A	VOI	WELEC
ONTARIO POWER 06	8.77	OPR	1908	HY	WAT	CONV	NONE	N/A	VOI	WELEC
ONTARIO POWER 07	8.77	OPR	1909	HY	WAT	CONV	NONE	N/A	VOI	WELEC
ONTARIO POWER 08	8.78	OPR	1910	HY	WAT	CONV	NONE	N/A	VOI	CGE
ONTARIO POWER 09	8.78	OPR	1911	HY	WAT	CONV	NONE	N/A	VOI	CGE
ONTARIO POWER 10	8.78	OPR	1911	HY	WAT	CONV	NONE	N/A	VOI	CGE
ONTARIO POWER 11	8.78	OPR	1913	HY	WAT	CONV	NONE	N/A	VOI	CGE
ONTARIO POWER 12	8.78	OPR	1913	HY	WAT	CONV	NONE	N/A	WSM	CGE
OTTER RAPIDS 1	46.00	OPR	1961	HY	WAT	CONV	NONE	N/A	ACC	CGE
OTTER RAPIDS 2	46.00	OPR	1961	HY	WAT	CONV	NONE	N/A	ACC	CGE
OTTER RAPIDS 3	46.00	OPR	1963	HY	WAT	CONV	NONE	N/A	ACC	CGE
OTTER RAPIDS 4	46.00	OPR	1963	HY	WAT	CONV	NONE	N/A	ACC	CGE
OTTO HOLDEN 1	27.00	OPR	1952	HY	WAT	CONV	NONE	N/A	ACC	WHC
OTTO HOLDEN 2	27.00	OPR	1952	HY	WAT	CONV	NONE	N/A	ACC	WHC
OTTO HOLDEN 3	27.00	OPR	1952	HY	WAT	CONV	NONE	N/A	ACC	WHC
OTTO HOLDEN 4	27.00	OPR	1952	HY	WAT	CONV	NONE	N/A	ACC	WHC
OTTO HOLDEN 5	27.00	OPR	1952	HY	WAT	CONV	NONE	N/A	JI	WHC
OTTO HOLDEN 6	27.00	OPR	1952	HY	WAT	CONV	NONE	N/A	JI	WHC
OTTO HOLDEN 7	27.00	OPR	1952	HY	WAT	CONV	NONE	N/A	JI	WHC
OTTO HOLDEN 8	27.00	OPR	1953	HY	WAT	CONV	NONE	N/A	JI	WHC
PICKERING A GT1	5.00	OPR	1970	GT	OIL	DIST	NONE	N/A	OREND	BRUSH
PICKERING A GT2	5.00	OPR	1970	GT	OIL	DIST	NONE	N/A	OREND	BRUSH

DIRECTORY OF POWER PLANTS IN CANADA
UDI/MCGRAW-HILL

PLANT/UNIT NAME	MW	STA	YEAR	TYPE	FUEL	FUEL TYPE	ALT FUEL	SSS MFR	TURB MFR	GEN MFR
OPERATOR: ONTARIO HYDRO										
PICKERING A GT3	5.00	OPR	1970	GT	OIL	DIST	NONE	N/A	OREND	BRUSH
PICKERING A GT4	5.00	OPR	1972	GT	OIL	DIST	NONE	N/A	OREND	BRUSH
PICKERING A GT5	5.00	OPR	1972	GT	OIL	DIST	NONE	N/A	OREND	BRUSH
PICKERING A GT6	5.00	OPR	1973	GT	OIL	DIST	NONE	N/A	OREND	BRUSH
PICKERING A1	542.00	OPR	1971	ST	UR	CANDU	NONE	AECL	PAR	PAR
PICKERING A2	542.00	OPR	1971	ST	UR	CANDU	NONE	AECL	PAR	PAR
PICKERING A3	542.00	OPR	1972	ST	UR	CANDU	NONE	AECL	PAR	PAR
PICKERING A4	542.00	OPR	1973	ST	UR	CANDU	NONE	AECL	PAR	PAR
PICKERING B GT1	7.00	OPR	1982	GT	OIL	DIST	NONE	N/A	OREND	BRUSH
PICKERING B GT2	7.00	OPR	1982	GT	OIL	DIST	NONE	N/A	OREND	BRUSH
PICKERING B GT3	7.00	OPR	1982	GT	OIL	DIST	NONE	N/A	OREND	BRUSH
PICKERING B GT4	7.00	OPR	1982	GT	OIL	DIST	NONE	N/A	OREND	BRUSH
PICKERING B GT5	7.00	OPR	1982	GT	OIL	DIST	NONE	N/A	OREND	BRUSH
PICKERING B GT6	7.00	OPR	1982	GT	OIL	DIST	NONE	N/A	OREND	BRUSH
PICKERING B GT7	2.50	OPR	1982	GT	OIL	DIST	NONE	N/A	SOLAR	BBC
PICKERING B GT8	2.50	OPR	1982	GT	OIL	DIST	NONE	N/A	SOLAR	BBC
PICKERING B1	540.00	OPR	1983	ST	UR	CANDU	NONE	AECL	PAR	PAR
PICKERING B2	540.00	OPR	1984	ST	UR	CANDU	NONE	AECL	PAR	PAR
PICKERING B3	540.00	OPR	1984	ST	UR	CANDU	NONE	AECL	PAR	PAR
PICKERING B4	540.00	OPR	1986	ST	UR	CANDU	NONE	AECL	PAR	PAR
PINE PORTAGE 1	33.00	OPR	1950	HY	WAT	CONV	NONE	N/A	ACC	WHC
PINE PORTAGE 2	33.00	OPR	1950	HY	WAT	CONV	NONE	N/A	ACC	WHC
PINE PORTAGE 3	38.50	OPR	1954	HY	WAT	CONV	NONE	N/A	SMS	WHC
PINE PORTAGE 4	38.50	OPR	1954	HY	WAT	CONV	NONE	N/A	SMS	WHC
QUEENSTON	1050.00	PLN	9999	HY	WAT	CONV	NONE	N/A		
RAGGED RAPIDS 1	4.50	OPR	1938	HY	WAT	CONV	NONE	N/A	SMS	WHC
RAGGED RAPIDS 2	4.50	OPR	1938	HY	WAT	CONV	NONE	N/A	SMS	WHC
RAWNEY FALLS 1	4.50	OPR	1922	HY	WAT	CONV	NONE	N/A	BV	CGE
RAWNEY FALLS 2	4.50	OPR	1922	HY	WAT	CONV	NONE	N/A	BV	CGE
RAWNEY FALLS 3	0.90	OPR	1926	HY	WAT	CONV	NONE	N/A	WH	CGE
RED ROCK FALLS 1	22.50	OPR	1960	HY	WAT	CONV	NONE	N/A	DEW	CGE

DIRECTORY OF POWER PLANTS IN CANADA
UDI/MCGRAW-HILL

PLANT/UNIT NAME	MW	STA	YEAR	TYPE	FUEL	FUEL TYPE	ALT FUEL	SSS MFR	TURB MFR	GEN MFR
OPERATOR: ONTARIO HYDRO										
RED ROCK FALLS 2	22.50	OPR	1961	HY	WAT	CONV	NONE	N/A	DEW	CGE
RICHARD L HEARN 1	100.00	STN	1951	ST	COAL	BIT	GAS	BWGM	PAR	PAR
RICHARD L HEARN 2	100.00	STN	1952	ST	COAL	BIT	GAS	BWGM	PAR	PAR
RICHARD L HEARN 3	100.00	STN	1952	ST	COAL	BIT	GAS	BWGM	PAR	PAR
RICHARD L HEARN 4	100.00	STN	1953	ST	COAL	BIT	GAS	BWGM	PAR	PAR
RICHARD L HEARN 5	200.00	STN	1959	ST	COAL	BIT	GAS	CE	PAR	PAR
RICHARD L HEARN 6	200.00	STN	1960	ST	COAL	BIT	GAS	BWGM	PAR	PAR
RICHARD L HEARN 7	200.00	STN	1960	ST	COAL	BIT	GAS	CE	PAR	PAR
RICHARD L HEARN 8	200.00	STN	1961	ST	COAL	BIT	GAS	BWGM	PAR	PAR
RICHARD L HEARN GT1	6.40	OPR	1967	GT	OIL	DIST	NONE	N/A	OREND	BRUSH
RICHARD L HEARN GT2	6.40	OPR	1967	GT	OIL	DIST	NONE	N/A	OREND	BRUSH
RICHARD L HEARN GT3	6.40	OPR	1967	GT	OIL	DIST	NONE	N/A	OREND	BRUSH
ROBERT H SAUNDERS 01	60.00	OPR	1958	HY	WAT	CONV	NONE	N/A	EE	CGE
ROBERT H SAUNDERS 02	60.00	OPR	1958	HY	WAT	CONV	NONE	N/A	EE	CGE
ROBERT H SAUNDERS 03	60.00	OPR	1958	HY	WAT	CONV	NONE	N/A	EE	WHC
ROBERT H SAUNDERS 04	60.00	OPR	1958	HY	WAT	CONV	NONE	N/A	EE	WHC
ROBERT H SAUNDERS 05	60.00	OPR	1958	HY	WAT	CONV	NONE	N/A	EE	CGE
ROBERT H SAUNDERS 06	60.00	OPR	1958	HY	WAT	CONV	NONE	N/A	EE	CGE
ROBERT H SAUNDERS 07	60.00	OPR	1958	HY	WAT	CONV	NONE	N/A	EE	WHC
ROBERT H SAUNDERS 08	60.00	OPR	1959	HY	WAT	CONV	NONE	N/A	EE	WHC
ROBERT H SAUNDERS 09	60.00	OPR	1959	HY	WAT	CONV	NONE	N/A	EE	CGE
ROBERT H SAUNDERS 10	60.00	OPR	1959	HY	WAT	CONV	NONE	N/A	EE	CGE
ROBERT H SAUNDERS 11	60.00	OPR	1959	HY	WAT	CONV	NONE	N/A	EE	WHC
ROBERT H SAUNDERS 12	60.00	OPR	1959	HY	WAT	CONV	NONE	N/A	EE	WHC
ROBERT H SAUNDERS 13	60.00	OPR	1959	HY	WAT	CONV	NONE	N/A	EE	CGE
ROBERT H SAUNDERS 14	60.00	OPR	1959	HY	WAT	CONV	NONE	N/A	EE	CGE
ROBERT H SAUNDERS 15	60.00	OPR	1959	HY	WAT	CONV	NONE	N/A	EE	WHC
ROBERT H SAUNDERS 16	60.00	OPR	1959	HY	WAT	CONV	NONE	N/A	EE	WHC
SANDY FALLS 1	0.95	OPR	1911	HY	WAT	CONV	NONE	N/A	SMS	WHC
SANDY FALLS 2	0.95	OPR	1911	HY	WAT	CONV	NONE	N/A	SMS	WHC
SANDY FALLS 3	1.60	OPR	1916	HY	WAT	CONV	NONE	N/A	IPM	CGE

DIRECTORY OF POWER PLANTS IN CANADA
UDI/MCGRAW-HILL

PLANT/UNIT NAME	MW	STA	YEAR	TYPE	FUEL	FUEL TYPE	ALT FUEL	SSS MFR	TURB MFR	GEN MFR
OPERATOR: ONTARIO HYDRO										
SEYMOUR 1	0.75	OPR	1909	HY	WAT	CONV	NONE	N/A	WK	CGE
SEYMOUR 2	0.60	OPR	1909	HY	WAT	CONV	NONE	N/A	WK	CGE
SEYMOUR 3	0.60	OPR	1910	HY	WAT	CONV	NONE	N/A	WK	CGE
SEYMOUR 4	0.60	OPR	1911	HY	WAT	CONV	NONE	N/A	WK	CGE
SEYMOUR 5	0.60	OPR	1911	HY	WAT	CONV	NONE	N/A	WK	CGE
SIDNEY 1	0.94	OPR	1911	HY	WAT	CONV	NONE	N/A	BV	CGE
SIDNEY 2	0.94	OPR	1911	HY	WAT	CONV	NONE	N/A	BV	CGE
SIDNEY 3	0.94	OPR	1911	HY	WAT	CONV	NONE	N/A	BV	CGE
SIDNEY 4	0.94	OPR	1911	HY	WAT	CONV	NONE	N/A	BV	CGE
SILLS ISLAND 1	1.35	OPR	1926	HY	WAT	CONV	NONE	N/A	SMS	CGE
SILLS ISLAND 2	1.20	OPR	1926	HY	WAT	CONV	NONE	N/A	SMS	CGE
SILVER FALLS 1	50.00	OPR	1959	HY	WAT	CONV	NONE	N/A	ACC	WHC
SIR ADAM BECK I 01	45.00	OPR	1921	HY	WAT	CONV	NONE	N/A	WSM	WHC
SIR ADAM BECK I 02	45.00	OPR	1921	HY	WAT	CONV	NONE	N/A	WSM	WHC
SIR ADAM BECK I 03	55.00	OPR	1924	HY	WAT	CONV	NONE	N/A	DEW	WHC
SIR ADAM BECK I 04	63.50	OPR	1924	HY	WAT	CONV	NONE	N/A	DEW	CGE
SIR ADAM BECK I 05	63.50	OPR	1921	HY	WAT	CONV	NONE	N/A	DEW	CGE
SIR ADAM BECK I 06	55.00	OPR	1930	HY	WAT	CONV	NONE	N/A	DEW	WHC
SIR ADAM BECK I 07	54.00	OPR	1924	HY	WAT	CONV	NONE	N/A	DEW	CGE
SIR ADAM BECK I 08	63.50	OPR	1921	HY	WAT	CONV	NONE	N/A	IPM	WHC
SIR ADAM BECK I 09	55.00	OPR	1921	HY	WAT	CONV	NONE	N/A	IPM	CGE
SIR ADAM BECK I 10	55.00	OPR	1921	HY	WAT	CONV	NONE	N/A	IPM	CGE
SIR ADAM BECK II 01	80.50	OPR	1954	HY	WAT	CONV	NONE	N/A	DEW	CGE
SIR ADAM BECK II 02	80.50	OPR	1954	HY	WAT	CONV	NONE	N/A	DEW	WHC
SIR ADAM BECK II 03	80.50	OPR	1954	HY	WAT	CONV	NONE	N/A	DEW	CGE
SIR ADAM BECK II 04	80.50	OPR	1954	HY	WAT	CONV	NONE	N/A	DEW	WHC
SIR ADAM BECK II 05	80.50	OPR	1954	HY	WAT	CONV	NONE	N/A	DEW	CGE
SIR ADAM BECK II 06	80.50	OPR	1954	HY	WAT	CONV	NONE	N/A	DEW	WHC
SIR ADAM BECK II 07	80.50	OPR	1954	HY	WAT	CONV	NONE	N/A	DEW	CGE
SIR ADAM BECK II 08	80.50	OPR	1955	HY	WAT	CONV	NONE	N/A	DEW	WHC
SIR ADAM BECK II 09	80.50	OPR	1955	HY	WAT	CONV	NONE	N/A	DEW	CGE

DIRECTORY OF POWER PLANTS IN CANADA
UDI/MCGRAW-HILL

PLANT/UNIT NAME	MW	STA	YEAR	TYPE	FUEL	FUEL TYPE	ALT FUEL	SSS MFR	TURB MFR	GEN MFR
OPERATOR: ONTARIO HYDRO										
SIR ADAM BECK II 10	80.50	OPR	1955	HY	WAT	CONV	NONE	N/A	DEW	WHC
SIR ADAM BECK II 11	80.50	OPR	1955	HY	WAT	CONV	NONE	N/A	DEW	CGE
SIR ADAM BECK II 12	80.50	OPR	1955	HY	WAT	CONV	NONE	N/A	DEW	WHC
SIR ADAM BECK II 13	80.50	OPR	1957	HY	WAT	CONV	NONE	N/A	DEW	CGE
SIR ADAM BECK II 14	80.50	OPR	1957	HY	WAT	CONV	NONE	N/A	DEW	WHC
SIR ADAM BECK II 15	80.50	OPR	1958	HY	WAT	CONV	NONE	N/A	DEW	CGE
SIR ADAM BECK II 16	80.50	OPR	1958	HY	WAT	CONV	NONE	N/A	DEW	WHC
SIR ADAM BECK III	550.00	PLN	9999	HY	WAT	CONV	NONE	N/A		
SIR ADAM BECK P&G 1	31.00	OPR	1957	HY	WAT	CONV	NONE	N/A	EE	WHC
SIR ADAM BECK P&G 2	31.00	OPR	1957	HY	WAT	CONV	NONE	N/A	EE	WHC
SIR ADAM BECK P&G 3	31.00	OPR	1957	HY	WAT	CONV	NONE	N/A	EE	WHC
SIR ADAM BECK P&G 4	31.00	OPR	1958	HY	WAT	CONV	NONE	N/A	EE	WHC
SIR ADAM BECK P&G 5	31.00	OPR	1958	HY	WAT	CONV	NONE	N/A	EE	WHC
SIR ADAM BECK P&G 6	31.00	OPR	1958	HY	WAT	CONV	NONE	N/A	EE	WHC
SOUTH FALLS 1	0.75	OPR	1916	HY	WAT	CONV	NONE	N/A	WH	CGE
SOUTH FALLS 2	2.00	OPR	1925	HY	WAT	CONV	NONE	N/A	WH	PEEB
SOUTH FALLS 3	2.00	OPR	1925	HY	WAT	CONV	NONE	N/A	WH	PEEB
STEWARTVILLE 1	24.00	OPR	1948	HY	WAT	CONV	NONE	N/A	ACC	CGE
STEWARTVILLE 2	24.00	OPR	1948	HY	WAT	CONV	NONE	N/A	ACC	CGE
STEWARTVILLE 3	24.00	OPR	1948	HY	WAT	CONV	NONE	N/A	ACC	CGE
STEWARTVILLE 4	51.00	OPR	1969	HY	WAT	CONV	NONE	N/A	ACC	CGE
STEWARTVILLE 5	51.00	OPR	1969	HY	WAT	CONV	NONE	N/A	ACC	CGE
STINSON 1	2.50	OPR	1925	HY	WAT	CONV	NONE	N/A	AC	CGE
STINSON 2	2.50	OPR	1925	HY	WAT	CONV	NONE	N/A	AC	CGE
THUNDER BAY 1	93.00	OPR	1963	ST	COAL	LIG	NONE	FW	EE	EE
THUNDER BAY 2	165.00	OPR	1981	ST	COAL	LIG	NONE	CE	BBC	BBC
THUNDER BAY 3	165.00	OPR	1981	ST	COAL	LIG	NONE	CE	BBC	BBC
THUNDER BAY GT1	11.60	OPR	1968	GT	OIL	DIST	NONE	N/A	RR	AEI
THUNDER BAY GT2	11.60	OPR	1968	GT	OIL	DIST	NONE	N/A	RR	AEI
TRETHEWEY FALLS 1	1.60	OPR	1929	HY	WAT	CONV	NONE	N/A	SMS	CGE
WAWAITIN 1	3.75	OPR	1912	HY	WAT	CONV	NONE	N/A	SMS	WHC

DIRECTORY OF POWER PLANTS IN CANADA
UDI/MCGRAW-HILL

PLANT/UNIT NAME	MW	STA	YEAR	TYPE	FUEL	FUEL TYPE	ALT FUEL	SSS MFR	TURB MFR	GEN MFR
OPERATOR: ONTARIO HYDRO										
WAWAITIN 2	3.75	OPR	1912	HY	WAT	CONV	NONE	N/A	SMS	WHC
WAWAITIN 3	2.50	OPR	1913	HY	WAT	CONV	NONE	N/A	SMS	WHC
WAWAITIN 4	2.50	OPR	1918	HY	WAT	CONV	NONE	N/A	SMS	WHC
WELLS 1	107.00	OPR	1970	HY	WAT	CONV	NONE	N/A	DEW	CGE
WELLS 2	107.00	OPR	1970	HY	WAT	CONV	NONE	N/A	DEW	CGE
WHITEDOG FALLS 1	24.00	OPR	1958	HY	WAT	CONV	NONE	N/A	DEW	WHC
WHITEDOG FALLS 2	24.00	OPR	1958	HY	WAT	CONV	NONE	N/A	DEW	WHC
WHITEDOG FALLS 3	24.00	OPR	1958	HY	WAT	CONV	NONE	N/A	DEW	WHC
OPERATOR: ORILLIA WATER LT & POWER COMM										
MATTHIAS 1	2.81	OPR	1950	HY	WAT	CONV	NONE	N/A	SMS	CGE
MINDEN 1	1.80	OPR	1935	HY	WAT	CONV	NONE	N/A	SMS	CGE
MINDEN 2	1.80	OPR	1935	HY	WAT	CONV	NONE	N/A	SMS	CGE
ORILLIA IC1	1.00	OPR	1947	IC	OIL	DIESEL	NONE	N/A	FM	EMC
ORILLIA IC2	1.14	OPR	1948	IC	OIL	DIESEL	NONE	N/A	FM	FM
SWIFT RAPIDS 1	2.70	OPR	1966	HY	WAT	CONV	NONE	N/A	ACC	CGE
SWIFT RAPIDS 2	2.70	OPR	1966	HY	WAT	CONV	NONE	N/A	ACC	CGE
SWIFT RAPIDS 3	2.70	OPR	1979	HY	WAT	CONV	NONE	N/A	BARBER	CGE
OPERATOR: OTTAWA HYDRO										
CHAUDIERE II 1	1.46	OPR	1908	HY	WAT	CONV	NONE	N/A	SMS	WHC
CHAUDIERE II 2	1.46	OPR	1908	HY	WAT	CONV	NONE	N/A	SMS	WHC
CHAUDIERE II 3	1.46	OPR	1984	HY	WAT	CONV	NONE	N/A	BARBER	WHC
CHAUDIERE IV 1	3.96	OPR	1931	HY	WAT	CONV	NONE	N/A	WH	CGE
CHAUDIERE IV 2	3.96	OPR	1985	HY	WAT	CONV	NONE	N/A	LF	CGE
OPERATOR: PACIFIC INTEGRATED WASTE SYS										
TROCHU	1.50	PLN		ST	TIRE		NONE			
OPERATOR: PAPIER JOURNAL DOMTAR LTEE										
BIRDS 1	1.92	OPR	1937	HY	WAT	CONV	NONE	N/A	DEW	WH

DIRECTORY OF POWER PLANTS IN CANADA
UDI/MCGRAW-HILL

PLANT/UNIT NAME	MW	STA	YEAR	TYPE	FUEL	FUEL TYPE	ALT FUEL	SSS MFR	TURB MFR	GEN MFR
OPERATOR: PAPIER JOURNAL DOMTAR LTEE										
MACDOUGALL 1	1.20	OPR	1925	HY	WAT	CONV	NONE	N/A	SMS	WH
MACDOUGALL 2	1.20	OPR	1927	HY	WAT	CONV	NONE	N/A	SMS	WH
OPERATOR: PARRY SOUND PUBLIC UTIL COMM										
PARRY SOUND 1	0.42	OPR	1919	HY	WAT	CONV	NONE	N/A	BV	CGE
PARRY SOUND 2	0.92	OPR	1919	HY	WAT	CONV	NONE	N/A	BV	WHC
OPERATOR: PEEL RESOURCE RECOVERY										
PEEL COUNTY 1	10.00	OPR	1992	ST	REF		NONE			
OPERATOR: PEMBROKE ELECTRIC LIGHT CO LTD										
WALTHAM 1	1.25	OPR	1917	HY	WAT	CONV	NONE	N/A	BV	WH
WALTHAM 2	1.53	OPR	1940	HY	WAT	CONV	NONE	N/A	LF	WH
WALTHAM 3	1.80	OPR	1944	HY	WAT	CONV	NONE	N/A	SMS	WH
WALTHAM 4	2.25	OPR	1950	HY	WAT	CONV	NONE	N/A	LF	WH
WALTHAM 5	2.25	OPR	1951	HY	WAT	CONV	NONE	N/A	LF	WH
OPERATOR: PEMBROKE HYDRO ELECTRIC COMM										
PEMBROKE IC1	0.93	OPR	1929	IC	OIL	DIESEL	NONE	N/A	BESS	WH
PEMBROKE IC2	0.68	OPR	1949	IC	OIL	DIESEL	NONE	N/A	GM	AC
OPERATOR: PETERBOROUGH UTILITIES COMM										
PETERBOROUGH 1	1.20	OPR	1950	HY	WAT	CONV	NONE	N/A	CVIC	WH
PETERBOROUGH 2	1.40	OPR	1950	HY	WAT	CONV	NONE	N/A	LF	CGE
PETERBOROUGH 3	1.50	OPR	1950	HY	WAT	CONV	NONE	N/A	WH	CGE
OPERATOR: PETRO CANADA										
TAYLOR 1	2.50	OPR	1957	ST	GAS		NONE	VVIW	GE	GE
TAYLOR 2	2.50	OPR	1957	ST	GAS		NONE	VVIW	GE	GE
TAYLOR 3	2.50	OPR	1957	ST	GAS		NONE	VVIW	GE	GE

DIRECTORY OF POWER PLANTS IN CANADA
UDI/MCGRAW-HILL

PLANT/UNIT NAME	MW	STA	YEAR	TYPE	FUEL	FUEL TYPE	ALT FUEL	SSS MFR	TURB MFR	GEN MFR
OPERATOR: PLACER DEVELOPMENT LTD										
ENDAKO MINES IC1	1.25	OPR	1964	IC	OIL	DIESEL	NONE	N/A	MIRR	BRUSH
ENDAKO MINES IC2	1.00	OPR	1964	IC	OIL	DIESEL	NONE	N/A	GM	EL
OPERATOR: POLSKY ENERGY CORP										
BROOKLYN ENERGY CENTRE 1	21.30	PLN	1995	ST/S	WOOD		NONE			
OPERATOR: POLYSAR LTD										
SARNIA (PL) 1	4.00	OPR	1943	ST/S	GAS		NONE	BW/CE	WHC	WH
SARNIA (PL) 2	5.00	OPR	1948	ST/S	GAS		NONE	BW/CE	WHC	WH
SARNIA (PL) 3	13.28	OPR	1956	ST/S	GAS		NONE	BW/CE	GE	GE
SARNIA (PL) 4	28.75	OPR	1983	ST/S	GAS		NONE	BW/CE	WHC	MHI
OPERATOR: PPG INDUSTRIES CANADA LTD										
RADIUM IC1	0.50	STN	1984	IC	OIL	DIESEL	NONE	N/A	DD	STM
OPERATOR: PROCTER & GAMBLE CELLULOSE LTD										
WAPITI RIVER 1	34.50	OPR	1973	ST	LIQ		NONE	CE/FW	STAL	STAL
OPERATOR: PROVINCIAL THERMAL UTILITIES										
SHIP POINT	8.80	PLN	1995	ST	REF		WOOD			
OPERATOR: PUBLIC WORKS CANADA										
GOOSE BAY 1	2.00	OPR	1953	ST	OIL	DIESEL	NONE	UN	WORT	EMC
GOOSE BAY 2	2.00	OPR	1953	ST	OIL	DIESEL	NONE	UN	WORT	EMC
GOOSE BAY 3	2.00	OPR	1953	ST	OIL	DIESEL	NONE	UN	WORT	EMC
GOOSE BAY 4	2.00	OPR	1953	ST	OIL	DIESEL	NONE	UN	WORT	EMC
OPERATOR: QUEEN CHARLOTTE FOREST PROD										
PORT CLEMENTS 1	21.00	PLN	9999	ST/S	WOOD		NONE			
OPERATOR: QUEEN CHARLOTTE POWER CORP										
MORESBY LAKE 1	6.00	OPR	1990	HY	WAT	CONV	NONE	N/A		

DIRECTORY OF POWER PLANTS IN CANADA
UDI/MCGRAW-HILL

PLANT/UNIT NAME	MW	STA	YEAR	TYPE	FUEL	FUEL TYPE	ALT FUEL	SSS MFR	TURB MFR	GEN MFR
OPERATOR: RAMSEN ENGINEERING										
DUNBRACK I	0.20	PLN	1994	HY	WAT	CONV	NONE	N/A		
TANGIER LAKE	2.20	PLN	1995	HY	WAT	CONV	NONE	N/A		
OPERATOR: REDPATH SUGARS LTD										
TORONTO 1	2.50	OPR	1959	ST/S	GAS		NONE	BW	GE	GE
OPERATOR: RENFREW HYDRO ELECTRIC COMM										
RENFREW I 1	0.27	OPR	1910	HY	WAT	CONV	NONE	N/A	SMS	CGE
RENFREW I 2	0.27	OPR	1911	HY	WAT	CONV	NONE	N/A	SMS	CGE
RENFREW I 3	0.48	OPR	1953	HY	WAT	CONV	NONE	N/A	COOP	EE
RENFREW II 1	0.58	OPR	1927	HY	WAT	CONV	NONE	N/A	COOP	CGE
RENFREW II 2	0.38	OPR	1936	HY	WAT	CONV	NONE	N/A	COOP	CGE
OPERATOR: ROBSON VALLEY POWER CORP										
PTARMIGAN CREEK 1	0.15	OPR	1989	HY	WAT	CONV	NONE	N/A		
PTARMIGAN CREEK 2	0.15	OPR	1989	HY	WAT	CONV	NONE	N/A		
OPERATOR: ROLLAND INC										
ST-JEROME		PLN	1994	CC	GAS		NONE			
OPERATOR: ROMAN CORPORATION LTD										
STRATHCONA (RC) 1	1.66	OPR	1955	ST/S	GAS		NONE	BW	GE	GE
STRATHCONA (RC) 2	1.66	OPR	1955	ST/S	GAS		NONE	BW	GE	GE
OPERATOR: SACKVILLE HYDRO POWER CORP										
SACKVILLE HYDRO	0.58	PLN	1994	HY	WAT	CONV	NONE	N/A		
OPERATOR: SASKPOWER										
AL COLE 1	10.00	STN	1929	ST	GAS		NONE	BW/FW	PAR	PAR
AL COLE 2	15.00	STN	1947	ST	GAS		NONE	BW/FW	PAR	PAR
AL COLE 3	25.00	STN	1953	ST	GAS		NONE	BW/FW	PAR	PAR
AL COLE 4	25.00	STN	1954	ST	GAS		NONE	BW/FW	PAR	PAR

DIRECTORY OF POWER PLANTS IN CANADA
UDI/MCGRAW-HILL

PLANT/UNIT NAME	MW	STA	YEAR	TYPE	FUEL	FUEL TYPE	ALT FUEL	SSS MFR	TURB MFR	GEN MFR
OPERATOR: SASKPOWER										
AL COLE 5	30.00	STN	1957	ST	GAS		NONE	CE	PAR	PAR
BOUNDARY DAM 1	66.00	OPR	1959	ST	COAL	LIG	NONE	BW	PAR	PAR
BOUNDARY DAM 2	66.00	OPR	1960	ST	COAL	LIG	NONE	CE	PAR	PAR
BOUNDARY DAM 3	150.00	OPR	1969	ST	COAL	LIG	NONE	CE	GE	GE
BOUNDARY DAM 4	150.00	OPR	1970	ST	COAL	LIG	NONE	CE	GE	GE
BOUNDARY DAM 5	150.00	OPR	1973	ST	COAL	LIG	NONE	CE	HT	HT
BOUNDARY DAM 6	292.50	OPR	1978	ST	COAL	LIG	NONE	CE	HT	HT
BRABANT LAKE IC1	0.10	OPR	1969	IC	OIL	DIESEL	NONE	N/A	CAT	TAMPER
BRABANT LAKE IC2	0.10	OPR	1975	IC	OIL	DIESEL	NONE	N/A	CAT	TAMPER
CHARLOT RIVER 1	5.13	OPR	1978	HY	WAT	CONV	NONE	N/A	DEW	CGE
CHARLOT RIVER 2	5.13	OPR	1978	HY	WAT	CONV	NONE	N/A	DEW	CGE
COTEAU CREEK 1	55.98	OPR	1968	HY	WAT	CONV	NONE	N/A	EE	WH
COTEAU CREEK 2	55.98	OPR	1968	HY	WAT	CONV	NONE	N/A	EE	WH
COTEAU CREEK 3	55.98	OPR	1968	HY	WAT	CONV	NONE	N/A	EE	WH
DESCHAMBEAULT IC1	0.15	OPR	1972	IC	OIL	DIESEL	NONE	N/A	CAT	TAMPER
DESCHAMBEAULT IC2	0.25	OPR	1978	IC	OIL	DIESEL	NONE	N/A	CAT	BBC
DESCHAMBEAULT IC3	0.25	OPR	1979	IC	OIL	DIESEL	NONE	N/A	CAT	BBC
EB CAMPBELL 1	33.75	OPR	1963	HY	WAT	CONV	NONE	N/A	JOHN	EE
EB CAMPBELL 2	33.75	OPR	1963	HY	WAT	CONV	NONE	N/A	JOHN	EE
EB CAMPBELL 3	33.75	OPR	1963	HY	WAT	CONV	NONE	N/A	JOHN	EE
EB CAMPBELL 4	33.75	OPR	1963	HY	WAT	CONV	NONE	N/A	JOHN	EE
EB CAMPBELL 5	33.75	OPR	1964	HY	WAT	CONV	NONE	N/A	JOHN	EE
EB CAMPBELL 6	33.75	OPR	1964	HY	WAT	CONV	NONE	N/A	JOHN	EE
EB CAMPBELL 7	38.70	OPR	1966	HY	WAT	CONV	NONE	N/A	AC	WH
EB CAMPBELL 8	38.70	OPR	1966	HY	WAT	CONV	NONE	N/A	AC	WH
ERMINE GT1	50.00	PLN	2000	GT	GAS		NONE	N/A		
ESTEVAN 1	15.00	OPR	1950	ST	COAL	LIG	NONE	CE/FW	PAR	PAR
ESTEVAN 2	20.00	OPR	1953	ST	COAL	LIG	NONE	CE/FW	PAR	PAR
ESTEVAN 3	30.00	OPR	1957	ST	COAL	LIG	NONE	CE/FW	MV	MV
FOND DU LAC IC1	0.25	OPR	1975	IC	OIL	DIESEL	NONE	N/A	CAT	TAMPER
FOND DU LAC IC2	0.25	OPR	1976	IC	OIL	DIESEL	NONE	N/A	CAT	TAMPER

DIRECTORY OF POWER PLANTS IN CANADA
UDI/MCGRAW-HILL

PLANT/UNIT NAME	MW	STA	YEAR	TYPE	FUEL	FUEL TYPE	ALT FUEL	SSS MFR	TURB MFR	GEN MFR
OPERATOR: SASKPOWER										
FOND DU LAC IC3	0.30	OPR	1977	IC	OIL	DIESEL	NONE	N/A	CAT	BBC
HALL LAKE IC1	0.05	OPR	1983	IC	OIL	DIESEL	NONE	N/A	CAT	TAMPER
HALL LAKE IC2	0.05	OPR	1983	IC	OIL	DIESEL	NONE	N/A	CAT	TAMPER
ISLAND FALLS 01	0.80	OPR	1928	HY	WAT	CONV	NONE	N/A	IPM	CGE
ISLAND FALLS 02	0.80	OPR	1928	HY	WAT	CONV	NONE	N/A	IPM	CGE
ISLAND FALLS 03	10.80	OPR	1930	HY	WAT	CONV	NONE	N/A	DEW	CGE
ISLAND FALLS 04	10.80	OPR	1930	HY	WAT	CONV	NONE	N/A	DEW	CGE
ISLAND FALLS 05	10.80	OPR	1930	HY	WAT	CONV	NONE	N/A	DEW	CGE
ISLAND FALLS 06	18.00	OPR	1937	HY	WAT	CONV	NONE	N/A	DEW	CGE
ISLAND FALLS 07	18.00	OPR	1939	HY	WAT	CONV	NONE	N/A	DEW	CGE
ISLAND FALLS 08	18.00	OPR	1948	HY	WAT	CONV	NONE	N/A	DEW	CGE
ISLAND FALLS 09	17.10	OPR	1959	HY	WAT	CONV	NONE	N/A	DEW	CGE
ISLAND FALLS 10	85.00	PLN	1995	HY	WAT	CONV	NONE	N/A		
KINOOSAO IC1	0.08	OPR	1970	IC	OIL	DIESEL	NONE	N/A	CAT	TAMPER
KINOOSAO IC2	0.10	OPR	1976	IC	OIL	DIESEL	NONE	N/A	CAT	TAMPER
LANDIS GT1	68.40	OPR	1975	GT	GAS		NONE	N/A	BBC	EMC
MEADOW LAKE GT1	51.00	OPR	1984	GT	GAS		NONE	N/A	JBE	BRUSH
NIPAWIN 1	85.00	OPR	1985	HY	WAT	CONV	NONE	N/A	HT	
NIPAWIN 2	85.00	OPR	1985	HY	WAT	CONV	NONE	N/A	HT	
NIPAWIN 3	85.00	OPR	1986	HY	WAT	CONV	NONE	N/A	HT	
POPLAR RIVER 1	294.00	OPR	1980	ST	COAL	LIG	NONE	BW	HT	HT
POPLAR RIVER 2	297.80	OPR	1983	ST	COAL	LIG	NONE	BW	HT	HT
QUEEN ELIZABETH 1	75.00	OPR	1958	ST	GAS		NONE	FW	BBC	BBC
QUEEN ELIZABETH 2	66.00	OPR	1959	ST	GAS		NONE	FW	EE	EE
QUEEN ELIZABETH 3	100.00	OPR	1972	ST	GAS		NONE	BW	HT	HT
SASKATOON GT1	50.00	PLN	2000	GT	GAS		NONE	N/A		
SHAND 1	300.00	OPR	1992	ST	COAL	LIG	NONE	BW		
SHAND 2	300.00	PLN	9999	ST	COAL	LIG	NONE	BW		
SOUTHEND IC1	0.20	OPR	1975	IC	OIL	DIESEL	NONE	N/A	CAT	TAMPER
SOUTHEND IC2	0.25	OPR	1978	IC	OIL	DIESEL	NONE	N/A	CAT	BBC
SOUTHEND IC3	0.25	OPR	1979	IC	OIL	DIESEL	NONE	N/A	CAT	BBC

UDI-033-92
December 1992

DIRECTORY OF POWER PLANTS IN CANADA
UDI/MCGRAW-HILL

PLANT/UNIT NAME	MW	STA	YEAR	TYPE	FUEL	FUEL TYPE	ALT FUEL	SSS MFR	TURB MFR	GEN MFR
OPERATOR: SASKPOWER										
STONY RAPIDS IC1	0.25	OPR	1976	IC	OIL	DIESEL	NONE	N/A	CAT	TAMPER
STONY RAPIDS IC2	0.25	OPR	1978	IC	OIL	DIESEL	NONE	N/A	CAT	BBC
STONY RAPIDS IC3	0.60	OPR	1981	IC	OIL	DIESEL	NONE	N/A	CAT	BBC
STONY RAPIDS IC4	0.60	OPR	1981	IC	OIL	DIESEL	NONE	N/A	CAT	BBC
SUCCESS GT1	11.84	OPR	1967	GT	GAS		NONE	N/A	ABBS	GE
SUCCESS GT2	11.84	OPR	1967	GT	GAS		NONE	N/A	ABBS	GE
SUCCESS GT3	11.84	OPR	1967	GT	GAS		NONE	N/A	ABBS	GE
WATERLOO 1	9.51	OPR	1961	HY	WAT	CONV	NONE	N/A		
WELLINGTON LAKE 1	2.40	OPR	1939	HY	WAT	CONV	NONE	N/A	AC	CGE
WELLINGTON LAKE 2	2.40	OPR	1959	HY	WAT	CONV	NONE	N/A	AC	CGE
WOLLASTON IC1	0.25	OPR	1978	IC	OIL	DIESEL	NONE	N/A	CAT	TAMPER
WOLLASTON IC2	0.25	OPR	1978	IC	OIL	DIESEL	NONE	N/A	CAT	BBC
WOLLASTON IC3	0.60	OPR	1981	IC	OIL	DIESEL	NONE	N/A	CAT	BBC
WOLLASTON IC4	0.60	OPR	1981	IC	OIL	DIESEL	NONE	N/A	CAT	BBC
OPERATOR: SCARFE LAKE ELECTRIC										
CHIBLOW DAM 1	1.60	CON	1992	HY	WAT	CONV	NONE	N/A	OSS	OSS
OPERATOR: SCOTT MARITIMES PULP LTD										
ABERCROMBIE POINT 1	18.75	OPR	1971	ST/S	WOOD		NONE	BW	WORT	EMC
OPERATOR: SERPENT RIVER POWER CORP										
SERPENT RIVER 1	6.50	OPR	1990	HY	WAT	CONV	NONE	N/A		
OPERATOR: SHERRITT-GORDON MINES LTD										
FORT SASKATCHEWAN 1	2.50	OPR	1954	ST	GAS		NONE	CE	BBC	BBC
FORT SASKATCHEWAN 2	2.50	OPR	1959	ST	GAS		NONE	CE	BBC	BBC
FORT SASKATCHEWAN GT1	2.80	OPR	1981	GT	GAS		NONE	N/A	SOLAR	IDEAL
OPERATOR: SIMM KAMEEN MINING										
SIMM KAMEEN 1	8.50	OPR	1974	HY	WAT	CONV	NONE	N/A		ABB

DIRECTORY OF POWER PLANTS IN CANADA
UDI/MCGRAW-HILL

PLANT/UNIT NAME	MW	STA	YEAR	TYPE	FUEL	FUEL TYPE	ALT FUEL	SSS MFR	TURB MFR	GEN MFR
OPERATOR: SIMM KAMEEN MINING										
SIMM KAMEEN 2	8.50	OPR	1974	HY	WAT	CONV	NONE	N/A		ABB
OPERATOR: SITHE ENERGIES GROUP										
INGLESIDE	130.00	PLN		CC	GAS		NONE			
OPERATOR: SITHE ENERGIES GROUP/HUSKY OIL										
CARDINAL (CASCO) GT1	106.00	CON	1994	GT/C	GAS		NONE	N/A	WH	WH
CARDINAL (CASCO) SC1	50.00	CON	1994	ST/C	WSTH		NONE	ZURN	WH	WH
OPERATOR: SKEENA CELLULOSE INC										
WATSON ISLAND (SKEENA) 1	7.50	OPR	1950	ST	LIQ		NONE	FW/BW	WORT	EMC
WATSON ISLAND (SKEENA) 2	34.50	OPR	1966	ST	LIQ		NONE	FW/BW	BBC	BBC
OPERATOR: SOUTHERN ALBERTA INST TECH										
SAIT POWER PLANT 1	0.60	OPR	1959	ST	GAS		NONE	FW/BW	BELLIS	MTPL
SAIT POWER PLANT IC1	0.50	OPR	1967	IC	GAS		NONE	N/A	WAU	TAMPER
OPERATOR: SOUTHVIEW ATHABASCA INC										
FLATBUSH 1	30.00	PLN	1995	ST	WOOD		NONE			
OPERATOR: SPRUCE FALLS POWER & PAPER CO										
KAPUSKASING MILL 1	12.50	OPR	1945	ST	GAS		NONE	MV/BW	GE	GE
KAPUSKASING MILL 2	9.10	OPR	1958	ST/S	GAS		NONE	MV/BW	PAR	PAR
KAPUSKASING MILL H1	1.80	OPR	1923	HY	WAT	CONV	NONE	N/A	DEW	CGE
SMOKY FALLS 1	13.20	OPR	1928	HY	WAT	CONV	NONE	N/A	AC	CGE
SMOKY FALLS 2	13.20	OPR	1928	HY	WAT	CONV	NONE	N/A	AC	CGE
SMOKY FALLS 3	13.20	OPR	1928	HY	WAT	CONV	NONE	N/A	AC	CGE
SMOKY FALLS 4	13.20	OPR	1931	HY	WAT	CONV	NONE	N/A	AC	CGE
OPERATOR: ST ANNE NACKAWIC PULP & PAPER										
NACKAWIC 1	25.00	OPR	1970	ST	OIL	HFO	NONE	BW	TERR	STAL

DIRECTORY OF POWER PLANTS IN CANADA
UDI/MCGRAW-HILL

PLANT/UNIT NAME	MW	STA	YEAR	TYPE	FUEL	FUEL TYPE	ALT FUEL	SSS MFR	TURB MFR	GEN MFR
OPERATOR: ST CATHERINES HYDRO COMM										
PORT DALHOUSIE 1	4.00	OPR	1988	HY	WAT	CONV	NONE	N/A	DBSZ	
PORT DALHOUSIE 2	4.00	OPR	1988	HY	WAT	CONV	NONE	N/A	DBSZ	
OPERATOR: ST GEORGE PULP & PAPER CO LTD										
ST GEORGE 1	0.70	OPR	1950	HY	WAT	CONV	NONE	N/A	BV	EE
ST GEORGE 2	0.70	OPR	1950	HY	WAT	CONV	NONE	N/A	BARBER	EE
ST GEORGE 3	1.50	OPR	1978	HY	WAT	CONV	NONE	N/A	BV	CGE
ST GEORGE 4	1.50	OPR	1978	HY	WAT	CONV	NONE	N/A	BV	CGE
OPERATOR: ST LAWRENCE SEAWAY AUTHORITY										
WELLAND 1	5.00	OPR	1932	HY	WAT	CONV	NONE	N/A	SMS	CGE
WELLAND 2	5.00	OPR	1932	HY	WAT	CONV	NONE	N/A	SMS	CGE
WELLAND 3	5.00	OPR	1932	HY	WAT	CONV	NONE	N/A	SMS	CGE
OPERATOR: ST REGIS (ALBERTA) LTD										
HINTON 1	21.96	OPR	1957	ST/S	GAS		NONE	FW/CE	GE	GE
HINTON 2	30.00	OPR	1989	ST/S	GAS		NONE		GE	GE
HINTON IC1	1.10	OPR	1956	IC	OIL	DIESEL	NONE	N/A	SCMK	EMC
HINTON IC2	1.00	OPR	1956	IC	OIL	DIESEL	NONE	N/A	GM	WH
OPERATOR: STELCO INC										
HAMILTON 1	4.00	OPR	1948	ST/S	BGAS		NONE	CE	MST	GE
HAMILTON 2	6.00	OPR	1959	ST	BGAS		NONE	CE	GE	GE
OPERATOR: STRATHCONA POWER LTD										
STRATHCONA PARK 1	0.08	OPR	1991	HY	WAT	CONV	NONE	N/A	PELT	KATO
OPERATOR: SUMMERSIDE ELECTRIC DEPT										
SUMMERSIDE IC1	0.20	OPR	1940	IC	OIL	DIESEL	NONE	N/A	FM	FM
SUMMERSIDE IC2	0.25	OPR	1940	IC	OIL	DIESEL	NONE	N/A	FM	FM
SUMMERSIDE IC3	0.25	OPR	1941	IC	OIL	DIESEL	NONE	N/A	FM	FM
SUMMERSIDE IC4	0.56	OPR	1947	IC	OIL	DIESEL	NONE	N/A	FM	FM

DIRECTORY OF POWER PLANTS IN CANADA
UDI/MCGRAW-HILL

PLANT/UNIT NAME	MW	STA	YEAR	TYPE	FUEL	FUEL TYPE	ALT FUEL	SSS MFR	TURB MFR	GEN MFR
OPERATOR: SUMMERSIDE ELECTRIC DEPT										
SUMMERSIDE IC5	1.14	OPR	1950	IC	OIL	DIESEL	NONE	N/A	FM	FM
SUMMERSIDE IC6	2.25	OPR	1960	IC	OIL	DIESEL	NONE	N/A	MBD	BRUSH
SUMMERSIDE IC7	2.25	OPR	1963	IC	OIL	DIESEL	NONE	N/A	MBD	BRUSH
SUMMERSIDE IC8	4.25	OPR	1983	IC	OIL	DIESEL	NONE	N/A	MIRR	BRUSH
OPERATOR: SUNCOR INC										
TAR ISLAND 1	32.50	OPR	1967	ST	COKE		NONE	FW/CE	GE	GE
TAR ISLAND 2	32.50	OPR	1967	ST	COKE		NONE	FW/CE	GE	GE
OPERATOR: SUNRIDGE POWER CORP										
DRYDEN 1	0.60	OPR	1912	HY	WAT	CONV	NONE	N/A	SMS	LANCS
DRYDEN GT1	37.70	OPR	1989	GT	GAS			N/A	WHC	BRUSH
EAGLE RIVER 1	1.76	OPR	1928	HY	WAT	CONV	NONE	N/A	SMS	CGE
MCKENZIE FALLS 1	1.12	OPR	1938	HY	WAT	CONV	NONE	N/A	SMS	CGE
WAINWRIGHT FALLS 1	1.00	OPR	1921	HY	WAT	CONV	NONE	N/A	SMS	WHC
OPERATOR: SYDNEY STEEL CORP										
SYDNEY 1	5.00	OPR	1919	ST/S	OIL	HFO	NONE	BWGM	GE	GE
SYDNEY 2	7.60	OPR	1937	ST/S	OIL	HFO	NONE	BWGM	BBC	BBC
SYDNEY 3	16.00	OPR	1943	ST/S	OIL	HFO	NONE	BWGM	PAR	PAR
OPERATOR: SYNEX INTERNATIONAL INC										
SURPRISE LAKE 1	3.55	PLN	9999	HY	WAT	CONV	NONE	N/A		
SURPRISE LAKE 2	3.55	PLN	9999	HY	WAT	CONV	NONE	N/A		
OPERATOR: TENNECO INC										
NORTH VANCOUVER	80.00	PLN		CC	GAS			H2	N/A	
OPERATOR: TRANSALTA RESOURCES CORP										
CROWSNEST	130.00	PLN	1995	CC	GAS			NONE		
MALTON (MCDONNELL) GT1	42.00	OPR	1992	GT/C	GAS			NONE	N/A	GE	GE
MALTON (MCDONNELL) GT2	42.00	OPR	1992	GT/C	GAS			NONE	N/A	GE	GE

DIRECTORY OF POWER PLANTS IN CANADA
UDI/MCGRAW-HILL

PLANT/UNIT NAME	MW	STA	YEAR	TYPE	FUEL	FUEL TYPE	ALT FUEL	SSS MFR	TURB MFR	GEN MFR
OPERATOR: TRANSALTA RESOURCES CORP										
MALTON (MCDONNELL) SC1	13.00	OPR	1992	ST/C	WSTH		NONE			
MALTON (MCDONNELL) SC2	13.00	OPR	1992	ST/C	WSTH		NONE			
OHSC PROJECT GT1	42.00	OPR	1992	GT/C	GAS		NONE	N/A	GE	GE
OHSC PROJECT SC1	13.00	OPR	1992	ST/C	WSTH		NONE			
OPERATOR: TRANSALTA UTILITIES CORP										
BARRIER 1	9.56	OPR	1947	HY	WAT	CONV	NONE	N/A	DEW	WHC
BEARSPAW 1	15.30	OPR	1954	HY	WAT	CONV	NONE	N/A	KMW	WHC
BIGHORN 1	59.00	OPR	1972	HY	WAT	CONV	NONE	N/A	DEW	EE
BIGHORN 2	59.00	OPR	1972	HY	WAT	CONV	NONE	N/A	DEW	EE
BRAZEAU 1	144.00	OPR	1965	HY	WAT	CONV	NONE	N/A	DEW	WHC
BRAZEAU 2	161.50	OPR	1967	HY	WAT	CONV	NONE	N/A	DEW	WHC
CASCADE 1	17.00	OPR	1942	HY	WAT	CONV	NONE	N/A	DEW	WHC
CASCADE 2	17.00	OPR	1957	HY	WAT	CONV	NONE	N/A	DEW	WHC
GHOST 1	12.75	OPR	1929	HY	WAT	CONV	NONE	N/A	DEW	WHC
GHOST 2	12.75	OPR	1929	HY	WAT	CONV	NONE	N/A	DEW	WHC
GHOST 3	21.15	OPR	1954	HY	WAT	CONV	NONE	N/A	EE	WHC
HORSESHOE 1	3.38	OPR	1911	HY	WAT	CONV	NONE	N/A	KMW	CGE
HORSESHOE 2	5.63	OPR	1911	HY	WAT	CONV	NONE	N/A	DEW	CGE
HORSESHOE 3	3.38	OPR	1911	HY	WAT	CONV	NONE	N/A	KMW	CGE
HORSESHOE 4	5.63	OPR	1911	HY	WAT	CONV	NONE	N/A	DEW	CGE
INTERLAKES 1	5.04	OPR	1955	HY	WAT	CONV	NONE	N/A	ACC	WHC
KANANASKIS 1	3.40	OPR	1913	HY	WAT	CONV	NONE	N/A	ACC	CGE
KANANASKIS 2	3.40	OPR	1913	HY	WAT	CONV	NONE	N/A	ACC	CGE
KANANASKIS 3	9.56	OPR	1951	HY	WAT	CONV	NONE	N/A	DEW	WHC
KEEPHILLS 1	403.20	OPR	1983	ST	COAL	SUB	NONE	CE	HT	HT
KEEPHILLS 2	403.20	OPR	1983	ST	COAL	SUB	NONE	CE	HT	HT
OUTLET WORKS 1	9.72	OPR	1965	HY	WAT	CONV	NONE	N/A	DEW	WHC
OUTLET WORKS 2	9.72	OPR	1967	HY	WAT	CONV	NONE	N/A	DEW	WHC
POCATERRA 1	13.50	OPR	1955	HY	WAT	CONV	NONE	N/A	ACC	WHC
RUNDLE 1	17.00	OPR	1951	HY	WAT	CONV	NONE	N/A	DEW	WHC

DIRECTORY OF POWER PLANTS IN CANADA
UDI/MCGRAW-HILL

PLANT/UNIT NAME	MW	STA	YEAR	TYPE	FUEL	FUEL TYPE	ALT FUEL	SSS MFR	TURB MFR	GEN MFR
OPERATOR: TRANSALTA UTILITIES CORP										
RUNDLE 2	29.75	OPR	1960	HY	WAT	CONV	NONE	N/A	DEW	WHC
SPRAY 1	40.40	OPR	1951	HY	WAT	CONV	NONE	N/A	DEW	WHC
SPRAY 2	40.40	OPR	1960	HY	WAT	CONV	NONE	N/A	DEW	WHC
SUNDANCE 1	300.00	OPR	1970	ST	COAL	SUB	NONE	CE	AEI	EE
SUNDANCE 2	300.00	OPR	1973	ST	COAL	SUB	NONE	CE	AEI	EE
SUNDANCE 3	400.00	OPR	1976	ST	COAL	SUB	NONE	CE	GEC	EE
SUNDANCE 4	400.00	OPR	1976	ST	COAL	SUB	NONE	CE	GEC	EE
SUNDANCE 5	400.00	OPR	1977	ST	COAL	SUB	NONE	CE	GEC	EE
SUNDANCE 6	400.00	OPR	1980	ST	COAL	SUB	NONE	CE	GEC	AEI
THREE SISTERS 1	3.40	OPR	1951	HY	WAT	CONV	NONE	N/A	DEW	WHC
WABAMUN 1	66.00	OPR	1956	ST	COAL	SUB	NONE	BWGM	MV	MV
WABAMUN 2	66.00	OPR	1958	ST	COAL	SUB	NONE	BWGM	MV	MV
WABAMUN 3	150.00	OPR	1962	ST	COAL	SUB	NONE	CE	AEI	MV
WABAMUN 4	300.00	OPR	1967	ST	COAL	SUB	NONE	CE	AEI	AEI/GE
OPERATOR: TRANSCANADA PIPELINES LTD										
KAPUSKASING	150.00	PLN	1995	CC	GAS		NONE		WH	WH
NIPIGON GT1	21.00	OPR	1992	GT/C	GAS		NONE	N/A	RUST	
NIPIGON GT2	21.00	OPR	1992	GT/C	GAS		NONE	N/A	RUST	
NIPIGON SC1	14.00	OPR	1992	ST/C	WSTH		NONE			
NORTH BAY	150.00	PLN	1995	CC	GAS		NONE		WH	WH
OPERATOR: TRENT UNIVERSITY										
NASSAU 1	0.36	OPR	1902	HY	WAT	CONV	NONE	N/A	WK	CGE
NASSAU 2	0.36	OPR	1902	HY	WAT	CONV	NONE	N/A	WK	CGE
NASSAU 3	1.50	OPR	1926	HY	WAT	CONV	NONE	N/A	MV	CGE
OPERATOR: TRICIL LTD										
SWARU 1	4.57	OPR	1982	ST/S	REF		NONE	BW	EL	EL

DIRECTORY OF POWER PLANTS IN CANADA
UDI/MCGRAW-HILL

PLANT/UNIT NAME	MW	STA	YEAR	TYPE	FUEL	FUEL TYPE	ALT FUEL	SSS MFR	TURB MFR	GEN MFR
OPERATOR: TWIN FALLS POWER CORP										
TWIN FALLS 1	45.00	STN		HY	WAT	CONV	NONE	N/A		
TWIN FALLS 2	45.00	STN		HY	WAT	CONV	NONE	N/A		
TWIN FALLS 3	45.00	STN		HY	WAT	CONV	NONE	N/A		
TWIN FALLS 4	45.00	STN		HY	WAT	CONV	NONE	N/A		
TWIN FALLS 5	45.00	STN		HY	WAT	CONV	NONE	N/A		
OPERATOR: UNIVERSITY OF OTTAWA										
UO ENERGY PLANT IC1	0.55	OPR	1985	IC	GAS		NONE	N/A		
OPERATOR: US WINDPOWER/PEIGAN BAND										
PINCHER CREEK I	9.00	PLN	1993	WTG	WIND		NONE	N/A		
OPERATOR: WAMINDJI BAND COUNCIL										
MAQUATUA	1.40	OPR	1985	HY	WAT	CONV	NONE	N/A	DBSZ	
OPERATOR: WELDWOOD OF CANADA LTD										
FLAVELLE 1	3.00	OPR	1915	ST	WOOD		NONE	BWGM	GE	GE
FLAVELLE 2	3.50	OPR	1941	ST	WOOD		NONE	BWGM	GE	GE
OPERATOR: WEST KOOTENAY POWER LTD										
CORRA LYNN 1	13.50	OPR	1932	HY	WAT	CONV	NONE	N/A	DEW	CGE
CORRA LYNN 2	13.50	OPR	1932	HY	WAT	CONV	NONE	N/A	DEW	CGE
CORRA LYNN 3	13.50	OPR	1932	HY	WAT	CONV	NONE	N/A	DEW	CGE
LOWER BONNINGTON 1	15.75	OPR	1925	HY	WAT	CONV	NONE	N/A	ACC	CGE
LOWER BONNINGTON 2	15.75	OPR	1926	HY	WAT	CONV	NONE	N/A	ACC	CGE
LOWER BONNINGTON 3	15.75	OPR	1971	HY	WAT	CONV	NONE	N/A	MHI	CGE
MOBILE (WK) IC1	0.20	OPR	1963	IC	OIL	DIESEL	NONE	N/A	GM	GE
SOUTH SLOCAN 1	15.75	OPR	1928	HY	WAT	CONV	NONE	N/A	ACC	CGE
SOUTH SLOCAN 2	15.75	OPR	1928	HY	WAT	CONV	NONE	N/A	ACC	CGE
SOUTH SLOCAN 3	15.75	OPR	1929	HY	WAT	CONV	NONE	N/A	ACC	CGE
UPPER BONNINGTON 1	5.06	OPR	1907	HY	WAT	CONV	NONE	N/A	IPM	CGE

DIRECTORY OF POWER PLANTS IN CANADA
UDI/MCGRAW-HILL

PLANT/UNIT NAME	MW	STA	YEAR	TYPE	FUEL	FUEL TYPE	ALT FUEL	SSS MFR	TURB MFR	GEN MFR
OPERATOR: WEST KOOTENAY POWER LTD										
UPPER BONNINGTON 2	5.06	OPR	1907	HY	WAT	CONV	NONE	N/A	IPM	CGE
UPPER BONNINGTON 3	6.75	OPR	1914	HY	WAT	CONV	NONE	N/A	ACC	CGE
UPPER BONNINGTON 4	6.75	OPR	1916	HY	WAT	CONV	NONE	N/A	ACC	CGE
UPPER BONNINGTON 5	15.75	OPR	1940	HY	WAT	CONV	NONE	N/A	ACC	WHC
UPPER BONNINGTON 6	15.75	OPR	1940	HY	WAT	CONV	NONE	N/A	ACC	WHC
OPERATOR: WESTAR LTD										
CELGAR PULP MILL 1	2.50	OPR	1963	ST	WOOD		NONE	CE/FW	GE	GE
OPERATOR: WESTCOAST ENERGY INC										
FORT FRANCES GT1	46.70	OPR	1990	GT/C	GAS		NONE	N/A	WHC	BRUSH
FORT FRANCES S1	59.00	OPR	1990	ST/C	WSTH		NONE	FW	WH	BRUSH
MCMAHON GT1	46.50	CON	1993	GT/S	GAS		NONE	N/A	WH	WH
MCMAHON GT2	46.50	CON	1993	GT/S	GAS		NONE	N/A	WH	WH
POWELL RIVER II	180.00	PLN		CC	GAS		NONE	N/A		
OPERATOR: WESTERN COOP FERTILIZER LTD										
MEDICINE HAT (WCF) 1	0.80	OPR	1956	ST	GAS		NONE	BW	GE	GE
OPERATOR: WESTERN PULP LP										
PORT ALICE 4	3.50	OPR	1949	ST	LIQ		NONE	CE	EL	EL
PORT ALICE 5	16.60	OPR	1976	ST	LIQ		NONE	CE	GE	GE
PORT ALICE H1	2.00	OPR	1953	HY	WAT	CONV	NONE	N/A	CVIC	EL
WOODFIBRE 1	2.00	OPR	1947	ST	LIQ		NONE	B/T/Z	EL	EL
WOODFIBRE 2	2.00	OPR	1947	ST	LIQ		NONE	B/T/Z	EL	EL
WOODFIBRE 3	3.00	OPR	1961	ST	LIQ		NONE	B/T/Z	GE	GE
WOODFIBRE H1	2.59	OPR	1947	HY	WAT	CONV	NONE	N/A	PWW	WHC
OPERATOR: WESTMINSTER RESOURCES LTD										
CAMPBELL RIVER IC1	0.75	OPR	1970	IC	OIL	DIESEL	NONE	N/A	GM	GE
CAMPBELL RIVER IC2	0.75	OPR	1970	IC	OIL	DIESEL	NONE	N/A	CAT	GE

DIRECTORY OF POWER PLANTS IN CANADA
UDI/MCGRAW-HILL

PLANT/UNIT NAME	MW	STA	YEAR	TYPE	FUEL	FUEL TYPE	ALT FUEL	SSS MFR	TURB MFR	GEN MFR

OPERATOR: *WESTMINSTER RESOURCES LTD*

PLANT/UNIT NAME	MW	STA	YEAR	TYPE	FUEL	FUEL TYPE	ALT FUEL	SSS MFR	TURB MFR	GEN MFR
CAMPBELL RIVER IC3	0.80	OPR	1971	IC	OIL	DIESEL	NONE	N/A	CAT	KATO
CAMPBELL RIVER IC4	0.80	OPR	1972	IC	OIL	DIESEL	NONE	N/A	GM	KATO
CAMPBELL RIVER IC5	0.75	OPR	1977	IC	OIL	DIESEL	NONE	N/A	CAT	WH
CAMPBELL RIVER IC6	0.80	OPR	1980	IC	OIL	DIESEL	NONE	N/A	CAT	KATO
CAMPBELL RIVER IC7	0.80	OPR	1980	IC	OIL	DIESEL	NONE	N/A	CAT	KATO
CAMPBELL RIVER IC8	0.80	OPR	1980	IC	OIL	DIESEL	NONE	N/A	CAT	KATO
CAMPBELL RIVER IC9	0.80	OPR	1983	IC	OIL	DIESEL	NONE	N/A	CAT	KATO
TENNANT LAKE 1	3.06	OPR	1966	HY	WAT	CONV	NONE	N/A	GGG	CGE
THELWOOD 1	8.20	OPR	1985	HY	WAT	CONV	NONE	N/A	GGG	TS

OPERATOR: *WEYERHAEUSER CANADA LTD*

PLANT/UNIT NAME	MW	STA	YEAR	TYPE	FUEL	FUEL TYPE	ALT FUEL	SSS MFR	TURB MFR	GEN MFR
KAMLOOPS 1	27.00	OPR	1972	ST	LIQ		NONE	BW/FW	STAL	STAL
KAMLOOPS 2	14.00	OPR	1972	ST	LIQ		NONE	BW/FW	STAL	STAL
PRINCE ALBERT 1	22.31	OPR	1968	ST/S	LIQ		NONE	BW	STAL	ASEA

OPERATOR: *WINDSOR COGENERATION/DOMTAR*

PLANT/UNIT NAME	MW	STA	YEAR	TYPE	FUEL	FUEL TYPE	ALT FUEL	SSS MFR	TURB MFR	GEN MFR
EASTERN TOWNSHIPS MILL	170.50	PLN		CC	GAS		NONE			

OPERATOR: *WINNIPEG HYDRO*

PLANT/UNIT NAME	MW	STA	YEAR	TYPE	FUEL	FUEL TYPE	ALT FUEL	SSS MFR	TURB MFR	GEN MFR
AMY STREET 1	5.00	OPR	1924	ST	COAL	LIG	NONE	JI/BW	HOW	PAR
AMY STREET 2	5.00	OPR	1924	ST	COAL	LIG	NONE	JI/BW	HOW	PAR
AMY STREET 4	25.00	OPR	1954	ST	COAL	LIG	NONE	JI/BW	BBC	BBC
POINTE DU BOIS 01	3.00	OPR	1911	HY	WAT	CONV	NONE	N/A	BV	MV
POINTE DU BOIS 02	3.00	OPR	1911	HY	WAT	CONV	NONE	N/A	BV	MV
POINTE DU BOIS 03	3.00	OPR	1911	HY	WAT	CONV	NONE	N/A	BV	MV
POINTE DU BOIS 04	3.00	OPR	1911	HY	WAT	CONV	NONE	N/A	BV	MV
POINTE DU BOIS 05	3.00	OPR	1911	HY	WAT	CONV	NONE	N/A	BV	MV
POINTE DU BOIS 06	4.00	OPR	1914	HY	WAT	CONV	NONE	N/A	EW	WHC
POINTE DU BOIS 07	4.00	OPR	1914	HY	WAT	CONV	NONE	N/A	EW	WHC
POINTE DU BOIS 08	4.00	OPR	1914	HY	WAT	CONV	NONE	N/A	EW	WHC
POINTE DU BOIS 09	5.20	OPR	1922	HY	WAT	CONV	NONE	N/A	BV	CGE

DIRECTORY OF POWER PLANTS IN CANADA
UDI/MCGRAW-HILL

PLANT/UNIT NAME	MW	STA	YEAR	TYPE	FUEL	FUEL TYPE	ALT FUEL	SSS MFR	TURB MFR	GEN MFR
OPERATOR: WINNIPEG HYDRO										
POINTE DU BOIS 10	5.20	OPR	1922	HY	WAT	CONV	NONE	N/A	BV	CGE
POINTE DU BOIS 11	5.20	OPR	1922	HY	WAT	CONV	NONE	N/A	BV	CGE
POINTE DU BOIS 12	5.20	OPR	1923	HY	WAT	CONV	NONE	N/A	CVIC	CGE
POINTE DU BOIS 13	5.20	OPR	1923	HY	WAT	CONV	NONE	N/A	CVIC	CGE
POINTE DU BOIS 14	5.20	OPR	1923	HY	WAT	CONV	NONE	N/A	CVIC	CGE
POINTE DU BOIS 15	5.20	OPR	1925	HY	WAT	CONV	NONE	N/A	BV	CGE
POINTE DU BOIS 16	5.20	OPR	1925	HY	WAT	CONV	NONE	N/A	BV	CGE
SLAVE FALLS 1	9.00	OPR	1931	HY	WAT	CONV	NONE	N/A	DEW	CGE
SLAVE FALLS 2	9.00	OPR	1931	HY	WAT	CONV	NONE	N/A	DEW	CGE
SLAVE FALLS 3	9.00	OPR	1936	HY	WAT	CONV	NONE	N/A	DEW	CGE
SLAVE FALLS 4	9.00	OPR	1936	HY	WAT	CONV	NONE	N/A	DEW	CGE
SLAVE FALLS 5	9.00	OPR	1946	HY	WAT	CONV	NONE	N/A	DEW	CGE
SLAVE FALLS 6	9.00	OPR	1946	HY	WAT	CONV	NONE	N/A	DEW	CGE
SLAVE FALLS 7	9.00	OPR	1948	HY	WAT	CONV	NONE	N/A	DEW	CGE
SLAVE FALLS 8	9.00	OPR	1948	HY	WAT	CONV	NONE	N/A	DEW	CGE
OPERATOR: YOHO POWER LTD										
FIELD IC1	0.15	OPR	1959	IC	OIL	DIESEL	NONE	N/A	MIRR	TERR
FIELD IC2	0.15	OPR	1959	IC	OIL	DIESEL	NONE	N/A	MIRR	TERR
FIELD IC3	0.10	OPR	1960	IC	OIL	DIESEL	NONE	N/A	MIRR	GE
FIELD IC4	0.25	OPR	1969	IC	OIL	DIESEL	NONE	N/A	LB	TAMPER
OPERATOR: YUKON ELECTRICAL CO LTD										
BEAVER CREEK IC1	0.25	OPR	1967	IC	OIL	DIESEL	NONE	N/A	CAT	TAMPER
BEAVER CREEK IC2	0.35	OPR	1967	IC	OIL	DIESEL	NONE	N/A	CAT	COEL
BEAVER CREEK IC3	0.25	OPR	1969	IC	OIL	DIESEL	NONE	N/A	CAT	KATO
CARMACKS IC1	0.35	OPR	1968	IC	OIL	DIESEL	NONE	N/A	CAT	COEL
DESTRUCTION BAY IC1	0.15	OPR	1962	IC	OIL	DIESEL	NONE	N/A	CAT	EMC
DESTRUCTION BAY IC2	0.25	OPR	1966	IC	OIL	DIESEL	NONE	N/A	CAT	TAMPER
DESTRUCTION BAY IC3	0.30	OPR	1973	IC	OIL	DIESEL	NONE	N/A	CAT	GE
HAINES JUNCTION IC1	0.10	OPR	1958	IC	OIL	DIESEL	NONE	N/A	VENG	COEL

DIRECTORY OF POWER PLANTS IN CANADA
UDI/MCGRAW-HILL

PLANT/UNIT NAME	MW	STA	YEAR	TYPE	FUEL	FUEL TYPE	ALT FUEL	SSS MFR	TURB MFR	GEN MFR
OPERATOR: YUKON ELECTRICAL CO LTD										
HAINES JUNCTION IC2	0.15	OPR	1963	IC	OIL	DIESEL	NONE	N/A	CAT	TAMPER
MCINTYRE CREEK 1	0.65	OPR	1955	HY	WAT	CONV	NONE	N/A	GGG	WH
OLD CROW IC1	0.20	OPR	1970	IC	OIL	DIESEL	NONE	N/A	CAT	EMC
OLD CROW IC2	0.15	OPR	1973	IC	OIL	DIESEL	NONE	N/A	CAT	KATO
OLD CROW IC3	0.23	OPR	1981	IC	OIL	DIESEL	NONE	N/A	CAT	KATO
PELLY RIVER CROSSING IC1	0.15	OPR	1963	IC	OIL	DIESEL	NONE	N/A	CAT	TAMPER
PELLY RIVER CROSSING IC2	0.20	OPR	1983	IC	OIL	DIESEL	NONE	N/A	VOLV	COEL
PELLY RIVER CROSSING IC3	0.25	OPR	1984	IC	OIL	DIESEL	NONE	N/A	VOLV	TAMPER
PORTER CREEK 1	0.30	OPR	1949	HY	WAT	CONV	NONE	N/A	PWW	CGE
PORTER CREEK 2	0.70	OPR	1952	HY	WAT	CONV	NONE	N/A	GGG	WH
ROSS RIVER IC1	0.35	OPR	1973	IC	OIL	DIESEL	NONE	N/A	CAT	KATO
STEWART CROSSING IC1	0.10	OPR	1958	IC	OIL	DIESEL	NONE	N/A	CAT	COEL
STEWART CROSSING IC2	0.06	OPR	1971	IC	OIL	DIESEL	NONE	N/A	CAT	CAT
STEWART CROSSING IC3	0.10	OPR	1985	IC	OIL	DIESEL	NONE	N/A	VOLV	TAMPER
SWIFT RIVER IC1	0.06	OPR	1965	IC	OIL	DIESEL	NONE	N/A	CAT	CAT
SWIFT RIVER IC2	0.10	OPR	1967	IC	OIL	DIESEL	NONE	N/A	CAT	COEL
SWIFT RIVER IC3	0.09	OPR	1976	IC	OIL	DIESEL	NONE	N/A	CAT	COEL
TESLIN IC1	0.35	OPR	1973	IC	OIL	DIESEL	NONE	N/A	CAT	KATO
TESLIN IC2	0.50	OPR	1983	IC	OIL	DIESEL	NONE	N/A	CAT	KATO
WATSON LAKE IC1	0.50	OPR	1966	IC	OIL	DIESEL	NONE	N/A	CAT	TAMPER
WATSON LAKE IC2	0.60	OPR	1967	IC	OIL	DIESEL	NONE	N/A	CAT	TAMPER
WATSON LAKE IC3	0.60	OPR	1970	IC	OIL	DIESEL	NONE	N/A	CAT	TAMPER
WATSON LAKE IC4	0.80	OPR	1976	IC	OIL	DIESEL	NONE	N/A	CAT	BBC
WATSON LAKE IC5	0.80	OPR	1978	IC	OIL	DIESEL	NONE	N/A	CAT	BBC
WATSON LAKE IC6	0.80	OPR	1985	IC	OIL	DIESEL	NONE	N/A	CAT	TAMPER
WATSON LAKE IC7	1.00	OPR	1985	IC	OIL	DIESEL	NONE	N/A	CAT	IDEAL
OPERATOR: YUKON ENERGY CORP										
ASIHIHIK 1	16.00	OPR	1975	HY	WAT	CONV	NONE	N/A	DEW	CGE
ASIHIHIK 2	16.00	OPR	1975	HY	WAT	CONV	NONE	N/A	DEW	CGE
MAYO RIVER 1	2.55	OPR	1952	HY	WAT	CONV	NONE	N/A	DEW	CGE

DIRECTORY OF POWER PLANTS IN CANADA
UDI/MCGRAW-HILL

PLANT/UNIT NAME	MW	STA	YEAR	TYPE	FUEL	FUEL TYPE	ALT FUEL	SSS MFR	TURB MFR	GEN MFR
OPERATOR: YUKON ENERGY CORP										
MAYO RIVER 2	2.55	OPR	1958	HY	WAT	CONV	NONE	N/A	GGG	CGE
WHITE HORSE RAPIDS 1	5.70	OPR	1958	HY	WAT	CONV	NONE	N/A	KMW	WHC
WHITE HORSE RAPIDS 2	5.70	OPR	1958	HY	WAT	CONV	NONE	N/A	KMW	WHC
WHITE HORSE RAPIDS 3	8.00	OPR	1969	HY	WAT	CONV	NONE	N/A	AC	CGE
WHITE HORSE RAPIDS 4	23.60	OPR	1984	HY	WAT	CONV	NONE	N/A	DEW	CGE
OPERATOR: ZZ/UNIDENTIFIED										
MARYS HARBOUR H1	0.18	OPR	1987	HY	WAT	CONV	NONE	N/A		
TROUT CREEK 1	0.60	OPR		HY	WAT	CONV	NONE	N/A	OSS	OSS

APPENDIX A

ELECTRIC POWER PLANT OPERATORS IN CANADA

Appendix A

Electric Power Plant Operators in Canada

AEC Power Ltd.
P.O. Bag 4009
Fort McMurray, AB T9H 3L1
CANADA

Abitibi-Price, Inc.
Box 500
Grand Falls, NF A2A 2K1
CANADA

Alberta Government Services

    Legislature Building Power Plant
    Alberta Government Services
    Government Center
    Edmonton, AB T5K 2C7
    CANADA

Alberta Hospital

    Edmonton Hospital Power Plant
    Alberta Hospital
    Box 307
    Edmonton, AB T5J 2J7
    CANADA

    Ponoka Hospital Power Plant
    Alberta Hospital
    Ponoka, AB T0C 2H0
    CANADA

Alberta Power Ltd.
10035-105th Street
Edmonton, AB T5J 2V6
CANADA
TEL: 403-420-7310
FAX: 403-441-4111

    Frank R. Bajc
    Plant Manager
    Battle River Station
    Alberta Power Ltd.
    P.O. Box 498
    Forestburg, AB T0B 1N0
    CANADA

    Graham B. Wilson
    Plant Manager
    H.R. Milner Station
    Alberta Power Ltd.
    P.O. Box 180
    Grande Cache, AB T0E 0Y0
    CANADA

    Richard R. Walthall
    Plant Manager
    Sheerness Station
    Alberta Power Ltd.
    P.O. Box 1540
    Hanna, AB T0J 1P0
    CANADA

Alberta Public Works Supply & Services

    Michener Centre South Power Plant
    Alberta Public Works Supply & Services
    Box 5002
    Red Deer, AB T4N 5Y5
    CANADA

Appendix A

Electric Power Plant Operators in Canada

Alberta Sugar Co.

    Taber Power Plant
    Alberta Sugar Co.
    Box 1209
    Taber, AB T0K 2G0
    CANADA

Albright e Wilson Amerique

    Buckingham Power Plant
    Albright e Wilson Amerique
    470 Erco Street
    Buckingham, PQ J8L 1E1
    CANADA

Alcan Smelters & Chemicals, Ltd.
Case Postale 1800
Jonquiere, PQ G7S 4R5
CANADA
TEL: 418-699-2131
FAX: 418-699-3357

    D. Timlick
    Plant Supervisor
    Kemano Hydroelectric Power Plant
    Alcan Smelters & Chemicals, Ltd.
    Box 1800
    Kitimat, BC V8C 2H2
    CANADA
    TEL: 604-639-8538
    FAX: 604-639-8602

Algoma Steel Corporation Ltd.

    Sault Ste. Marie Power Plant
    Algoma Steel Corporation Ltd.
    Queen Street West
    Sault Ste. Marie, ON P6A 5P2
    CANADA

Allied Chemicals Canada Ltd.

    Amherstburg Power Plant
    Allied Chemicals Canada Ltd.
    Box 2000
    Amherstburg, ON N9V 2Z6
    CANADA

Appendix A

Electric Power Plant Operators in Canada

Almonte Public Utilities Commission
28 Mill Street
Box 179
Almonte, ON K0A 1A0
CANADA

Amoco Canada Petroleum Ltd.
6815 - 8th Street, N.E.
Calgary, AB T2E 7H7
CANADA

Atlantic Sugar Ltd.

    Saint John Power Plant
    Atlantic Sugar Ltd.
    Box 2800
    Saint John, NB E2L 4L4
    CANADA

BC Forest Products Ltd.
371 Gorge Road
Victoria, BC V8W 2N5
CANADA

    Cowichan Power Plant
    BC Forest Products Ltd.
    Box 2800
    Youbou, BC V0R 3E0
    CANADA

    Crofton Power Plant
    BC Forest Products Ltd.
    Box 70
    Crofton, BC V0R 1R0
    CANADA

    Mackenzie Power Plant
    BC Forest Products Ltd.
    Box 310
    Mackenzie, BC V0J 2C0
    CANADA

BC Hydro
970 Burrard Street
Vancouver, BC V6Z 1Y3
CANADA
TEL: 604-663-2114
FAX: 604-663-3515

    A.E. Zink
    Plant Manager
    Burrard Power Plant
    BC Hydro
    Box 1, Site 7
    RR 1
    Port Moody, BC V3H 3C8
    CANADA
    TEL: 604-469-6100
    FAX: 604-469-8622

    Gordon M. Shrum Hydroelectric Plant
    BC Hydro
    Hudson Hope, BC
    CANADA

    A.S. Woodruff
    Plant Manager
    Lake Buntzen Hydroelectric Plant
    BC Hydro
    Box 6, Site 3
    RR 1
    Port Moody, BC V3H 3C8
    CANADA
    TEL: 604-293-5870
    FAX: 604-293-5874

    J.R. Irvine
    Plant Manager
    Walter Hardman Hydroelectric Plant
    BC Hydro
    Bag 5700
    Revelstoke, BC
    CANADA
    TEL: 604-837-6211
    FAX: 604-837-5624

Appendix A

Electric Power Plant Operators in Canada

BC Packers Ltd.

    Namue Power Plant
    BC Packers Ltd.
    P.O. Box 5000
    Vancouver, BC V6B 4A8
    CANADA

BC Sugar
Box 2150
Vancouver, BC V6B 3V2
CANADA

BC Sugar Refining Company Ltd.

    Fort Garry Power Plant
    BC Sugar Refining Company Ltd.
    Manitoba Sugar Company Division
    555 Hervo Street
    Winnipeg, MB R3T 3L6
    CANADA

BJ Hartgrove Ltd.

    Hargrove Power Plant
    BJ Hartgrove Ltd.
    Box 128
    Bath, NB E0J 1E0
    CANADA

BPCO Incorporated
Box 576
Edmonton, AB T5J 2K8
CANADA

Belleterre Hydro Electric Commission
CP 130
Belleterre, PQ J0Z 1L0
CANADA

Boise Cascade Canada Ltd.
145 Third Street West
Fort Frances, ON P9A 3N2
CANADA

    Kenora Power Plant
    Boise Cascade Canada Ltd.
    504 - 9th Street South
    P.O. Box 5000
    Kenora, ON P9N 3Y1
    CANADA

Appendix A

Electric Power Plant Operators in Canada

Bowaters Mersey Paper Company Ltd.
Box 1150
Liverpool, NS B0T 1K0
CANADA

Bracebridge Hydro
Box 1749
Bracebridge, ON P0B 1C0
CANADA

CIP Inc.

    Gold River Pulp Mill Power Plant
    CIP Inc.
    P.O. Box 1000
    Gold River, BC V0P 1G0
    CANADA

Calgary City Waterworks
Box 2100
Location 35
Calgary, AB T2P 2M5
CANADA

Campbellford Hydro
113 Front Street North
Box 1055
Campbellford, ON K0L 1L0
CANADA

Canadian Forest Products Ltd.

    Englewood Power Plant
    Canadian Forest Products Ltd.
    Woss, BC V0N 3P0
    CANADA

Canadian General Electric Co.

    Peterborough Power Plant
    Canadian General Electric Co.
    107 Park Street North
    Peterborough, ON K9J 7B5
    CANADA

Canadian Niagara Power Company Ltd.
1130 Bertie Street
P.O. Box 1218
Fort Erie, ON L2A 5Y2
CANADA
TEL: 416-871-0330

    Jim Fretz
    Vice President, Production
    Rankine Power Plant
    Canadian Niagara Power Company Ltd.
    Box 118
    Queen Victoria Park
    Niagara Falls, ON L2E 6S8
    CANADA
    TEL: 416-354-1641
    FAX: 416-354-6597

Appendix A

Electric Power Plant Operators in Canada

Canadian Salt Company Ltd.

    Lindbergh Power Plant
    Canadian Salt Company Ltd.
    Box 480
    Elk Point, AB T0A 1A0
    CANADA

Cariboo Pulp & Paper Co.

    Quesnel Power Plant
    Cariboo Pulp & Paper Co.
    Box 7500
    Quesnel, BC V2J 3J6
    CANADA

Cassaiar Mining Corp.

    Cassiar Resources Division Power Plant
    Cassaiar Mining Corp.
    Cassiar, BC V0C 1E0
    CANADA

Celanese Canada Ltd.

    Clover Bar Power Plant   Drummondville Power Plant
    Celanese Canada Ltd.   **Celanese Canada Ltd.**
    Box 99   2575 Blvd. St. Joseph
    Edmonton, AB T5J 2H7   Drummondville, PQ J2B 7V4
    CANADA   CANADA

Central Coast Power Corp.

    Ocean Falls Power Plant
    Central Coast Power Corp.
    General Delivery
    Ocean Falls, BC V0T 1P0
    CANADA

Centrale SPC Ltd.

    Chicoutimi Power Plant
    Centrale SPC Ltd.
    2020 Chemin de la Reserve
    Chicoutimi, PQ G7H 5B7
    CANADA

Appendix A

Electric Power Plant Operators in Canada

Churchill Falls (Labrador) Corp.
P.O. Box 12500
St. John's, NF A1B 3T5
CANADA
TEL: 709-737-1450

    Churchill Falls Power Plant
    Churchill Falls (Labrador) Corp.
    P.O. Box 310
    Churchill Falls, NF A1B 3T5
    CANADA
    TEL: 709-925-3311
    FAX: 709-925-8220

Coaticook Electric Dept.
150 Child Rue
Coaticook, PQ J1A 2B3
CANADA
TEL: 819-849-6333
FAX: 819-849-9472

    S. Morin
    Electrical Engineer
    Coaticook Hydroelectric Plant
    Coaticook Electric Dept.
    St. Paul Street
    Coaticook, PQ J1A 2B3
    CANADA

Cominco Ltd.
BC Group
81 Rock Island Highway
Trail, BC V1R 4L4
CANADA

Consolidated-Bathurst Ltd.
891 Main Street
Bathurst, NB E2A 4A3
CANADA

    Grand Baie One Power Plant
    Consolidated-Bathurst Ltd.
    Box 69
    Montreal, PQ H3C 2R5
    CANADA

Corner Brook Pulp & Paper Ltd.

    Corner Brook Power Plant
    Corner Brook Pulp & Paper Ltd.
    P.O. Box 2001
    Corner Brook, NF A2H 6J4
    CANADA

Appendix A

Electric Power Plant Operators in Canada

Crestbrook Pulp & Paper Ltd.

    Skookmunchuck Power Plant
    Crestbrook Pulp & Paper Ltd.
    P.O. Box 4600
    Crestbrook, BC V1C 4J7
    CANADA

Crown Forest Industries Ltd.
700-815 West Hastings Street
Vancouver, BC V6C 2Y4
CANADA

Deer Lake Power Co.
Box 2000
Deer Lake, NF A0K 2E0
CANADA
TEL: 709-635-2125
FAX: 709-635-5569

Dominion Textile Inc.

    Magog Power Plant
    Dominion Textile Inc.
    Box 6250
    Montreal, PQ H3C 3L1
    CANADA

Domtar Chemicals Group

    Unity Power Plant
    Domtar Chemicals Group
    Box 98
    Unity, SK S0K 4L0
    CANADA

Dow Chemical of Canada Ltd.

    Dow Alberta Power Plant   Sarnia Power Plant
    Dow Chemical of Canada Ltd.   **Dow Chemical of Canada Ltd.**
    P.O. Box 759   South Vidal Street
    Building 211   Box 3030
    Fort Saskatchewan, AB T8L 2P4   Sarnia, ON N7T 7M1
    CANADA   CANADA

E.B. Eddy Forest Products Ltd.
Box 600
Hull, PQ J8X 3Y7
CANADA

    Espanola Power Plant
    E.B. Eddy Forest Products Ltd.
    Espanola, ON P0P 1C0
    CANADA

Appendix A

Electric Power Plant Operators in Canada

Eastern Power Developers Inc.
2200 Lakeshore Blvd. West
Suite 206
Toronto, ON M8V 1A4
CANADA

Edmonton Power
10065 Jasper Avenue
Capitol Square
Edmonton, AB T5J 3B1
CANADA
TEL: 403-448-3193

    John J. Mulka
    Plant Manager
    Clover Bar Generating Station
    Edmonton Power
    1515-130th Avenue
    Edmonton, AB T5J 2R7
    CANADA
    TEL: 403-428-3179
    FAX: 403-428-4663

    Ken Warren
    Plant Manager
    Genesee Generating Station
    Edmonton Power
    Secondary Road 770
    Warburg, AB T0C 2T0
    CANADA
    TEL: 403-963-4192

    Rossdale Generating Station
    Edmonton Power
    10155-96th Avenue
    Edmonton, AB
    CANADA

Edmunston Electric Department
5 - 31st Avenue
Edmunston, NB E3V 1X7
CANADA
TEL: 506-739-2106
FAX: 506-739-6323

Esso Resources Canada Ltd.

    Norman Wells Power Plant
    Esso Resources Canada Ltd.
    Bag Service 5000
    Norman Wells, NT X0E 0V0
    CANADA

Evans Products Company Ltd.

    Golden Power Plant
    Evans Products Company Ltd.
    Box 170
    Golden, BC V0A 1H0
    CANADA

Appendix A

Electric Power Plant Operators in Canada

Fer et Titane du Quebec Inc.

>Havre Ste Pierre Power Plant
Fer et Titane du Quebec Inc.
1625 Route Marie Victorin
CP 560
Sorel, PQ J3P 5P6
CANADA

Foothills Hospital

>Foothills Hospital Power Plant
Foothills Hospital
1403 - 29th Street NW
Calgary, AB T1Y 1M7
CANADA

Fraser Ltd.
27 Rice Street
Edmunston, NB E3V 1S9
CANADA

Gananoque Light & Power Ltd.
5 King Street
Box 548
Gananoque, ON K7G 2V1
CANADA
TEL: 613-382-2118

Great Lakes Forest Products

>Fort William Power Plant
Great Lakes Forest Products
Box 430
Station F
Thunder Bay, ON P7C 4W3
CANADA

Great Lakes Power Company Ltd.
122 East Street
Box 100
Sault Ste Marie, ON P6A 5L4
CANADA
TEL: 705-759-7600
FAX: 705-759-7633

Gulf Canada Resources Ltd.

>Rimbey Power Plant
Gulf Canada Resources Ltd.
Box 530
Rimbey, AB T0C 2J0
CANADA

Appendix A

Electric Power Plant Operators in Canada

Hiram Walker & Son Ltd.

    Walkerville Power Plant
    Hiram Walker & Son Ltd.
    2072 Riverside Drive East
    P.O. Box 2518
    Walkerville, ON N8Y 4S5
    CANADA

Howe Sound Pulp & Paper

    Port Mellon Power Plant
    Howe Sound Pulp & Paper
    Port Mellon, BC V0N 2S0
    CANADA

Hudson Bay Mining & Smelting

    Flin Flon Power Plant
    Hudson Bay Mining & Smelting
    Box 1500
    Flin Flon, MB R8A 1N9
    CANADA

Hydro-Quebec
75 Rene-Levesque West
Montreal, PQ H2Z 1A4
CANADA
TEL: 514-289-2211
FAX: 514-843-3163

Hydro-Sherbrooke
1800 Rue Roy
Sherbrooke, PQ J1K 1B6
CANADA
TEL: 819-821-5727
FAX: 819-822-6085

Hydromega Development Inc.

    Mont Laurier Power Plant
    Hydromega Development Inc.
    701 Rue Iberville
    CP 150
    Mont Laurier, PQ J9L 3G9
    CANADA

Inco Metals Co.
Copper Cliff, ON P0M 1N0
CANADA

Appendix A

Electric Power Plant Operators in Canada

Iron Ore Company of Canada
CP 1000
Sept-Iles, PQ G4R 3E1
CANADA

Irving Pulp & Paper Ltd.

    Saint John Power Plant
    Irving Pulp & Paper Ltd.
    Box 3007
    Postal Station B
    Saint John, NB E2M 3H1
    CANADA

Jonquiere Electric Service
1710 Rue Ste Famille
Jonquiere, PQ G7X 7W7
CANADA
TEL: 418-695-7725
FAX: 418-548-6560

Kalium Chemicals

    Belle Plaine Power Plant
    Kalium Chemicals
    1801 Hamilton Street
    Regina, SK S4P 4B5
    CANADA

La Cie Gaspesia Ltee.

    Chandler Power Plant
    La Cie Gaspesia Ltee.
    CP 3000
    Chandler, PQ G0C 1K0
    CANADA

La Cie Hydroelectrique Manicouagan

    McCormick Dam Power Plant
    La Cie Hydroelectrique Manicouagan
    20 Marquette
    Baie Comeau, PQ G4Z 1K6
    CANADA

La Cie Price Ltee.
3750 Rue de Champlain
Jonquiere, PQ G7X 7W5
CANADA

MacLaren Forest Products

    Gaspe Power Plant
    MacLaren Forest Products
    Murdockville, PQ G0E 1W0
    CANADA

Appendix A

Electric Power Plant Operators in Canada

Maclaren-Quebec Power Co.
Buckingham, PQ J8L 2X3
CANADA
TEL: 819-986-5045
FAX: 819-986-5045

    High Falls Hydroelectric Plant
    Maclaren-Quebec Power Co.
    RR 2
    Val des Bois, PQ
    CANADA

    Ronald Lean
    System Superintendent
    Masson Hydroelectric Plant
    Maclaren-Quebec Power Co.
    10 MacLaren East
    Masson, PQ J8L 2X3
    CANADA
    TEL: 819-986-3345
    FAX: 819-986-8367

Macmillan Bloedel Ltd.
6270 Yew Street
Powell River, BC V8A 4Z9
CANADA

    Sturgeon Falls Power Plant
    Macmillan Bloedel Ltd.
    Sturgeon Falls, ON P0H 2G0
    CANADA

Magog Electric Department
7 Rue Principale Est
Magog, PQ J1X 1Y4
CANADA
TEL: 819-843-6501

Maine-New Brunswick Electric Power Co.
209 State Street
Presque Isle, ME 04769
USA
TEL: 207-768-5811
FAX: 207-768-8511

Malette Kraft Pulp & Paper

    Smooth Rock Falls Power Plant
    Malette Kraft Pulp & Paper
    P.O. Box 310
    Smooth Rock Falls, ON P0L 2B0
    CANADA

Manitoba Forestry Resources

    The Pas Power Plant
    Manitoba Forestry Resources
    Box 2189
    The Pas, MB R9A 1L9
    CANADA

Appendix A

Electric Power Plant Operators in Canada

Manitoba Hydro
820 Taylor Avenue
Box 815
Winnipeg, MB R3C 2P4
CANADA
TEL: 204-474-3311
FAX: 204-475-9044

Jack Stothard
Superintendent
Brandon Generating Station
Manitoba Hydro
Box 850
Brandon, MB R7A 5Z8
CANADA

Great Falls Generating Station
Manitoba Hydro
Great Falls, MB R0E 0V0
CANADA

Kelsey Generating Station
Manitoba Hydro
Box 280
Gillam, MB R0B 0L0
CANADA

Limestone Generating Station
Manitoba Hydro
Box 280
Gillam, MB R0B 0L0
CANADA

McArthur Falls Generating Station
Manitoba Hydro
Great Falls, MB R0E 0V0
CANADA

Jim Nicholl
Superintendent
Selkirk Generating Station
Manitoba Hydro
Box 307
Selkirk, MB R1A 2B3
CANADA

Bill Sallows
Superintendent
Grand Rapids Generating Station
Manitoba Hydro
Box 555
Grand Rapids, MB R0C 1E0
CANADA

Darrell D. McKay
Superintendent
Jenpeg Generating Station
Manitoba Hydro
Box 815
Winnipeg, MB R3C 2P4
CANADA

Kettle Generating Station
Manitoba Hydro
Box 280
Gillam, MB R0B 0L0
CANADA

Horst A. Salewski
Superintendent
Long Spruce Generating Station
Manitoba Hydro
Box 280
Gillam, MB R0B 0L0
CANADA

Pine Falls Generating Station
Manitoba Hydro
Great Falls, MB R0E 0V0
CANADA

Jim LeRoye
Superintendent
Seven Sisters Generating Station
Manitoba Hydro
Great Falls, MB R0E 0V0
CANADA

Appendix A

Electric Power Plant Operators in Canada

Maritime Electric Company Ltd.
Court Tower
134 Kent Street
Charlottetown, PE C1A 7N2
CANADA
TEL: 902-566-1599
FAX: 902-566-2692

    W. Lynds
    Plant Superintendent
    Charlottetown Steam-Electric Plant
    Maritime Electric Company Ltd.
    Cumberland Street
    Charlottetown, PE
    CANADA

Medicine Hat Electric Utility
580 First Street S.E.
Medicine Hat, AB T1A 8E6
CANADA
TEL: 403-529-8259

    H. Kollross
    Plant Manager
    Medicine Hat Power Plant
    Medicine Hat Electric Utility
    Power House Road
    Medicine Hat, AB T1A 8E6
    CANADA

Minas Basin Pulp & Paper Co.
Hantsport, NS B0P 1P0
CANADA

Miramichi Pulp & Paper Ltd.
Box 1020
Newcastle, NB E1V 3M8
CANADA

NBIP Forest Products Ltd.

    Dalhousie Power Plant
    NBIP Forest Products Ltd.
    William Street
    Dalhousie, NB E0K 1B0
    CANADA

NERCO Consolidated Mine Ltd.
Box 2000
Yellowknife, NT X1A 2M1
CANADA

Appendix A

Electric Power Plant Operators in Canada

NWT Power Corp.
Box 5700
Edmonton, AB T6C 4J8
CANADA

Nelson Light Department
502 Vernon Street
Nelson, BC V1L 4E8
CANADA
TEL: 604-352-5511
FAX: 604-352-2131

New Brunswick Electric Power Commission
527 King Street
Fredricton, NB E38 4X1
CANADA
TEL: 506-458-4444
FAX: 506-458-4249

James L. Cummings
Station Manager
Beechwood Hydraulic Power Plant
New Brunswick Electric Power Commission
R.R. #3
Bath, NB E0J 1E0
CANADA
TEL: 506-278-3291
FAX: 506-278-5013

Donald MacDonald
Station Manager
Chatham Generating Station
New Brunswick Electric Power Commission
P.O. Box 279
Chatham, NB E1N 3A6
CANADA
TEL: 506-773-9486
FAX: 506-773-5378

Bob Brown
Station Manager
Courtenay Bay Generating Station
New Brunswick Electric Power Commission
509 Prom. Bayside Street
Saint John, NB E2J 1B4
CANADA
TEL: 506-635-7695
FAX: 506-633-1765

Richard Shaw
Station Manager
Grand Falls Hydraulic Power Plant
New Brunswick Electric Power Commission
P.O. Box 1260
Grand Falls, NB E0J 1M0
CANADA
TEL: 506-473-6050
FAX: 506-473-5236

Rod White
Station Manager
Belledune Generating Station
New Brunswick Electric Power Commission
Belledune, NB E0B 1G0
CANADA
TEL: 506-522-2500
FAX: 506-522-5313

Jim Brogan
Station Manager
Coleson Cove Generating Station
New Brunswick Electric Power Commission
P.O. Box 3600
Station "B"
Saint John, NB E2M 4Y2
CANADA
TEL: 506-635-8225
FAX: 506-635-8651

Blair Kennedy
Station Manager
Dalhousie Generating Station
New Brunswick Electric Power Commission
P.O. Box 520
Dalhousie, NB E0K 1B0
CANADA
TEL: 506-684-3387
FAX: 506-684-3060

Carl Bailey
Station Manager
Grand Lake Generating Station
New Brunswick Electric Power Commission
P.O. Box 18
Site 23
Minto, NB E0E 1J0
CANADA
TEL: 506-327-3317
FAX: 506-327-4319

Appendix A

Electric Power Plant Operators in Canada

Gene S. Guptill
Station Manager
Grand Manan Power Plant
New Brunswick Electric Power Commission
Grand Harbor, NB E0G 1X0
CANADA
TEL: 506-662-3223
FAX: 506-662-3798

Don MacDonald
Station Manager
Millbank Gas Turbine Power Plant
New Brunswick Electric Power Commission
P.O. Box 118
Douglastown, NB E0C 1H0
CANADA
TEL: 506-778-7000
FAX: 506-773-7284

Musquash Power Plant
New Brunswick Electric Power Commission
R.R. #1
Lepreau, NB E0G 1X0
CANADA

Don MacDonald
Station Manager
Ste-Rose Gas Turbine Power Plant
New Brunswick Electric Power Commission
General Delivery
Tracadie, NB E0C 2B0
CANADA
TEL: 506-395-7800
FAX: 506-395-7818

Newfoundland & Labrador Hydro
Hydro Place, Columbus Drive
P.O. Box 12400
St. John's, NF A1B 4K7
CANADA
TEL: 709-737-1400
FAX: 709-737-1231

T.R. Vatcher
Manager, Hydro Plant Operations
Bay d'Espoir Power Plant
Newfoundland & Labrador Hydro
P.O. Box 100
Milltown, NF A0H 1W0
CANADA
TEL: 709-882-2551
FAX: 709-882-2707

Bob A. Washburn
Station Manager
Mactaquac Hydraulic Power Plant
New Brunswick Electric Power Commission
Mouth of Keswick, NB E0H 1N0
CANADA
TEL: 506-363-3093
FAX: 506-363-3826

Robert Morrison
Station Manager
Milltown Hydraulic Power Plant
New Brunswick Electric Power Commission
P.O. Box 336
St. Stephen, NB E0G 2K0
CANADA
TEL: 506-466-5411

A.R. Johnson
Station Manager
Point Lepreau Nuclear Generating Station
New Brunswick Electric Power Commission
P.O. Box 10
Lepreau, NB E0G 2H0
CANADA
TEL: 506-458-3120
FAX: 506-659-2703

Happy Valley GT Power Plant
Newfoundland & Labrador Hydro
Bag 3017
Happy Valley, NF A0P 1E0
CANADA
TEL: 709-896-2993
FAX: 709-896-8948

Appendix A

Electric Power Plant Operators in Canada

Trevor J. Atkin
Manager, Thermal Operations
Holyrood Power Plant
Newfoundland & Labrador Hydro
P.O. Box 29
Holyrood, NF A0A 2R0
CANADA
TEL: 709-737-1340
FAX: 709-229-7894

Stephenville GT Power Plant
Newfoundland & Labrador Hydro
P.O. Box 234
Stephenville, NF A2N 2Z4
CANADA
TEL: 709-643-2158
FAX: 709-643-4620

Roddickton Thermal Plant
Newfoundland & Labrador Hydro
Roddickton, NF A0K 4P0
CANADA
TEL: 709-457-2759
FAX: 709-457-2671

Newfoundland Light & Power Company Ltd.
55 Kenmount Road
P.O. Box 8910
St. John's, NF A1B 3P6
CANADA
TEL: 709-737-5600
FAX: 709-737-5832

Noranda Mines Ltd.

Murdochville Power Plant
Noranda Mines Ltd.
Murdochville, PQ G0E 1W0
CANADA

Noranda Smelter Power Plant
Noranda Mines Ltd.
CP 4000
Noranda, PQ J9X 5B6
CANADA

Northland Utilities (NWT) Ltd.
77 Woodland Road
Box 1248
Hay River, NT X0E 0R0
CANADA
TEL: 403-874-6879
FAX: 403-874-6829

Northwest Territories Power Corp.
3 Capitol Road
Bag 6000
Hay River, NT X0E 0R0
CANADA
TEL: 403-465-3377

Northwood Pulp & Paper Ltd.

Fraser Flats Power Plant
Northwood Pulp & Paper Ltd.
Box 9000
Prince George, BC V2L 4W2
CANADA

Appendix A

Electric Power Plant Operators in Canada

Nova Scotia Forest Industries Ltd.

    Port Hawkesbury Power Plant
    Nova Scotia Forest Industries Ltd.
    Box 59
    Port Hawkesbury, NS B0E 2V0
    CANADA

Nova Scotia Power Corp.
Box 910
Halifax, NS B3J 2W5
CANADA
TEL: 902-424-6230
FAX: 902-428-6112

    Dickie Brook Hydroelectric Plant
    Nova Scotia Power Corp.
    Box 129
    Guysborough, NS
    CANADA
    TEL: 902-533-2646
    FAX: 902-533-2031

    Glace Bay Generating Station
    Nova Scotia Power Corp.
    P.O. Box 69
    Glace Bay, NS B1A 5V1
    CANADA

    Lingan Generating Station
    Nova Scotia Power Corp.
    P.O. Box 183
    New Waterford, NS B1H 4N9
    CANADA
    TEL: 902-862-6422
    FAX: 902-862-6087

    Maccan Generating Station
    Nova Scotia Power Corp.
    Maccan, NS B0L 1B0
    CANADA

    Point Tupper Generating Station
    Nova Scotia Power Corp.
    P.O. Box 2004
    Point Hawkesbury, NS B0E 2V0
    CANADA

    Tidewater Hydroelectric Plant
    Nova Scotia Power Corp.
    P.O. Box 70
    Halifax, NS B0J 3J0
    CANADA
    TEL: 902-826-2396

    Trenton Generating Station
    Nova Scotia Power Corp.
    P.O. Box 190
    Trenton, NS B0K 1X0
    CANADA

    Tufts Cove Generating Station
    Nova Scotia Power Corp.
    315 Windmill Road
    Dartmouth, NS B3J 2W5
    CANADA
    TEL: 902-428-7639
    FAX: 902-428-7699

    Weymouth Falls Hydroelectric Plant
    Nova Scotia Power Corp.
    P.O. Box 150
    Bear River, NS B0S 1B0
    CANADA
    TEL: 902-467-3406
    FAX: 902-467-4182

    Wreck Cove Hydroelectric Plant
    Nova Scotia Power Corp.
    RR 1
    Englishtown, NS B0C 1H0
    CANADA
    TEL: 902-929-2637

Appendix A

Electric Power Plant Operators in Canada

Ontario Hydro
700 University Avenue
Toronto, ON M5G 1X6
CANADA
TEL: 416-592-5111
FAX: 416-590-2284

D.W. Taylor
Station Manager
Atikokan Thermal Generating Station
Ontario Hydro
Box 1900
Atitokan, ON P0T 1C0
CANADA
TEL: 807-597-1110

R.E. Lewis
Station Manager
Bruce B Nuclear Generating Station
Ontario Hydro
Box 4000
Tiverton, ON N0G 2T0
CANADA
TEL: 519-368-7031

Cameron Falls Hydraulic Gen. Station
Ontario Hydro
Box 1060
Nipigon, ON P0T 2J0
CANADA

G.R. Childerhose
Plant Manager
Darlington Nuclear Generating Station
Ontario Hydro
Box 4000
Bowmanville, ON L1C 3W2
CANADA
TEL: 416-623-6606

R.R. Wheatley
Production Manager
J. Clark Keith Thermal Gen. Station
Ontario Hydro
4460 Sandwich Street West
Box 7160
Windsor, ON N9C 3Z1
CANADA
TEL: 519-969-7533

D.G. Shelton
Station Manager
Lambton Thermal Generating Station
Ontario Hydro
Box 2100
Courtright, ON N0N 1H0
CANADA
TEL: 519-867-2663

A.G. Holt
Station Manager
Bruce A Nuclear Generating Station
Ontario Hydro
Box 3000
Tiverton, ON N0G 2T0
CANADA
TEL: 519-368-7031

G.D. Davidson
Plant Manager
Bruce Heavy Water Plant
Ontario Hydro
Box 2000
Tiverton, ON N0G 2T0
CANADA
TEL: 519-368-7031

Chats Falls Hydraulic Generating Station
Ontario Hydro
106 Colonnade Road North
Suite 210
Nepean, ON K2E 7L6
CANADA

Ear Falls Hydraulic Generating Station
Ontario Hydro
Box 220
Ear Falls, ON P0V 1T0
CANADA

T.W.B. MacFarlane
Station Manager
Lakeview Thermal Generating Station
Ontario Hydro
Box 369 (Port Credit)
Mississauga, ON L5G 4M1
CANADA
TEL: 416-274-3461
FAX: 416-274-0733

R.J. Forrest
Station Manager
Lennox Thermal Generating Station
Ontario Hydro
Highway 33
Box 1000
Bath, ON K0H 1G0
CANADA
TEL: 613-352-3525

Appendix A

Electric Power Plant Operators in Canada

H.J. Kirwin
Station Manager
Nanticoke Thermal Generating Station
Ontario Hydro
Box 2000
Nanticoke, ON N0A 1L0
CANADA
TEL: 519-587-2203
FAX: 519-587-5366

Pickering B Nuclear Generating Station
Ontario Hydro
813 Brock Road
Box 160
Pickering, ON L1V 2R5
CANADA
TEL: 416-839-1151

Robert H. Saunders Hydraulic Gen. Sta.
Ontario Hydro
2nd Street West
Box 999
Cornwall, ON K6H 5P6
CANADA

R.R. Wheatley
Station Manager
Thunder Bay Thermal Generating Station
Ontario Hydro
108th Avenue, Mission Island
Box 816
Thunder Bay, ON P7C 4X7
CANADA
TEL: 807-623-2701

Orillia Water, Light & Power Commission
360 West Street South
Box 398
Orillia, ON L3V 6J9
CANADA

Ottawa Hydro
3025 Albion Road
Box 8700
Ottawa, ON K1G 3S4
CANADA

Papeterie Reed Ltee.

Forestville Power Plant
Papeterie Reed Ltee.
Box 600
Forestville, PQ G0T 1E0
CANADA

R.O. Schuelke
Station Manager
Pickering A Nuclear Generating Station
Ontario Hydro
813 Brock Road
Box 160
Pickering, ON L1V 2R5
CANADA
TEL: 416-839-1151

R.F. Gibson
Station Manager
Richard L. Hearn Thermal Gen. Station
Ontario Hydro
440 Unwin Avenue
Toronto, ON M4M 3B9
CANADA
TEL: 416-592-6286
FAX: 416-465-2236

Sir Adam Beck #1 Hydraulic Gen. Station
Ontario Hydro
Niagara River Parkway
Box 219
Queenston, ON L0S
CANADA

Appendix A

Electric Power Plant Operators in Canada

Papier Journal Domatar Ltee.

    MacDougal Power Plant
    Papier Journal Domatar Ltee.
    CP 370
    Donnacona, PQ G0A 1T0
    CANADA

Parry Sound Public Utilities Commission
41 Gibson Street
Parry Sound, ON P2A 1W5
CANADA

Peel Resource Recovery

    Peel Resource Recovery Facility
    Peel Resource Recovery
    17 Dean Street
    Brampton, ON
    CANADA

Pembroke Electric Light Company Ltd.
P.O. Box 217
Pembroke, ON K8A 6X3
CANADA

Pembroke Hydro Electric Commission
283 Pembroke Street West
Pembroke, ON K8A 5N5
CANADA

Peterborough Utilities Commission
1867 Ashburnham Avenue
Peterborough, ON K9L 1P8
CANADA

Petro Canada

    Taylor Power Plant
    Petro Canada
    Box 270
    Taylor, BC V0C 2K0
    CANADA

Placer Dome Ltd.
Box 49330
Bentall Postal Station
Vancouver, BC V7X 1P1
CANADA

Polysar Ltd.

    Sarnia Power Plant
    Polysar Ltd.
    Videl Street South
    Sarnia, ON N7T 7H7
    CANADA

Appendix A

Electric Power Plant Operators in Canada

Procter & Gamble Cellulose Ltd.

    Wapiti River Power Plant
    Procter & Gamble Cellulose Ltd.
    Postal Bag 1020
    Grand Prairie, AB T8V 3A9
    CANADA

Public Works Canada

    Goose Bay Power Plant
    Public Works Canada
    Box 520
    Goose Bay, NF A0O 1S0
    CANADA

Redpath Sugars Ltd.

    Toronto Power Plant
    Redpath Sugars Ltd.
    95 Queen's Quay East
    Toronto, ON M5E 1A3
    CANADA

Renfrew Hydro Electric Commission
29 Bridge Avenue West
Renfrew, ON K7V 3R3
CANADA

SASKPOWER
2025 Victoria Avenue
Regina, SK S4P 0S1
CANADA
TEL: 306-566-2121

    B.A. McLellan
    Plant Manager
    Boundary Dam Power Station
    SASKPOWER
    Box 790
    Estevan, SK S4A 2A6
    CANADA
    TEL: 306-634-1300

    D.R. Dagg
    Superintendent
    Charlet River Hydro Station
    SASKPOWER
    Box 277
    Uranium City, SK S0J 2W0
    CANADA
    TEL: 306-498-3377

    H.H. Jeske
    Station Electrician
    Couteau Creek Hydro Station
    SASKPOWER
    Box 940
    Outlook, SK S0L 2N0
    CANADA
    TEL: 306-857-2024

    T.B.T. Werrett
    Plant Superintendent
    E.B. Campbell Hydro Station
    SASKPOWER
    Box 550
    Nipawin, SK S0E 1E0
    CANADA
    TEL: 306-862-3808

    F. Corley
    Plant Manager
    Estevan Generating Station
    SASKPOWER
    Box 399
    Estevan, SK S4A 2A4
    CANADA
    TEL: 306-634-4734

    R.R. Glombowski
    Plant Superintendent
    Island Falls Hydro Station
    SASKPOWER
    Box 26
    Sandy Bay, SK S0P 0G0
    CANADA
    TEL: 306-754-2101

Appendix A

Electric Power Plant Operators in Canada

G.E. Callahan
Plant Superintendent
Nipawin Hydro Station
SASKPOWER
Box 2045
Nipawin, SK S0E 1E0
CANADA
TEL: 306-862-3148
FAX: 306-862-3065

H.H. Hubenig
Plant Manager
Queen Elizabeth Power Station
SASKPOWER
Box 1748
Saskatoon, SK S7K 3S1
CANADA
TEL: 306-934-7994

W.A. Mitchell
Plant Manager
Poplar River Power Station
SASKPOWER
Box 360
Coronach, SK S0E 0Z0
CANADA
TEL: 306-267-3397
FAX: 306-267-4919

T. Scott
Plant Manager
Shand Power Station
SASKPOWER
234 - 5th Street
Estevan, SK S4A 0X8
CANADA
TEL: 306-634-9778

Scott Maritime Pulp Ltd.

Abercrombie Point Power Plant
Scott Maritime Pulp Ltd.
Box 549D
New Glasgow, NS B2H 5E8
CANADA

Sherritt-Gordon Mines Ltd.

Fort Saskatchewan Power Plant
Sherritt-Gordon Mines Ltd.
Fort Saskatchewan, AB T8L 2P2
CANADA

Skeena Cellulose Inc.

Watson Island (Skeena) Power Plant
Skeena Cellulose Inc.
P.O. Box 1000
Prince Rupert, BC V8J 3S2
CANADA

Southern Alberta Institute of Technology

SAIT Power Plant
Southern Alberta Institute of Technology
1301 - 16th Avenue West
Calgary, AB T2M 0I2
CANADA

Appendix A

Electric Power Plant Operators in Canada

Spruce Falls Power & Paper Company Ltd.
Box 100
Kapuskasing, ON P5N 2Y2
CANADA

St. Anne Nackawic Pulp & Paper

>Nackawic Power Plant
>**St. Anne Nackawic Pulp & Paper**
>Box 1000
>Nackawic, NB E0H 1P0
>CANADA

St. George Pulp & Paper Company Ltd.

>St. George Power Plant
>**St. George Pulp & Paper Company Ltd.**
>284 Union Street
>Saint John, NB E2L 1B7
>CANADA

St. Lawrence Seaway Authority
508 Glendale Avenue
Box 370
St. Catherines, ON L2R 6V8
CANADA

St. Regis (Alberta) Ltd.

>Hinton Power Plant
>**St. Regis (Alberta) Ltd.**
>Hinton, AB T0E 1B0
>CANADA

Stelco Ltd.

>Hamilton Power Plant
>**Stelco Ltd.**
>P.O. Box 2030
>Hamilton, ON L8N 3T1
>CANADA

Summerside Electric Department
Box 1510
Summerside, PE C1N 4K4
CANADA
TEL: 902-436-4222
FAX: 902-436-4255

Suncor Ltd.

>Tar Island Power Plant
>**Suncor Ltd.**
>Oil Sands Group
>P.O. Box 4001
>Fort McMurray, AB T9H 3E3
>CANADA
>FAX: 403-743-6419

Appendix A

Electric Power Plant Operators in Canada

Sunridge Power Corp.
Duke Street
Box 3001
Dryden, ON P8N 2Z7
CANADA

    Dryden Power Plant
    Sunridge Power Corp.
    Duke Street
    Box 3001
    Dryden, ON P8N 2Z7
    CANADA

Teck Corporation Ltd.

    Beaverdell Power Plant
    Teck Corporation Ltd.
    Beaverdell, BC V0H 1A0
    CANADA

TransAlta Utilities Corp.
202-215 12th Avenue SE
Calgary, AB T2P 2M1
CANADA
TEL: 403-267-7110
FAX: 403-267-3630

    W. Ketchin
    Plant Manager
    Bighorn Dam Power Plant
    TransAlta Utilities Corp.
    Box 1325
    Rocky Mountain House, AB T0M 1T0
    CANADA
    TEL: 403-721-3952
    FAX: 403-721-2297

    H.G. Olsen
    Plant Manager
    Brazeau Hydroelectric Plant
    TransAlta Utilities Corp.
    Box 128
    Lodgepole, AB
    CANADA
    TEL: 403-894-3594

    S.E. Wylie
    Plant Manager
    Keephills Generating Plant
    TransAlta Utilities Corp.
    RR #1
    Site 3, Compartment #1
    Duffield, AB T0E 0N0
    CANADA
    TEL: 403-731-3756
    FAX: 403-731-2121

    W. Shiman
    Plant Manager
    Pocaterra Hydroelectric Plant
    TransAlta Utilities Corp.
    General Delivery
    Seebe, AB T0L 1X0
    CANADA

    F.A. Mulligan
    Plant Manager
    Sundance Generating Plant
    TransAlta Utilities Corp.
    RR #1
    Site 3, Compartment #2
    Duffield, AB T0E 0N0
    CANADA

    L.D. Krause
    Plant Manager
    Wabamun Generating Plant
    TransAlta Utilities Corp.
    Box 120
    Wabamun, AB T0E 2K0
    CANADA

Appendix A

Electric Power Plant Operators in Canada

Trent University

    Nassau Power Plant
    Trent University
    Box 4800
    Peterborough, ON K9J 7B8
    CANADA

Tricil Ltd.
Resource Recovery Division
89 Queensway West
Mississauga, ON L5B 2V2
CANADA

    Swaru Power Plant
    Tricil Ltd.
    470 Kenora Avenue North
    Hamilton, ON L8E 3X8
    CANADA

Weldwood of Canada Ltd.

    Flavelle Power Plant
    Weldwood of Canada Ltd.
    2400 Murray Street
    Port Moody, BC V3H 4H6
    CANADA

West Kootenay Power & Light Co.
8100 Rock Island Highway
Trail, BC V1P 4L4
CANADA
TEL: 604-368-3321
FAX: 604-364-1270

Westar Ltd.

    Celgar Pulp Mill Power Plant
    Westar Ltd.
    Box 1000
    Castlegar, BC V1N 3H9
    CANADA

Western Cooperative Fertilizer Ltd.

    Medicine Hat Power Plant
    Western Cooperative Fertilizer Ltd.
    Box 110
    Medicine Hat, AB T1A 7L6
    CANADA

Appendix A

Electric Power Plant Operators in Canada

Western Pulp Limited Partnership
Box 5000
Squamish, BC V0N 2N0
CANADA

Westminster Resources Ltd.

    Campbell River Power Plant
    Westminster Resources Ltd.
    Box 8000
    Campbell River, BC V9W 5E2
    CANADA

Weyerhaeuser Canada Ltd.

    Kamloops Power Plant
    Weyerhaeuser Canada Ltd.
    Box 800
    Kamloops, BC V2C 5M7
    CANADA

    Prince Albert Power Plant
    Weyerhaeuser Canada Ltd.
    Box 3001
    Prince Albert, SK S6V 5T5
    CANADA

Winnipeg Hydro
1315 Notre Dame Avenue
Winnipeg, MB R38 1C1
CANADA
TEL: 204-986-2270
FAX: 204-942-7804

Yukon Electrical Company Ltd.
P.O. Box 4190
Whitehorse, YK Y1A 3T4
CANADA
TEL: 403-668-5211
FAX: 403-668-3965

APPENDIX B

CANADA POWER PLANT DATA BASE

DATA FILE DIRECTORY

UDI-033-92   December 1992

APPENDIX B

UDI CANADA Power Plant Data Base
Data File Directory

Field No.	Field Name	Field Type/Length	Description
1	UNIT	A 25	Name of generating unit (set). If unit-specific data are available, each unit at site has its own record in the data base. If the PLANT name is repeated as the UNIT name, it indicates that the unit numbering scheme is unknown at this time. The operating utility's numbering schemes are used wherever possible.

A block designation in parentheses [*e.g.* "(A)"] indicates that while the precise unit-specific data are unknown at this time, the amount of capacity so labeled is known to be separate from other blocks at the site.

"**GT**" is usually used to indicate gas/combustion turbine units; "**IC**" is used to designate internal combustion units; "**SC**" is used to designate steam turbines in combined-cycle; "**A**" is often used to designate auxiliary units at hydraulic sites [*e.g.* A1]. |
2	PLANT	A 25	Name of power plant site.
3	OPERATOR	A 30	Name of operating utility or autoproducer company.
4	STATUS	A 3	Current unit status: CAN = cancelled CON = under construction DAC = deactivated/mothballed DEF = indefinitely deferred

B-1

UDI-033-92   December 1992

			DEL =	delayed after construction start
			PLN =	planned
			OPR =	Operating
			RET =	Retired
			STN =	Shutdown or economic reserve
			UNK =	Unknown status

5   MW   N 8.3   Installed capacity of plant/unit. (MW)

6   YEAR   N 4   Year unit entered or is scheduled to enter commercial operation. 9999 may be used to indicate "unscheduled". Blank indicates unknown.

7   PTYPE   A 4   Type of plant/unit:

CC	=	combined-cycle
CC/S	=	combined-cycle with steam sendout
IC	=	internal combustion (diesel)
GT	=	gas/combustion turbine
GT/C	=	GT in combined-cycle
HY	=	hydroelectric
ST	=	steam turbine
ST/C	=	steam turbine in combined-cycle
ST/S	=	steam turbine with steam sendout
WTG	=	wind turbine/generator
TGT	=	topping gas turbine (in combined-cycle)

8   FUEL   A 4   Primary plant/unit fuel type: COAL; OIL; GAS; GEO; WAT; WIND; UR; etc. UNK = unknown. See Abbreviations Used in UDI Data Bases, *(Appendix C)*.

9   FUELTYPE A 6   Coal-fired units may have secondary fuel designation of ANTH (anthracite); BIT (bituminous); SUB (subbituminous); or LIG (lignite, brown coal). Hydroelectric units are designated as either CONV

B-2

				(conventional); PS (pumped storage); Nuclear units are generally BWR (boiling water reactor) or PWR (pressurized water reactor). Oil-fired units may show DIST (distillate [light] oil), etc. See Abbreviations Used in UDI Data Bases, *(Appendix C)*.
10	ALTFUEL	A	4	Alternate fuel(s) for plant/unit. See Abbreviations Used in UDI Data Bases, *(Appendix C)*.
11	SSSMFR	A	6	Steam supply system (boiler or reactor) manufacturer. See Abbreviations Used in UDI Data Bases, *(Appendix C)*. N/A for GT, hydraulic, IC, and geothermal units.
12	TURBMFR	A	6	Steam, gas/combustion, or hydraulic turbine manufacturer. If IC, engine manufacturer is shown. See Abbreviations Used in UDI Data Bases, *(Appendix C)*.
13	GENMFR	A	6	Generator manufacturer. See Abbreviations Used in UDI Data Bases, *(Appendix C)*.
14	TURBTYPE	A	10	Turbine type. **For steam turbines:** single cylinder, tandem, cross-compound, etc. Steam turbine designation may have numbers showing number of flows and last-stage blade length (in inches or mm). **For gas turbines:** vendor model designation. **For hydro units:** runner type. See Abbreviations Used in UDI Data Bases, *(Appendix C)*.
15	SFLOW	N	7.2	Steam flow (kg/second).
16	SPRESS	N	3	Steam pressure at the turbine (bar).
17	TEMP	N	3	Steam temperature at the turbine (°C).
18	REHEAT1	N	3	First reheat temperature, if applicable (°C).

19	REHEAT2	N	3	Second reheat temperature, if applicable (°C).
20	PARTCTL	A	6	Type of particulate control, see Abbreviations Used in UDI Data Bases, *(Appendix C)*. May be N/A.
21	PARTMFR	A	6	Manufacturer of particulate control device for unit, see Abbreviations Used in UDI Data Bases, *(Appendix C)*. May be N/A.
22	SO2CTL	A	6	Type of SO_2 control, see Abbreviations Used in UDI Data Bases, *(Appendix C)*. May be N/A.
23	FGDMFR	A	6	Manufacturer of FGD scrubber system if applicable, see Abbreviations Used in UDI Data Bases, *(Appendix C)*. May be N/A.
24	NOXCTL	A	6	Type of NO_x control, see Abbreviations Used in UDI Data Bases, *(Appendix C)*. May by N/A.
25	NOXMFR	A	6	Manufacturer of NO_x control system if applicable, see Abbreviations Used in UDI Data Bases, *(Appendix C)*.
26	COOL	A	6	Type of main condenser cooling system used by unit:

        OTF = Once through, fresh water
        OTB = Once through, brackish water
        OTS = Once through, sea water
        MDT = Mechanical draft cooling tower
        NDT = Natural draft cooling tower

Also see Abbreviations Used in UDI Data Bases, *(Appendix C)*. May be N/A.

27	AE	A	6	Architect/engineering firm: UTIL if utility did design. Also see Abbreviations Used in UDI Data Bases, *(Appendix C)*.
28	CONSTRUCT	A	6	Primary construction contractor: UTIL if utility managed construction. Also see Abbreviations Used in UDI Data Bases, *(Appendix C)*.
29	COUNTRY	A	25	Country in which plant is located.
30	AREA	A	10	Geographic area descriptor.
31	RETIRE	N	4	Year unit was retired, if applicable.
32	OPERTYPE	A	1	Power plant operator type: A = autoproducer (industrial or other non-utility); U = utility May be blank.
33	PROVINCE	A	4	Province in which plant is located: AB = Alberta; BC = British Columbia; MB = Manitoba; NB = New Brunswick; NF = Newfoundland; NT = Northwest Territories; NS = Nova Scotia; ON = Ontario; PE = Prince Edward Island; PQ = Quebec; SK = Saskatchewan; YK = Yukon Territory

APPENDIX C

ABBREVIATIONS USED IN

THE UDI UTILITY DATA BASE

Appendix C
Abbreviations Used in UDI Data Bases

A&A	Amos & Andrews Inc.	AEG/MN	AEG/MAN
A-G	Atlantic-Gulf Co. (subsidiary of Zurn Industries Inc.)	AEGK	AEG-Kanis (Germany) ** USE AEG **
A-J	Alsthom Jeumont (France)	AEI	Associated Electric Industries Ltd. (now merged into GEC) [England]
A-N	Anderson-Nichols	AEI/GE	AEI/General Electric
A/B	Anthracite and bituminous coal	AEI/PR	AEI/Parsons
A/T	Ansaldo/TIBB	AEP	American Electric Power Service Corp.
AA	Acres American Inc.	AERO	Aerotec
AAF	American Air Filter Company Inc.	AFUDC	Allowance for funds used during construction
AALBOR	Aalborg Ciserve International (Denmark)	AG WST	Agricultural waste ** USE AGWS **
AAO	A. Ahlstrom Oy (Finland)	AGK	Amme Giesecke and Konegen
ABB	ASEA Brown Boveri	AGR	Advanced gas-cooled reactor
ABBES	ABB Environmental Systems	AGWS	Agricultural waste
ABBRR	ABB Resource Recovery Systems	AHLST	Ahlstrom (Finland)
ABBS	ABB Stal (Sweden)	AHP	American Hydro Power Co.
ABC	Anglo Belgian Corp. (IC manufacturer)	AI	Acres International Corp.
ABCO	ABCO Industries Inc.	AIR	Air-cooled condenser
ABWR	Advanced boiling water reactor	AJ	Axel Johnson Engineering Corp.
AC	Allis-Chalmers	AJCC	Al Johnson Construction Co.
AC/SA	Allis-Chalmers/Siemens-Allis	AJCD	Axel Johnson (Canada) Ltd. (hydro turbines)
ACB	Allis-Chalmers Bullock (Canada)	AJM	Alsthom Jeumont (France) ** USE A-J **
ACBI	Arciero-Birtcher International Inc.	AL	Louis Allis (hydro generators) ** USE LA **
ACC	Allis-Chalmers Canada	AL&N	Ayers Lewis & Norris
ACEC	Ateliers de Constructions Electricques de Charleroi S.A. (now ABB ACEC) [Belgium]	ALB	American Line Builders
		ALCAN	Alcan Aluminum Co.
ACG	AC Generating Inc.	ALCO	ALCO
ACLF	ACEC/Cockeril/Westinghouse	ALCOA	Alcoa Aluminum Co.
ACPS	Allis Chalmers Power Systems Inc. (JV of Allis-Chalmers and Kraftwerk Union)	ALKO	ALKO (turbine manufacturer)
		ALL	Allison (gas turbine manufacturing subsidiary of General Motors Corp.)
ACSR	Aluminum conductor steel reinforced	ALL/PS	Allco/Peabody Sturtevant
AD	Abner Doble (hydro turbines)	ALMON	Almon & Associates
ADL/CE	Arthur D. Little/CEA	AM/BU	American Standard/Buell
ADM	Archer-Daniels-Midland	AM/CB	American Standard/Carborundum
ADVAN	Advanced Energy Systems Co.	AM/PB	American Standard/Peabody
AE	Associated Electric Industries Ltd. (England) ** USE AEI **	AM/WST	American Standard/Western
AE/GEC	Associated Electric Industries/General Electric UK (UK)	AMEN	American Energy Co.
AEC	Associated Electric Industries Ltd. (England) ** USE AEI **	AMGC	Association Momentanee de Genie Civil (Belgium)
AECC	American Energy Conservation Consultants	AMPAI	Ampai (A/E) [South Korea]
AECL	Atomic Energy of Canada	AMRG	Amergy Corp.
AEE	Atomenergoexport (USSR)	AMST	American Standard
AEG	AEG Telefunken AG or AEG Kanis (Germany)	AMT	Algar-Martin
AEG/BB	AEG/Brown Boveri	AN	Ansaldo SpA (Italy)

Revised December 1992

AN/EM	Ansaldo/Ercole Marelli
AN/SAV	Ansaldo/Savigliano
AN/T	Ansaldo/Tosi
ANCHOR	Anchor Metals
ANR	ANR Venture Management Co.
ANTH	Anthracite coal
AOS	AO Smith Harvestore
AP	American Pole (subsidiary of ABB Combustion Engineering)
AP&C	Air Products & Chemicals Inc.
APC	Atomic Power Construction Ltd. (England)
APC	Austin Power Company
APP	American Private Power Inc.
APW	American Pelton Wheel Co. (hydraulic turbine manufacturer)
APW/IR	American Pelton/Ingersoll Rand
ARFC	American Ref-Fuel Co. (JV of Air Products & Chemicals and Browning Ferris Industries)
ARG	Armstrong (turbine manufacturer) [England]
ARGE	Dyckerhoff & Widman AG/Wakyss & Freitag AG/Hegdkamp (Germany)
ARR	American Resource Recovery
AS	ASEA Electric Co. (Sweden) ** USE ASEA **
ASB	A.S. Bugshan (Saudi Arabia)
ASBW	ASEA/B&W
ASC	American Shack Corp.
ASE	Associated Southern Engineering Co.
ASEA	ASEA (Sweden)
ASH	Allen-Sherman-Hoff
ASK	AskCorp.
ASPT	Asphaltines
AST	ASEA-Stal (now ABB Stal)
ASTAL	Astaldi (Italy)
AT	Alsthom (France)
AT/ACE	Alsthom/ACEC
AT/BZD	Alsthom/BZD (China)
AT/R	Alsthom/Rateau or ART
ATI	Atomics International
ATK	Guy F. Atkinson Co.
ATLAS	Atlas Copco
ATR	Advanced thermal reactor
ATS	Applied Thermal Systems
AUST	Austin Building Co.
AVON	Avondale Industries
AW	Andy Weiser
AW&S	August Winter & Sons
AWS	American Windshark
B&B	Barnard and Burk Inc.
B&C	Brown & Caldwell
B&L	Barton & Loguidice P.C.
B&R	Burns & Roe Inc.
B&RT	Brown & Root Inc. (subsidiary of Halliburton)
B&V	Black & Veatch
B-H	Babcock-Hitachi K.K. (Japan)
B-J	Brayton-James
B-N	Barber-Nichols Engineering Inc.
B/A	Bituminous coal and anthracite coal
B/C	Babcock & Wilcox/Combustion Engineering
B/L	Bituminous coal and lignite (brown coal)
B/MCA	Bechtel/MacAlpine
B/PC	Bituminous coal and petroleum coke
B/RDF	Bituminous coal and RDF
B/S	Bituminous and subbituminous coal
B/T/Z	B&W/Trane/Zurn
BA	Besha Associates
BAG	Bagasse
BAL	Balfour Beatty & Co. (England)
BALD	Baldwin Associates
BALL	Joe Ball
BAM	Bataafsche Aannerning Maatschappij (Netherlands)
BARB	Barber Hydraulic Turbines (Canada)
BARK	Tree bark (waste wood)
BART	Bart Associates
BASIC	Basic Energy Ltd. (England)
BAY	Baylor
BB/TH	Balfour Beatty/Trafalgar House (England)
BBC	Brown Boveri & Cie. (Switzerland)
BBEC	Big Bend Engineering Company Inc.
BBR	Babcock-Brown Boveri Reaktor GmbH (Germany)
BC	Belmont Constructors
BE&C	BE&C Engineers Inc. (subsidiary of Boeing)
BEA	Braun Engineering Associates
BEAL	Fred Beal
BEC	Brandt Environmental Corp. (particulate control equipment)
BECH	Bechtel Power Corp.
BECH/U	Bechtel/Utility
BECK	R.W. Beck
BECO	Technogroup Beco SpA (Italy)
BECON	Becon Construction Co. (subsidiary of Bechtel Group)
BEE	Basic Environmental Engineering Inc.
BEI	Beard Engineering Co.
BEIC	Business Energy Improvements Corp.
BEL	Babcock Energy Ltd. (England)
BELCO	Belco Pollution Control Corp. (subsidiary of Foster Wheeler)
BELELL	Belelli (Italy)

BELL	Bell Engineering Corp.
BELLIS	Bellis and Morcom (turbine manufacturer) [Canada]
BEM	Brush Electric Machines (England) ** USE BRUSH **
BEN	Benham Holway Power Group
BENZ	Mercedes Benz (Germany)
BESS	Bessemer
BF	Buffalo Forge
BFB	Bubbling fluidized bed
BFCG	Blast-furnace gas/coke oven gas
BFG	Blast-furnace gas
BFGAS	Blast-furnace gas ** USE BFG **
BFI	Browning-Ferris Industries
BGAS	Biogas (unspecified)
BH	Baghouse
BH/KH	Babcock-Hitachi K.K./Kawasaki Heavy Industries
BHEL	Bharat Heavy Electricals Inc. (India)
BHM	Beijing Heavy Machinery (China)
BI	Bines
BIBB	Bibb and Associates Inc.
BIG	Bigelow/Liptak
BINARY	Binary geothermal plants
BIO	Biomass (not conventional wood or wood waste)
BIOGAS	Biogas of Colorado
BIP	Bingham Pump Co.
BIR	Birwelco
BIT	Bituminous coal
BIT/PC	Bituminous coal/petroleum coke ** USE B/PC **
BIT/SB	Bituminous coal/subbituminous coal/petroleum coke ** USE B/S/PC **
BITM	Bitumen (Orimulsion)
BIW	Bingham-Williamette Co.
BJ	Byron Jackson Pump Inc.
BLACK	Blackstone (IC manufacturer) now Mirrless Blackstone) [England]
BLAKE	Blake Construction
BLD	Baldwin Locomotive Works (hydraulic turbine manufacturer)
BLH	Baldwin-Lima-Hamilton (hydraulic turbine manufacturer)
BLT	Blount International Ltd.
BLV/AB	Blohm + Voss (steam turbine manufacturer) [Germany]
BLV/AB	Blohm + Voss/ABB
BM	Buhler Miag
BMAIER	B-Maier (generator manufacturer)
BMB	Birlesmis Muhendis Burosu (Turkey)
BMCD	Burns & McDonnell Engineering Company Inc.
BOE	Boeing Aerospace
BOUMA	Bouma WindEnergie (Netherlands)
BOUV	Bouvier Hydropower Inc.
BOUY	Bouygues SA (France)
BOV	Bovay Engineers Inc.
BOV/EE	Boving/English Electric
BOVING	Boving (hydroelectric turbine manufacturer) [England]
BP	Back-pressure steam turbine
BPC	Bonneville Pacific Corp.
BPENG	BP Energy
BPL	Babcock Power Ltd. (England)
BPS	Beloit Power Systems
BR	Brush Electric Machines Ltd. (England) ** USE BRUSH **
BRACK	Bracknell (Canada)
BRAU	C.F. Braun
BREDA	Backer en Rueb Breda BV (Netherlands)
BRIN	Brinderson Corp.
BRONS	Bronsweurk (boiler manufacturer)
BRUG	Kvaener Brug A/S (Norway)
BRUNN	Brunn & Sorenson
BRUSH	Brush Electric Machines Ltd. (England)
BRWN	Brown Engineering Co.
BS	Bristol Sidderley (England)
BSH	Buttner-Schilde-Haas AG (Germany)
BSW	Baldwin Southworks (hydro)
BTH	British Thompson Houston (England)
BTM	Ball tube mill
BU/AMS	Buell/American Standard
BU/RC	Buell/Research Cottrell
BU/WAL	Buell/Walther
BUDA	Buda (turbine manufacturer)
BUELL	Buell Emission Control Division of Envirotech Corp. (now GEES)
BUG	Brooklyn Union Gas Co.
BULB	Bulb-type hydroelectric turbine
BUNGE	Bunge
BUREC	U.S. Bureau of Reclamation
BURT	Burt & Mallow
BUT	Butane
BV	Boving (England)
BV/BM	Bovay/Burns & McDonald
BV/BR	Boving/Brush
BV/EE	Boving/English Electric
BV/GEC	Boving/GEC
BW	Babcock & Wilcox
BW/CE	Babcock & Wilcox/Combustion Engineering
BW/DUR	Babcock/Durr
BW/FW	Babcock & Wilcox/Foster Wheeler
BW/TIW	Babcock & Wilcox/Toronto Iron Works
BW/ULT	Babcock & Wilcox/Ultrapower
BW/VUI	Babcock & Wilcox/Vulcan Iron Works

BW/ZN	Babcock & Wilcox/Zurn
BWC	Babcock & Wilcox Canada
BWC	Borg Warner Corp.
BWE	Burmeister & Wain Energie or A/S Burmeister & Wain (boiler manufacturer) [Denmark]
BWG	Benjamin Woodhouse & Guenther
BWGM	Babcock & Wilcox Goldie McCulloch
BWK	Babcockwerke (Germany) ** USE DBAG **
BWR	Boiling water reactor
C&H	C&H Combustion
C&S	Carlson & Sweatt
C&T	Wm. F. Cosulich & I.C. Thomasson
C-B	Cleaver-Brooks
C-G	Co-Gen (micro cogen model)
C-K	Clark-Kenith Inc.
C/D	Crude and distillate fuel oil
C/O	Cablec/Okonite
C/P/G	Coal/peat/gas
C/PS	Conventional and pumped storage operation (reversible)
C/R	Coal and refuse
C/SB/B	Chantiers Modernes/Sainrapt et Brice/Ballot
CA	Cogeneration Associates (Canada)
CA	Commonwealth Associates
CAD	Cadoux
CADC	Cogeneration Acquisition & Development Co.
CAES	Compressed air energy storage system
CAHYD	Calcium hydroxide injection FGD scrubber
CALPIN	Calpine Corp.
CAMP	Campenon-Bernard (France)
CAN	Cancelled
CANAT	Canatom (Canada)
CAND	Canada
CANR	Canron
CANSLV	CANSOLV (regenerable FGD system)
CANY	Canyon Industries
CAP	Capitol Cogeneration Co.
CARB	Carborundum Environmental Systems Inc. (unit of Kennecott Corp.)
CARB	Dry aqueous carbonate FGD scrubber
CARSON	Carson Engineers
CART	Carter Energy Systems Inc.
CARTER	Carter Wind Systems Inc.
CAT	Caterpillar (IC manufacturer)
CATL	Catalytic converters
CB	Coal blending
CB	Cooper-Bessemer
CBFGD	Circulating-bed FGD scrubber
CBR	Construtores Reunidos SA (Brazil)
CC	Cogenic Energy Systems Inc.
CC	Combined-cycle
CC	Cooling canal
CC/D	Combined-cycle for desalzinization
CC/H	Combined-cycle for district heating (CHP)
CC/S	Combined-cycle with steam sendout (CHP)
CC4F	Cross-compound 4-flow steam-turbine
CCL	Clarke Chapman Ltd. [Northern Engineering Industries Ltd.] (England)
CCL/JB	Clarke Chapman/John Brown
CCL/JT	CCL/John Thompson
CCTX	Cogentrix Inc.
CCW	Century Contractors West Inc.
CCYC	Combined cycle
CDB	Campbell DeBoe & Associates
CDC	Cogeneration Development Corp.
CDG	Campbell DeBoe Geiss & Webster
CDK	zz/unknown
CDLA	Chantiers de l'Atlantique (IC manufacturer) [France]
CE	Combustion Engineering Inc. (now ABB Combustion Engineering)
CE/BW	Combustion Engineering/Babcock & Wilcox
CE/FW	Combustion Engineering/Foster Wheeler
CE/LUR	Combustion Engineering/Lurgi
CE/N	Combustion Engineering/Neyrpic
CE/RS	Combustion Engineering/Riley Stoker
CE/SZ	Combustion Engineering/Sulzer Brothers
CE/TOS	Combustion Engineering/Tosi
CEA	Combustion Equipment Associates Inc.
CEA	Commissariat a l'Energie Atomique (France)
CEAM	CE American
CEC	Combustion Engineering Canada Ltd.
CEDC	Catalyst Energy Development Corp.
CEDG	California Energy Design Group Inc.
CEG	Cegelec
CEI	Compagnia Elettrotecnica Italiana SpA (Italy)
CEM	Compagnie Electro-Mechanique (France)
CEMAR	Combustion Engineering Canada/Maurbeni
CEMIN	Cemindia (subsidiary of Trafalgar House) [India]
CENT	Centrifugal (hydro pump turbine) ** USE CENTRIF **
CENT	Century Design Inc. (WTG manufacturer)
CENTRIF	Centrifugal hydro pump
CENTX	Centrax Ltd. (GT packager) [England]
CES	Cogenic Energy Systems Inc.
CET	Cogeneration Energy Technology
CF	Compliance fuel
CFB	Circulating fluidized bed boiler atmospheric

Revised December 1992

C-4

CFB/PC	Circulating fluidized bed and pulverized-coal
CFB/RH	Circulating fluidized bed with reheat
CFBS	Circulating fluidized-bed FGD scrubber
CFE	Cie. d'Enterprises CFE S.A. (Belgium)
CGAS	Coal gas (from coal gasification)
CGE	Canadian General Electric (Canada)
CGN	Cogeneration/backpressure operation
CH	Chungking Water Works (hydro turbines) [China]
CH2M	CH2M Hill
CHAP	Chapman " " USE CCL " "
CHEM	Chemico Air Pollution Control Division of Envirotech Corp
CHENG	Cheng cycle
CHERNE	Cherne Construction
CHEV	Chevrolet
CHEVRN	Chevron Corp.
CHIC	Chicago Pneumatic (IC engine manufacturer)
CHIN	Chinese (made in PRC)
CHIY	Chiyoda Chemical Engineering & Construction Co. Ltd. (Japan) " " USE cyclone collector)
CHY " "	
CHL	CHL Energy Conversion Systems
CHP	Combined heat and power
CHRON	Chronar Corp.
CHW	C.H. Wheeler
CHY	Chiyoda Chemical Engineering & Construction Company Ltd. (Japan)
CIAN	Cianbro Corp.
CIE	CIE (Italy)
CIMI	Nuova Cimimontubi (Italy)
CITE	CITE Associates
CIWEC	China International Water & Electric Corp. (China)
CJ	Centric-Jones Co.
CK/C	Coke and coal
CK/G	Coke and gas
CK/O	Coke and oil
CL	Clarage
CL	Cooling lake or cooling pond
CLAR	Clarion Power Constructors (JV of Fru-Con and ABB Turbine Inc.)
CLAY	Clayton
CLI	CoGen Lyondell (subsidiary of PSE Inc.)
CLOUGH	Clough Engineering (Australia)
CLRK	Clark Brothers
CLYDE	Clyde Engineering (New Zealand)
CM	Chantiers Modernes (France)
CM	Coppus Murray
CMB	Combination cooling system
CMI	Cockerill Mechanical Industries (Belgium)
CMS	CMS Generation Co.

CNF	CNF Constructors Inc.
CNNC	China National Nuclear Corp.
CNTR	Centaur " " USE SOLAR " "
CO-TH	Co-Thermic Power Systems Inc.
COEL	Columbia Electric
COG	Cogentech
COG	Coke oven gas
COGE	Cogefar (Italy)
COGES	Cogeneration Energy Systems
COGNC	Cogenic Inc.
COGWST	Cogen West
COHN	Cohn Company
COKE	Petroleum coke
COL	Solar collector
COM	Coal-oil mixture
COM	Commonwealth Construction Company
COMB	Combination particulate control (usually ESP preceded by multiclones or cyclone collector)
COMP	Comprimo BV (Netherlands)
COMSIP	Comsip (generator manufacturer)
CON	Under construction
CONS	Consumat Systems Inc.
CONV	Conventional hydroelectric plant
CONVER	Conversion Technologies Inc.
COOP	Cooper Industries (IC manufacturer)
COP	Coppus Engineering Inc.
COPE	Cope
CORNEL	Cornell Pump Co. (hydroelectric turbine manufacturer)
CORTEA	Agroman/Entrecanales (Spain)
COTT	Cottrell
CP	Cogen Power
CP	Cooling pond
CPC	Combustion Power Co. (subsidiary of GWF Energy Systems)
CPE&A	CPE & Associates Inc.
CPI	Co-Power Inc.
CPINT	C.P. International
CPLND	Copeland
CPSB	Claus Pyle Schomer Burns & DeHuno
CR	Cooper Rolls
CR	Custodio Roe & Associates
CR/KHI	Cooper Rolls/Kawasaki Heavy Industries
CREUS	Creusot-Loire (hydraulic turbine manufacturer) [France]
CROCK	Crocker Wheeler (generator manufacturer)
CROSS	Cross-flow hydro turbines
CROUSE	Crouse Recovery Systems Inc.
CRS	CRS Sirrine

CRS/DR	CRS Sirrine/Dravo Corp.	
CRUDE	Crude oil	
CRUZ	F.F. Cruz Inc. (Philippines)	
CRY	Cryogenic	
CSE	Cold side ESP	
CSSR	Unspecified Czech supplier	
CSW	Central & South West Services	
CT	Cooling tower	
CT&A	Consoer Townsend & Associates	
CTA	Cockeril Ougree-Providence/Franco Tosi SpA/Ateliers de Constructions	
	Electricques de Charleroi SpA	
CTI	Concept Technology Inc.	
CUF/JN	CUF/John	
CULM	Anthracite coal waste	
CUM	Cummins Engine	
CV	Conveyor fuel delivery	
CV	Copes-Vulcan	
CVIC	Canadaian Vickers	
CVP	Centrifugal volute pump	
CW	Coal washing	
CW	Curtiss-Wright (gas turbine licensing associate of Pratt & Whitney)	
CW/O	Coal-water mixture/oil	
CWM	Coal-water mixture	
CWT	Chilled water	
CYCL	Cyclone particulate removal	
D&A	Donohue & Associates Inc.	
D&D	Dewberry & Davis	
D&Z	Day & Zimmerman	
D-A	Dillingham-Atkinson (Dillingham Construction Corp./Guy F. Atkinson Co.)	
D/FD	Duke/Fluor Daniel	
D/H	Distillate and heavy fuel oil	
D/R	D/R Hydro Company	
DAC	Deactivated/mothballed	
DAIHAT	Daihatsu (IC manufacturer) [Japan]	
DALE	Dale Electric (turbine manufacturer)	
DALK	Double alkali/dual akali FGD scrubber	
DAN	Daniel International (subsidiary of Fluor Corp.)	
DANE	Daneco Inc.	
DANKS	Danks (boiler manufacturer) [England]	
DAQ	Dry aqueous carbonate FGD scrubber	
DAV	Daverman Associates	
DAVY	Davy Powergas Inc.	
DB	Dominion Bridge (Canada)	
DBA	Deutsche Babcock AG (Germany)	
DBAG	Deutsche Babcock AG (Germany) ** USE DBAG **	
DBS/EW	Dominion Bridge-Sulzer/Escher Wyss	
DBS/GE	Dominion Bridge-Sulzer/General Electric	
DBS/SI	Dominion Bridge-Sulzer/Siemens	
DBSZ	Dominion Bridge-Sulzer Ltd. (Canada)	
DC	D. Connelley (boiler manufacturer)	
DD	Detroit Diesel	
DDR	Unspecified East German supplier	
DEC	Donfong Electric Co. (China)	
DEC	Dow Engineering Co.	
DECK	Decker Energy International	
DEERE	John Deere	
DEF	Deferred without construction start	
DEL	Delayed after construction start	
DELCO	Delco (generator manufacturer)	
DELT	Deltak Corp. (HRSG manufacturer)	
DENGE	Denge Power Projects (England)	
DES	Double Energy System Inc.	
DESC	Developers of Energy Systems Corp.	
DESTEC	Destec Energy Inc.	
DET	Detroit Stoker Company	
DEUTZ	Deutz MWM (Germany)	
DEVL	Developer (when serving as own A/E or constructor)	
DEW	Dominion Engineering Works Ltd. (Canada)	
DFBW	Dongfang Boiler Works (China)	
DFGD	Unspecified dry FGD scrubber	
DFGW	Dongfang Generator Works (China)	
DFLASH	Double flash geothermal plant	
DFTW	Dongfang Turbine Works (China)	
DGAS	Digester gas	
DGI	Dayton Globe Ironworks	
DHS	District heating (CHP) system	
DIAG	Diagonal hydroelectric turbine	
DICK	Dick Corp.	
DIESEL	Diesel oil	
DIG	Digester	
DIST	Distillate oil (also #2 oil and light fuel oil)	
DITT	Ditt SA (France)	
DK	Dick-Kerr	
DK/FL	Duke/Fluor	
DL	DeLaval Industries Inc.	
DL	Dry lime FGD scrubber	
DLL	DeLaval Ljungstrom	
DLNC	Dry low NOX combustors	
DMK	Davy McKee Corp.	
DMOG	Double motion overthrust grates	
DNG	Danregn Vindkraft A/S (WTG manufacturer) [Denmark]	
DNOX	Unspecified NOX removal equipment	

DOB	Doble ** USE AD **
DODDS	Riley Dodds (boiler manufacturer) [Australia]
DOM	Domestic
DOM	Dominion Construction Co.
DORMAN	Dorman Diesel (England)
DOUG	Douglas Energy
DOW	Dow Chemical Co.
DP	Diamond Power
DP/TIB	DePretto-Escher Wyss/TIBB
DPEW	DePretto-Escher Wyss SpA (Italy)
DPT	Diagonal pump turbine
DRAV	Dravo Corp.
DRES	Dresser Industries
DRY	Dry cooling
DRY ST	Dry steam geothermal plants
DSAL	Desaliniznation
DSCRB	Dry scrubber
DSOX	Unknown SO2 control system (Japan)
DT	Dominion Turbine (Canada)
DU/BO	Durr/Borsig
DUCT	Duct burner
DUKE	Duke Engineering & Services Inc.
DUM	Dumont S.A. (France)
DUMEZ	zz/unknown
DURR	Durr (boiler manufacturer [Germany]
DUTC	Dutch oven boiler
DUTCH	Dutch Industries (Canada)
DWL	Dengyosha Works Ltd. (AC licensee) [Japan]
DWT	Danish Wind Technology A/S (WTG manufacturer) [Denmark]
DYN	Dynergy (WTG manufacturer)
E&G	Ebaugh & Goethe Inc.
E-H	Enka Insaat ue Sanayi AS (Turkey)/Hydrogranja (Yugoslavia)
E/BR	Ebasco/Brinderson
E/EMC	Elliot/Electric Machinery Corp.
E/G&H	Electrowatt Engineering Services Ltd. (Switzerland)/Gibbs & Hill
EA	Empresarios Agurapados (Spain)
EA	Energy Alternatives
EA/SEN	EA/Sener
EB	Electrobel (Belgium)
EB/MEN	Ebara/Mendancha (Japan)
EBARA	EBARA Manufacturing Co. (AC and Elliot licensee) [Japan]
EBAS	Ebasco Services Inc. (subsidiary of ENSERCH Corp.)
EBP	EB Power (hydro generators)
EBT	English Electric/Babcock Power/Taylor Woodrow
EBY	Eby Corp.
EC	Econotherm Corp.
ECC	Electrical Construction Co. (England)
ECII	Energy Conversion Industries Inc.
ECO	Eco Energy Inc. (subsidiary of Outokumpu Group [Finland])
ECOL	Ecolaire Process Management Co.
ECP	Environmental Control Products
ECS	Enviro-Chem Systems
ED	Energy Dynamics Inc.
EDF	Electricite de France International
EDRI	Energy Design Research Inc.
EDS	Engineering Design Services
EE	English Electric Company (merged with GEC) [England]
EE/AS	English Electric/ASEA
EE/BPL	English Electric/Babcock Power Ltd.
EE/BW	English Electric/BW
EEC	Environmental Elements Corp.
EF	Energy Factors Inc.
EGT	European Gas Turbine Co.
EI	Erectors Inc.
EL	Elliott Co. (turbine manufacturer)
ELC	Electric Construction
ELEX	Elex AG (Switzerland)
ELIN	Elin (Austria) or American Elin Corp.
ELL	Ellerbe & Co.
ELLIS	Ellis-Don (constructors) [Canada]
ELPK	Elsamprojekt A/S (Denmark)
ELSL	Electrosila (generators) [USSR]
EM	Edgemoor (boiler manufacturer)
EMC	Electric Machinery Manufacturing Co.
EMC/WT	Electric Machinery/Worthington
EMCON	EMCON Associates
EMCS	Emission Control Systems Inc.
EME	EME (T/G manufacturer) [USSR]
EME	Emerson Electric Co.
EMERY	Emery Marker & Campbell
EMSTAR	Emstar (subsidiary of Shell UK) [England]
EN/LOS	Enserch Corp. (USA)/Losinger (Switzerland)
ENB	Empresa Nacional Bazan SA (Spain)
ENB	Entrecanales y Tavora (Spain)
ENER	Enercorp (HRSG manufacturer)
ENERST	Energy Strategies Inc.
ENG	Engler Electric (Canada)
ENGLE	Englehard (SCR supplier)
ENKA	ENKA (Turkey)
ENRON	Enron Corp.
ENS	ENSERCH
ENT	Enterprise

C-7

Revised December 1992

ENTC	Enertech Corp. (WTG manufacturer)
ENTEC	Entec
ENTECH	ENTECH Inc.
ENTHRM	Entherm
ENVT	Envirotech
EP	Electric Products Co.
EP	Ewbank Preece Ltd. (England)
EPC	EPC International Inc.
EPCOR	EPCOR Systems Inc.
EPDC	Electric Power Development Corp. (Japan)
EPI	Energy Products of Idaho (boiler manufacturer)
EPI/GE	Energy Products of Idaho/General Electric
EPS	Energy Productivity Services Inc.
ER	Erie City (now Zurn Industries Inc.)
ER&EC	Exxon Research & Engineering Co.
ERG	Energy Resources Group Inc.
ERI	Energy Recovery Inc.
ES	Energy Sciences (WTG manufacturer)
ESA	Energy Systems Associates Inc.
ESAC	ESAC (hydro) [France]
ESE	Energy Systems Engineers
ESI	Energy Services Inc.
ESL	Energy Systems Ltd.
ESP	Unspecified type of ESP
ESP/BH	ESP/Baghouse
ESX	Essex (hydro turbine manufacturer)
ET	Energy Technology Inc.
ETA	ET Archer Corp.
ETAC	Energy Tactics Inc.
EV/BO	EVT/Borsig
EVB	EV (low-NOX) burners
EVT	EVT Energie-und Verfahrenstechnik GmbH (Germany)
EW	Escher Wyss (Switzerland)
EW/BBC	Escher Wyss/BBC
EW/OK	Escher Wyss/Oerlikon
EW/SS	Escher Wyss/SSW
EXCEL	Excell (WTG manufacturer)
EXT	Extraction steam turbine (single or double)
EXXON	Exxon Corp.
EYT	Entrecanales y Tavora (Spain)
F&K	Flack & Kurtz EMC
F-C	Fru-Con Construction Corp.
F-P	Forsgren-Perkins Engineering
F/E	Franki/Engema
F/K	Francis/kaplan turbines
F/N	Flakt/Niro
FA	Fosgren Associates
FAY	Fayette Manufacturing Co.
FB&D	Ford Bacon & Davis (subsidiary of Deutsche Babcock AG)
FBC	Fludized bed combustion
FBR	Fast breeder reactor
FC	Fraser & Chalmers (England)
FC	Fuel cell
FEG	Fegles Power
FGD	Unspecified type of FGD scrubber
FGD/FB	FGD/Fluidized bed
FGR	Flue gas recirculation (particulate and NOX control)
FHN	FHN Energy Inc.
FIAT	Fiat TTG SpA (Italy)
FICH	Fichtner Consulting Engineers (Germany)
FINCAN	Fincantieri (Italy)
FINN	Finning Tractor (turbine manufacturer)
FISCHR	Fischer Energy Inc.
FISH	L.A. Fish Engineering Company Inc. or Fish Power Group
FIX	Fixed grate boiler
FJ/KH	Fuji/Korea Heavy Industries
FK/BHE	Flakt/BHEL
FKI-B	FKI-Babcock (Germany)
FLKT	Flakt Inc. (now ABB Flakt)
FLO	FloWind Corp.
FLS	F.L. Smidth (Denmark)
FLUD	Fluidyne Engineering Corp.
FLUO	Fluor Daniel Corp. (also corporate predecessors)
FLYGT	ITT Flygt Corp. (hydro turbine manufacturer)
FM	Fairbanks Morse division of Coltec (IC manufacturer)
FM/WAU	Fairbanks Morse/Waukesha
FMC	FMC Air Quality Control division of FMC Corp.
FO2	Fuel Oil No. 2/distillate fuel oil
FO6	Fuel Oil No. 6/heavy fuel oil
FO6/RES	Fuel oil 6/residual
FOCHI	Fochi Group (Italy)
FOR	Foreign
FORD	Ford Motor Co.
FORTH	Forth Energy Ltd. (subsidiary of Merz & McLellan) [England]
FOST	Foster-Miller
FOSTER	Foster Lau Inc.
FOUG	Fougerolle (France)
FPG	Fish Power Group ** USE FISH **
FPT	Francis-type pump turbine
FRAM	Framatome (France)
FRAN	Francis type hydraulic turbine
FRMACE	Framatome/ACEC/Cockerill

C-8

Revised December 1992

FRT	Front-fired boilers
FS	Frank Stevens Electric
FT	Firetube boiler
FT/NU	Fiat/Nuovo Pignone
FT/SUS	Firetube boiler with suspension grate
FU	Fujitsu (Japan)
FUJI	Fuji Heavy Industries (Japan)
FUJI	Fuji Heavy Industries/Fuji Denko Seizo K.K. (Voith licensee) [Japan]
FURST	Furst Energy Inc.
FW	Foster Wheeler Energy Corp.
FW/BW	Foster Wheeler/Babcock & Wilcox
FW/CE	Foster Wheeler/Combustion Engineering
FW/JBE	Foster Wheeler Power Ltd./John Brown Engineering
FW/MAG	FW/MAGUE
FW/MV	Foster Wheeler/Metropolitan Vickers
FW/TIW	Foster Wheeler/Toronto Iron Works
FW/VKE	Foster Wheeler/Vickers Keeler
FWC	Florida Wire & Cable
FWCD	Foster Wheeler Ltd. (Canada)
G&H	Gibbs & Hill Inc. (subsidiary of Dravo Corp.)
G&HT	Gibson & Hart
G&O	Gurries & Okamoto Inc.
G-C	Gilbert/Commonwealth
G/LN	Natural gas and LNG
G/O	Natural gas and oil
G/P	Natural gas and propane
G/PW	Natural gas and paper waste
G/RE	Natural gas and refuse
G/TB	Granger/TB
G2S/HA	G2S Constructors/Harbert Corp.
GA	General Atomics Corp.
GAR	Garrett (gas turbine manufacturer)
GAS	Natural gas
GASCGN	Gascogen (Australia)
GASF	Gasifier
GATOR	Gator Power (partnership of Florida Power Corp. and Babcock & Wilcox)
GBG	Granite Ball & Groves
GC	Groupe Constructeurs Francaise (France)
GCHWR	Gas-cooled heavy water reactor
GCR	Gas-cooled reactor
GDV	Generadores de Vapor (Foster Wheeler licensee) [Spain]
GE	General Electric Co. (USA)
GE/HT	General Electric/Hitachi
GE/KH	GE/Korea Heavy Industries & Construction
GE/STA	GE/Stal
GE/TD	General Electric/Turbodyne
GE/UTL	GE/Utility
GE/WH	General Electric/Westinghouse
GEA	zz/unknown
GEC	General Electric Co. (England)
GEC/PR	GEC/Parsons
GEES	General Electric Environmental Services (formerly Buell)
GEN	Genbach
GENT	Generation Technologies
GEO	Geothermal
GER	General Energy Resources Inc.
GERCED	Gerced Diesel
GES	Geselski
GFCC	Gannett Fleming Corddry & Carpenter Inc.
GG	Gulf General
GGG	Gilbert Gilkes & Gordon (England)
GGG/PR	GGG/Parsons
GHDR	Gibbs Hill Durham & Richardson
GIA	Green International Affiliates
GIE	Gruppo Industria Eletto Meccanische per Impianti All'Estero (Italy)
GIL	Gilbert Associates Inc.
GILK	Gilkes Pumps
GK	Gilkes (hydro turbines) [England]
GLOBAL	Global Boiler Mechanical
GM	General Motors (IC manufacturer)
GM	Grummi-Mayer KG (Germany)
GMT	Grandi Motori Trieste (Italy)
GOB	Gob (bituminous mining waste)
GOEB	Steve Goebel Construction
GOLD	Goldfire (IC engine manufacturer)
GORF	Gorf Contracting (Canada)
GOT	Gotaverken Energy Systems (Sweden)
GP	Gould Pumps Inc.
GPC	Geothermal Power Co.
GRANGR	Granger Renewable Resources Inc.
GRAY	Gray & Osborne
GREEN	Green Construction Co.
GRIF	Griffin Industries Inc.
GSF	GSF Energy Inc.
GT	Gas/combustion turbine
GT/C	Gas turbine in combined-cycle
GT/H	Gas turbine with heat recovery
GT/HR	Gas turbine and heat recovery boilers
GT/S	Gas turbine with steam sendout
GTM	Grand Trauaux de Marseille (France)
GULF	Gulf General
GULFD	Gulf Design & Engineering Inc.

Revised December 1992

C-9

GW	Gebrueder Weiss KG (Germany)
GWF	GWF Power Systems (subsidiary of Signal Co.)
GZ	Ganz Mavag
GZ/ELN	Ganz Mavag/Elin
H&A	Horiuchi and Associates Inc.
H&M	H&M Engineering Inc.
H&S	Hirschfeld & Stone Consulting Engineers
H&W	Harlan & Wolf (IC manufacturer)
H-C	Hydra-Co Enterprises Inc. (subsidiary of Niagara Mohawk Power Corp.)
H-RI	H-R International Inc.
H-W	Hayden-Wegman
H-Z	Hitachi-Zosen (Japan)
H/CSE	Hot and cold side ESPs
H/K	Hitachi/Kajima (Japan)
H/T	Harbert/Triga
H/T	Haz/Takenaka (Japan)
H/V	Horizontal and vertical hydro turbines
H2	Hydrogen
HAD	Hadson Power Systems (now owned by LG&E Energy Systems)
HALDOR	Haldor Topsoe A/S (Denmark)
HALE	Hale Energy Conversion Systems
HALL	Hall Machine Co. (WTG manufacturer)
HAM	Hamilton Brothers
HAMIL	Hamilton Standard (WTG manufacturer)
HAMMER	Hammermill
HANGZ	Hangzhou (China)
HANSON	James Hanson & Associates
HAR	Harbin ** USE HARBIN **
HARB	Harbert
HARBIN	Harbin Electrical Equipment Co. (China)
HARDIN	Hardin Energy Systems
HARK	Harker
HARPER	Harper Construction Co.
HARRIS	Harris Mechanical Contracting Co.
HARZ	Harza Engineering Co.
HAWK	Hawker Siddeley Power Engineering Inc.
HAWT	Horizontal axis wind turbine
HBW	Ha'erbin Boiler Works (China)
HBZ	H.B. Zachary Co.
HC	Hollow conductor (pumps)
HCCM	Huaxing (PRC)/China Construction Engineering Co./Campenon Bernard (France)/Maeda (Japan)
HCG	Hollandse Constructie Groep (NEM Boilers) ** USE NEM **
HCG	Hollandse Constructie Groep (Netherlands)
HDR	Henningson Durham & Richardson
HDRE	HDR Engineering
HDRI	HDR Infrastructure Inc.
HDRT	HDR Techserv
HEC	HEC Energy Corp.
HEIL	Heil Engineering Systems
HEMET	Hemeter Development Corp.
HERC	Hercules (IC engine manufacturer)
HES	Hawthorne Engine Systems
HEURT	Heurtey Petrochemical (HRSG manufacturer) [France]
HFO	Heavy fuel oil (#6 oil)
HFO/C	Heavy fuel oil/crude oil
HFO/D	Heavy fuel oil and distillate fuel oil
HGAS	Hydro gas
HGI	Harris Group Inc. (formerly Schuchart & Associates Inc.)
HHHH	Hoch/Hammers/Heitkamp/Philipp Holzmann AG (Germany)
HIC	Harrison International Corp.
HICK	Hickam Industries
HIN	A.E. Hinman
HIPP	Hipp Engineering Ltd.
HKG	Haynie Kallman & Gray
HLM	Hansen Lind Meyer
HMC	Henkels & McCoy
HN	Halliwell Nowe
HOCH	Hochtief AG (Germany)
HOK	Holyoke Machine Co. (hydroelectric turbine manufacturer)
HOLE	Holewinski
HOLL	Hollansen
HOLT	Ben Holt Co.
HONDA	Honda (IC engine manufacturer)
HOPENN	H.O. Penn Machinery
HOR	Hoven-Owens-Renschler (hydraulic turbine manufacturer)
HORNSB	Hornsby (IC manufacturer)
HORZ	Horizontal axis hydro turbine
HOSEY	Hosey & Associates
HOW	James Howden & Company Ltd. (Scotland)
HP	Howden Parsons (Canada)
HPC	Hidrotechnica Portugesa Consultores Estudos & Projects Lda. (Portugal)
HRB	Hochtemperatur Reaktorbau (Germany)
HRCC	Hydro Resources & Contractors Corp. (Philippines)
HRH	H.R. Hadlow
HS	Turbines Hispano Ogasco Inc. (Spain)
HSE	Hot side ESP
HT	Hitachi Ltd. (Japan)
HT/FT	Horizontal boiler/firetube boiler
HT/KH	Hitachi/Korea Heavy Industries
HTC	Hydraulic Turbine Co. (hydroelectric turbine manufacturer)
HTR	High temperature gas-cooled reactor

HU	Rodney Hunt
HUDSN	Hudson Engineering
HUGHES	Hughes Brothers
HUR	Hurst Boiler Co.
HW	Hot water
HWC	H. W. Campbell & Associates
HWR	Heavy water reactor
HY	Hydroelectric turbine generator
HYA/AN	Hydroart/Ansaldo
HYA/EM	Hydroart/Ercole Marelli
HYC	Hydraulic coupling (pumps)
HYDA	Hydroart SpA (Italy)
HYDROL	Hydrolec (hydroelectric turbine manufacturer) now Neypric Minihydro Inc. [Canada]
HYUN	Hyundai (ROK)
HYWS	Hydro West (hydroelectric turbine manufacturer)
HZ	Horizontal boiler
HZDWST	Hazardous waste
HZFT	Horizontal firetube boiler
I-S	Ishikawajima Heavy Industries - Shibaura Works (Japan)
I/SEMT	IHI/SEMT Pielstick
IAI	Industrieen Lagen (Germany)
IBC	Industrial Boiler Co.
IBM	International Business Machines Inc.
IC	Internal combustion (reciprocating engine or diesel engine)
ICA	Industrial Clean Air Co.
ICC	International Cogeneration Corp.
ICCT	ICC Technologies Inc.
ICL	International Combustion Ltd. (now NEI International Combustion) [England]
ICT	I.C. Thomasson Associates
IDC	Interest during construction
IDEAL	Ideal Electric (generator manufacturer)
IEC	International Engineering Co.
ECO	IECO (USA)
IEI	Interface Engineering Inc.
IEM	Isby Enico LE Myers
IEMSA	Iemsa (Italy)
IESA	IESA (A/E) [Brazil]
IESC	International Environmental Systems Corp. (subsidiary of UOP Inc.)
IH	International Harvester
IHI	Ishikawajima-Harima Heavy Industries (Japan)
IMP	Impell Corp. (subsidiary of Combustion Engineering)
IMP	Impulse hydroelectric turbine
IMPPWR	Impell Power Projects
IMPSA	IMPSA (hydroelectric turbine manufacturer) [Argentina]
INB	Internationale Natrium Brutreaktobrau GmbH (Germany)
INC	Incinerator
INCL	Inclined hydroelectric turbine
INDAR	Generator manufacturer (Spain)
INDATM	zz/unknown
INDECK	Indeck Energy Services Inc.
INDWST	Industrial waste
INENCO	Inenco Energy Ltd. (England)
INOR	INOR (France)
INT	Interel Corp.
INTEG	Integrated Total Energy Inc.
INTELL	Intellicon Inc.
INTER	Interchem (NA) Industries Inc.
INTI	PT Inti Karya Persada Teknik (Indonesia)
IPC	Integrated Power Corp.
IPM	I.P. Morris Division of The Wm. Cramp & Sons S & E.B. Company (hydraulic turbine manufacturer)
IPT	International Power Technology Inc.
IR	Ingersoll-Rand Co.
IR	Italstrade-Recchi SpA (Italy)
IRBY	Irby Construction Co.
ISI	International Systems Inc.
IVO	IVO Consulting Engineers (Finland)
IWST	Industrial waste
J&M	Jackson & Moreland (division of United Engineers & Constructors)
J-M	Johnson-Mathey
J-S	Jeumont-Schneider (France)
J-U-B	J-U-B Engineers Inc.
J/N	Joy/Niro
JAC	Jacobs Engineering Group Inc.
JACB	Jacobs Wind Electric
JAJ	J.A. Jones Construction Company (subsidiary of Phillip Holzmann AG)
JAPAN	Unspecified Japanese supplier
JAT	J.A. Trent
JBE	John Brown Engineering Ltd. (Scotland)
JELC	Jelco Inc. (division of Towsend & Bottum Inc.)
JEN	Jenbach
JET	Jet fuel
JFP	J.F. Pritchard & Co.
JG	Jeumont-Schneider (France)
JI	John Inglis
JI/BW	John Inglis/Babcock & Wilcox
JINAN	Jinan (China)
JKF	J. Kenneth Fraser & Associates
JL	John Laing Construction Ltd. (UK)
JM	Jenkes Machine

Revised December 1992

JM/RMP	JV of James M. Montgomery Consulting Engineers Inc. and Ralph M. Parsons Inc.)
JMM	James M. Montgomery Consulting Engineers
JOHN	Johnson
JORD	E.C. Jordan Co.
JOY	Joy Industrial Equipment Co.
JP	Johnson Pump Co.
JT	John Thompson Ltd. [Northern Engineering Industries Ltd.] (England)
JUG	Jugoturbina (Yugoslavia)
JW	JW Operating Co.
K&L	Koepf & Lang
K&S	Krieger & Stewart
K-S	Katy-Seghers
K/E/N	KRC/EVT/Noell
K/H/K	Kajima/Hazama-Gumi/Kumagai-Gumi (Japan)
K/O	Kum/Obay (Japan)
K/S/O	Kajima/Shimizu/Okumura (Japan)
K/T/O	Kum/Takenaka/Obay (Japan)
KA	Kaiser Aluminum
KAB	Kablitz Stoker (boiler manufacturer)
KAHN	Albert Kahn
KAIS	Kaiser Engineers Inc.
KAJI	Kajima (Japan)
KAL	Kalina cycle
KAM	Kamine Engineering & Mechanical Contracting Inc.
KAP	Kaplan type hydraulic turbine
KATO	Kato Engineering Division of Reliance Electric Co.
KAW	Kaweah Construction Co.
KC	Kelley Co.
KCMIL	Thousand circular mills (= MCM)
KDO	Keeler/Dorr-Oliver (now subsidiary of Tampella)
KDP	Kongsberg Dresser Power
KEEL	E. Keeler Co.
KELL	M.W. Kellogg Co. (subsidiary of Signal Co.)
KEN	William Kennedy
KERO	Kerosene
KGAS	Coke oven gas ** USE COG **
KH/CE	Korea Heavy Industries/Combustion Engineering
KH/GE	Korea Heavy Industries/General Electric
KH/HT	Korea Heavy Industries/Hitachi
KH/NG	Korea Heavy Industries/Niigata
KHI	Kawasaki Heavy Industries (Japan)
KHIC	Korea Heavy Industries & Construction (South Korea)
KIE	Peter Kiewit & Sons Co. ** USE KIEW **
KIER	Kier Construction (England)
KIEW	Peter Kiewit Inc. or Peter Kiewit & Sons Co.
KIKAI	Tsukishima Kikai
KIME	Jim Kime
KIP	Kipper & Sons
KIRK	A. C. Kirkwood & Associates
KISHA	Kisha Seizo (boiler manufacturer) [Japan]
KLEIN	Kleinschmidt Associates
KLOEP	Kloepfer Inc.
KMW	Karlstads Mekaniska Werkstand
KOHLER	Kohler
KONGS	Kongsberg
KOP/LC	Koppers/Lodge Cottrell
KOPP	Koppers Co.
KPEC	Korea Power Engineering Corp. (South Korea)
KRC	KRC Umwelttechnik GmbH (Germany)
KRC/EV	KRC/EVT
KREB	Krebs
KSG	zz/unknown [Germany]
KSG/DR	KSG/Durr
KSK	Kawasaki Heavy Industries (Japan) ** USE KHI **
KST	KST Hydroelectric Engineers (Canada)
KU	Kraftwerk Union (subsidiary of Siemens AG) [Germany]
KUL	Kuljian Corporation
KUN	Max J Kuney Co.
KUNMNG	Kunming Electrical Machinery Works (China)
KV	Kvaerner Hydro Power Inc. (subsidiary of Kvaerner Group) [Norway]
KV/ID	Kvaerner/Ideal
KV/MEP	Kvaerner/MEP (Brazil)
KV/VOI	Kvaerner/Voith
KVAER	Kvaerner Hydro Power Inc. ** USE KV **
KW	KW Energy Systems Inc.
KWU	Kraftwerk Union (unit of Siemens AG) [Germany]
KYO	Kyocera Inc.
L&A	Lockman & Associates
L&M	Lutz & May
L&T	Larson & Toubro (India)
L-A	Louis-Allis ** USE LA **
L-B	Lardet-Babcock
L-C	Lummus-Crest Inc. (subsidiary of ABB Combustion Engineering)
L/B	Lignite and bituminous coal
L/S	Lignite and subbituminous coal
LA	Louis Allis Co.
LACNTY	Los Angeles County Sanitation District
LAH	Lahmeyer International GmbH [subsidiary of RWE] (Germany)
LAI	Longardner & Associates Inc.
LAIDLW	Laidlaw Gas Recovery Systems
LAJET	LaJet Energy Co.

Revised December 1992

C-12

LAN	Lockwood Andres & Newnam
LANCS	Lancashire Dynamo & Motor
LANE	Lane
LAQ	Laquidara Inc.
LAW	Lawrence Scott
LB	Laurent Bouillet
LB	Lister Blackstone
LBO	Layne & Bowler Co.
LC	Lodge Cottrell Division of Dresser Industries Inc.
LD&B	Lutz Daily & Brain
LDP	Laramore Douglas & Popham
LEIT	Leittel
LF	James Leffel Co. (hydraulic turbine manufacturer)
LG/NOI	Lurgi/Noyes
LGM	zz/unknown
LGMP	L.G. Mochel & Partners
LGX	Liangxiang (China)
LIFAC	Limestone injection into furnace with CAO activation
LIG	Lignite (brown coal)
LIMA	Lima
LIQ	Pulping liquor
LIST	Lister
LIT	Litwin Engineers & Constructors Inc. (unit of AMCA International Ltd.)
LJ/RK	Litostroj-Blex/Rade Koncar-Biex
LLOYD	Lloyd Power Corp.
LM IJ	Limestone injection
LMBR	Liquid metal breeder reactor
LMFBR	Liquid metal fast breeder reactor
LMZ	Leningrad Machine Works (USSR)
LN	Leeds & Northrup Instruments (unit of General Signal)
LNB	Lo-NOX burners
LNG	Liquified natural gas
LOCK	Lockwood Greene Engineers Inc.
LOLL	Lolland (WTG manufacturer)
LOTTE	Lotte
LP/O	LPG and oil
LPG	Liquified petroleum gas (propane)
LQ	Liquid geothermal technology
LS	Leroy Somer (hyroelectric generators)
LSW	LSW Engineers
LURG	Lurgi Corp. (Germany)
LUZ	Luz International
LYD	Lydig Construction Inc.
M&A	Meyer & Associates
M&D	M & Del (hydro)
M&E	Metcalf & Eddy
M&H	Mead & Hunt
M-B	Mitchell-Banki (hydroelectric turbine manufacturer)
M-L	Minnesota-Moline (IC engine manufacturer) "" USE SMS ""
M-S	Morgan-Smith (hydraulic turbine manufacturer) (subsidiary of Brown & Root)
M-V	Mid-Valley Constructors Inc. (subsidiary of Brown & Root)
M-Z	Mitsubishi-Zosen (Japan)
M/H/T	Maeda/Haz/Taisei
M/K	Maeda/Kim (Japan)
M/K/O	Maeda/Kum/Obay (Japan)
M/T/P	Marubeni/Toshiba/Pritchard
MAACHI	Maachi (HRSG manufacturer) [Italy]
MACK	zz/unknown
MAG	MAGNOX gas-cooled reactor
MAG OX	Magnesium oxide FGD scrubber
MAGUE	Mague (Foster Wheeler licensee) [Portugal]
MAIN	Chas T. Main Inc.
MAM	M. A. Mortenson Company Inc.
MAN	MAN Gutehoffnungshutte GmbH (Germany)
MAN/BB	MAN/BBC
MAN/KW	MAN/KWU
MAN/SI	MAN/Siemens
MANBW	MAN B&W Diesel AG (Germany)
MANURE	Manure fuel
MAPI	Mitsubishi Atomic Power Industries Inc. (Japan)
MAR	Ercole Marelli Nuova SpA (Italy)
MAR	Marubeni Corp. (Japan)
MARIYA	Mariya (WTG manufacturer)
MARRON	Marron Associates
MATHER	Mather & Platt
MAW	Montreal Armature Works
MAX	MAXIM
MAXN	Maxon Construction Company Inc.
MAY	Mayfair
MB	Morse Boulger
MBBC (Tomano)	Brown Boveri & Cie. design built by Mitsui Shipbuilding & Engineering [Japan]
MBC	Michigan Boiler Co.
MBD	Mirrless Bickton & Daye
MC	Magnetic coupling (pumps)
MC	Medical Cogen (JV of Kaiser Engineers and General Electric)
MCA	zz/unknown
MCALP	Sir Robert McAlpine & Sons Ltd. (England)
MCB	McBurney
MCLARN	McLaren (IC manufacturer)
MCLEOD	R.J. McLeod (Scotland)
MCM	Thousand circular mills (use KCMIL)

Revised December 1992

C-13

MCS	Michigan Cogen Systems
MDT	Mechanical draft tower
MDT/D	Dry type mechanical draft cooling tower
ME	Mitsubishi Electric Corp. (generators) [Japan]
MEC	May Engineering Co.
MECH	Mechanical particulate control device
MEDWST	Medical waste
MEGA	Mega Renewables
MENDES	Mendes Junior (constructor) [Brazil]
MEP	Mecanica Pesada (hydroelectric turbine manufacturer) [Brazil]
MES/PC	Midwesco Energy Systems Inc./Paschen Constructors Inc.
MET	Metric Constructors Inc. (subsidiary of J.A. Jones Construction Co.) ** USE METRIC **
METAL	Metalurgical Corp. (China)
METH	Methane (landfill gas or sewage gas) see also DGAS
METRIC	Metric Constructors Inc. (subsidiary of J.A. Jones Construction Co.)
MEY	Meyer
MFO	Medium fuel oil
MGAS	Mine gas (low-BTU waste gas from coal mines)
MHI	Mitsubishi Heavy Industries (Japan)
MHI/BR	MHI/Brush
MICH	Michener & Associates
MICON	Micon (WTG manufacturer) [Denmark]
MICRO	Micro Cogen Systems Inc.
MID	Midwesco Inc.
MIL	Marine Industrie Ltee/Division Hydro-Electricque (Canada)
MILK	MILK/Wallace
MILL	Milled peat
MINE	Mine mouth
MIRR	Mirrlees Blackstone Inc. or Mirrlee Diesel Engineering
MIT	Mitchell (boiler manufacturer) [England]
MITCH	Mitchell Construction (England)
MITSUI	Mitsui & Co. (Japan)
MIW	Murray Iron Works Co. (boiler manufacturer)
MJT	Mike J Thiel Inc.
MK	Morrison-Knudsen Company Inc. or M-K Ferguson
MK/CB	Morrison-Knudsen/Clement Brothers Co.
MLB	MLB Industries
MLSE	M.L. Smith Environmental
MLW	Montreal Locomotive Works (Canada)
MM	Merz & McLellan
MM	Mixed mode cooling system
MMB	Moorhead Machinery & Boiler Co.
MMMC	Mitsui Mjike Machinery Co. (Japan)
MMR	MMR Power & Industrial
MOD	Modular incinerator
MON	Montreal Engineering (Canada)
MONE	Moneco Consultants (Canada)
MONS	Monsanto Enviro-Chem
MONT	Montenay Power Corp.
MOT	Motores SA (Colombia)
MOWLEM	Mowlem Engineering (England)
MP	Mechanica de la Pena (Spain)
MP	Modular Products
MP&L	Minnesota Power & Light Co.
MPA	Multipower Associates (JV of Pyropower Black & Veatch and Marubeni)
MPEC	Mission Power Engineering Co. (subsidiary of SCECorp.)
MR	Marathon
MRT	Josef Martin Feuerungsbau GMBH (boiler manufacturer) [Germany]
MS	Myer Schiller
MST	Moore Steam Turbine
MSW	Municipal solid waste ** USE REF **
MT	Mitsubishi Heavy Industries Ltd. (Japan) ** USE MHI **
MTPL	Mather & Platt
MTS	Mitsui & Co. (Japan) ** USE MITSUI **
MUL/WS	Multiclone/wet scrubber
MULT	Multiple contractors or vendors
MULTI	Multiclone particulate collector
MV	Metropolitan Vickers (England)
MV/BW	Metropolitan Vickers/Babcok & Wilcox
MVA	Megavoltampere
MVC	Missouri Valley Constructors Inc.
MVI	Missouri Valley Inc.
N-P	NEI Parsons
N/A	Not applicable
N/ASEA	NOHAB (Sweden)/ASEA (Sweden)
N/CG	Neypric/Cegelec
N/MDT	Natural and mechanical draft cooling towers
N/SIE	Neypric/Siemens
NAC	North American Cogeneration Inc.
NAH	North American Hydro Inc.
NAL	NALCO
NANJIN	Nanjing Electric Machinery (China)
NANQI	Nanqi (China)
NAP	Naptha
NAT	Natkin
NB	Nebraska Boiler
NBG	Nordberg
NCP	NEI Crossley-Pielstik (IC manufacturer)
NCPA	Northern California Power Agency
NCRRA	North County Resource Recovery Associates (JV of Thermo Electron and Brown & Root)

Revised December 1992

ND	No decision (no equipment specified)
NDF/GE	Neyrpic-Duro Felguera/General Electrica Espanola (Spain)
NDT	Natural draft cooling tower
NDT/D	Dry type natural draft cooling tower
NEA	Northeast Engineering Associates
NEC	National Energy Constructors
NEI	Northern Engineering Inc.
NEIC	NEI/CMI
NEIP	NEI Projects (England)
NEM	NEM Boilers & Process Equipment (JV of Hollandse Construcite Groep and L&C Steinmuller) [Netherlands]
NEPCO	National Energy Production Co. (Zurn subsidiary)
NES	National Energy Systems
NEWEST	Newest Inc.
NEWT	Newton Associates
NEY	Neypric (hydraulic turbine manufacturer) [France]
NEY/JG	Neypric/Jugoturbina
NH	Nassax Hemsley Inc.
NH3 IJ	Ammonia injection
NI	National Industries
NIC	National Industrial Contractors
NIIGAT	Niigata (IC manufacturer) [Japan]
NIPPON	Nippon Kokan KK (Rateau licensee) [Japan]
NISSHO	Nissho-Iwa ** USE NISSHO **
NM	National Machine
NN	Newport News Ship & Dry Dock (hydraulic turbine manufacturer)
NNC	National Nuclear Corp. (England)
NOE/K	NOELL/KRC
NOELL	NOELL GmbH (Germany)
NOHAB	NOHAB (Sweden)
NOOTER	Nooter-Eriksen (HRSG manufacturer)
NORD	Nordtank Energy Group (WTG manufacturer) [Denmark]
NOREN	NORENCO (subsidiary of Northern States Power Co.)
NORTON	Norton Chemical Process Products (SCR manufacturer)
NORWES	NorWest Energy Systems (Canada)
NOVA	Novatome (France)
NOYES	Noyes (ESP manufacturer)
NPCIL	Nuclear Power Corp. of India Ltd.
NRDBRG	Nordberg ** USE NBG **
NS	Nissho-Iwa ** USE NISSHO **
NSP	Northern States Power Co.
NTEMW	Nanjing ** USE NANJIN **
NUCLEN	Nuclebras Engenaria SA (Brazil)
NUOVO	Nuovo Pignone SpA (Italy)
NUTEC	NUTEC (IC engine manufacturer)
NYS	New York State
O/BG	Oil and blast-furnace gas
O/C	Oil and coal
O/G	Oil and natural gas
O/N	Oil and naptha
O/R	Oil and refuse
O/S/T	Obay/Shimizu/Takenaka (Japan)
O/T/K	Obayashi/Taisei/Kajima (Japan)
OB	Ohio Brass
OBER	Obermeyer (hydraulic turbine manufacturer)
OBES	O'Brien Energy Systems
OBMC	O'Brien Machinery Co.
OC	O'Connor Combustor Division of Westinghouse Electric Corp.
OC	Oscillating combustor
ODOM	J.M. Odom Co.
OE	zz/unknown
OERL	Oerlikon
OFAG	OFAG AG (Switzerland)
OFF	Refinery off-gas (= REFGAS) ** USE RGAS **
OFFSH	Offshore
OGD	Ogden Martin Systems Inc.
OIL	Fuel oil
OL	Oscar Larson
ON	Onan
ONSI	ONSI (fuel-cell manufacturer)
OPR	In commercial operation
OPS	Offshore Power Systems
OR	Ormat Turbines Ltd. (Israel)
OREN	Orenco Engineering Co.
OREND	Orenda
ORI	Oriemulsion
ORMAT	Ormat Energy Systems Inc. (Israel)
OSS	Ossberger Turbines Inc. (hydraulic turbine manufacturer)
OT	Once through cooling
OTB	Once through cooling using brackish water
OTF	Once through cooling using fresh water
OTG	Once through cooling using ground water
OTM	Once through cooling using municipal water
OTS	Once through cooling using saline water
OTSK	Obay/Takenaka/Shimizu/Kim (Japan)
OTTO	Otto-cycle engine
OUTO	Outokumpu (boiler manufacturer) [Finland] or Outokumpu EcoEnergy
P&K	Poole & Kent
P&P	Plarr & Pelton (hydraulic turbine manufacturer)
P&S	Pfeifer & Schultz Inc.
P&W	Parsons & Whittemore
P-BLH	Pelton-Baldwin Lima Hamilton (hydraulic turbine manufacturer)

Revised December 1992

C-15

P-D	Pelton-Doble (hydraulic turbine manufacturer)
P-J	Parsons-Jurden Corp.
PACIF	Pacific Energy
PACNRG	Pacific Energen Inc.
PAGE	Page Engineering
PALM	Palmer Electric
PAPER	Paper mill sludge
PAR	C.A. Parsons & Company Ltd (now NEI Parsons) [England]
PAR/AE	C.A. Parsons & Co./Associated Electric Industries Ltd. (UK)
PAR/MA	C.A. Parsons & Co./C.T. Main
PAS	Power ascension stage
PAX	Davey Paxman or Ruston Paxman (IC manufacturer)
PAYNE	Payne & Keller
PBEA	Palm Beach Energy Associates (JV of B&W and Bechtel)
PBQD	Parsons Brinckerhoff Quade & Douglas Inc.
PC	Pulverized coal boiler
PC/ARC	Pulverized coal boiler arc-fired
PCC	Pennsylvania Crusher Corp.
PCE	Pacific Coast Engineering
PCFB	Pressurized circulating fluidized-bed boiler ** USE PFBC **
PDBG	PowerDesignBuild Group (subsidiary of ElectriCorp) [New Zealand]
PDEV	Power Development Systems Inc.
PDS	PDS Engineers & Constructors
PE	Power Enterprises
PEAB	Peabody Process Systems Division (subsidiary of ABB Flakt)
PEAT	Peat
PEC	Pennsylvania Engineering Corp.
PEEB	A. Peebles (generator manufacturer)
PELT	Pelton Water Wheel Co. (hydraulic turbine manufacturer) ** USE PWW **
PELT	Pelton type hydraulic turbine
PEN	Power Engineers
PENSKE	Penske Power Systems
PEPS	Pacific Energy Production Systems Inc.
PERC	Pennsylvania Energy Resources Co.
PEREN	Perennial Energy Inc.
PET	Petroleum coke ** USE COKE **
PFBC	Pressurized fluidized-bed combustor
PG&E	Pacific Gas & Electric Co.
PGA	Paul L. Geiringer & Associates
PGC	Philadelphia Gear Corp.
PGE-B	PG&E-Bechtel Generating Co.
PGS	Petrolane Gas Service
PHWR	Pressurized heavy water reactor
PI	Platt Iron Works (hydraulic turbine manufacturer)
PII	Philippine Infastructures Inc. (Philippines)

PILE	Pile-type combustor
PINTO	Pinto Basto (Portugal)
PION	Pioneer Services & Engineering (Fluor)
PIPE	Pipeline
PIRNA	VEB Pirna (Germany)
PIT	Pit-type hydroelectric turbine
PIWS	Pacific Integrated Waste Systems Inc.
PIZZA	Pizzagalli Construction Co.
PJD	P.J. Dick Contracting Inc.
PK	Pullman Kellogg (division of Wheelabrator-Frye Inc.)
PLAS	Plastic waste
PLN	Planned and still in design
PM	Parsons Main
PMR	Pro Mass Recovery
PO	Pacific Oerlikon
POL	Polenko Holek Inc. (Holland)
POPE	Pope Engineers
POR	Portec Inc.
PP	Pacific Pumps Division of Dresser Industries Inc.
PPC	Phillips Petroleum Corp.
PPI	Power Projects Inc.
PPP	PWR Power Projects (JV of Westinghouse and NNC) [England]
PPT	Pelton-type pump hydraulic turbine
PR	Pyrite rejection system
PRC	Power Recovery Systems Inc.
PRDC	Power Reactor Development Co.
PRE	Preformed
PREP	Precipitair Pollution Control
PRIT	Pritchard Corp. (subsidiary of Black & Veatch)
PROCE	Procedair Industrie
PROCON	Procon Inc. (subsidiary of UOP Inc.)
PROMON	Promon (A/E) [Brazil]
PROP	Propane (also see LPG)
PROP	Propeller hydroelectric turbine
PROTOS	Protos Engineering (India)
PS	Pumped storage hydroelectric plant
PSE	Power Systems Engineering Inc. (now owned by Destec Energy Inc.)
PSL	Peabody Sturtevant Ltd. (England)
PT	Peter Brotherhood Ltd. (England) ** USE PETER **
PTCH	Pitch
PVI	Pacific Ventures Inc.
PW	Pratt & Whitney (AKA Turbo Power & Marine Systems Inc. unit of United Technologies)
PW/W	Pratt & Whitney engine with Worthington free turbine
PWR	Power Products Corporation
PWR	Pressurized water reactor

Revised December 1992

C-16

PWST	Paper mill waste and sludges
PWW	Pelton Water Wheel Co. (hydraulic turbine manufacturer)
PYR	Pyropower Corp. (boiler manufacturer)
QINGDA	Qingdao (China)
R-B	Riley Beird
R-F	Rist-Frost
R-H	Rodney-Hunt (hydaulic turbine manufacturer)
R-S	Rogers-Schmidt Engineering Co.
R/AN	Rateau/Ansaldo
R/B	Rail and barge fuel delivery
R/B/T	Rail and barge and truck fuel delivery
R/C	Residual and crude oil
R/M	Rateau Schneider (France/Ercole Marelli (Italy)
R/SHIP	Rail and ship fuel delivery
R/T	Rail and truck fuel delivery
RAD	Radial hydroelectric turbine
RADOS	Steve P. Rados Corp.
RAF	RAFACO (boiler manufacturer) [Poland]
RAIL	Rail fuel delivery
RANK	Rankine cycle engine
RAT	Rateau (France) ** USE RATEAU **
RATEAU	Rateau (turbine manufacturer) [France]
RAUMA	Rauma-Repola Witermo (Finland)
RAY	Raymond Engineers Pty Ltd. (Australia)
RC	Regulator control (pumps)
RC	Research-Cottrell Inc.
RC/BU	Research Cottrell/Buell
RC/KOP	RC/Koppers
RC/UOP	RC/UOP
RC/WAL	RC/Walther
RCA	Radio Corporation of America Inc.
RCP	Reciprocating grate boiler
RCP/WW	Waterwall reciprocating grate boiler
RDF	Refuse-derived fuel
RDT	R.D. Thomas
READY	Ready Power (generator manufacturer)
RECIP	Reciprocating Engine
RECIR	Gas recirculation
REF	Refuse (unprocessed municipal solid waste)
REFGAS	Refinery offgas ** USE RGAS **
REFR	Refractory furnace ** USE RW **
REG	Regulator (pumps)
REL	Reliance Electric Co.
RENDEL	Rendel Palmer & Tritton (A/E) [England]
REPCO	Repco (IC manufacturer)
RESID	Residual oil
RET	Retired
REY	Reynolds
RF	Refrigeration
RG	Roller grate boiler
RGAS	Refinery off-gas
RH	Reheat (boilers)
RHL	Russel-Hipwell Lister
RHM	Rodney Hunt
RHN	Ruston & Hornsby
RICE	Rice (biomass fuel)
RIM	Rim drive (hydro turbine)
RIV	Rivers Engineering
RIVA	Riva Calzoni (turbine manufacturer) [Italy]
RKB	Rade Koncar-Biex
RME	Rocky Mountain Engineering
RMP	Ralph M. Parsons Company
ROCK	Rockwell International
ROGER	Roger Construction Co.
ROM	PB Rombough Construction Ltd. (Canada)
ROTH	Rothemuehler [Babcock & Wilcox] (Germany)
RP	Rhone-Poulenc (France)
RPT	Reversible pump turbine
RPTT	Tube type reversible pump/turbine
RPX	Ruston Paxman
RR	Rolls Royce (England)
RRS	Resource Recovery Systems
RS	Riley Stoker Corp. (subsidiary of Deutsche Babcock AG)
RS/E	Riley Stoker/Environeering
RS/ER	Riley Stoker/Erie City
RSH	Reynolds-Smith-Hills
RTD	Rotary drum combustor
RTH	Rotary hearth combustor
RTK	Rotary kiln combustor
RUST	Rust International Corp.
RUSTON	Ruston Gas Turbines Inc.
RW	Refractory wall combustors
RW	Richardsons Westgarth Ltd. (UK)
RWST	Refinery wastes
RWT	R. W. Taylor Steel Co.
RYAN	Ryan Construction
S&L	Sargent & Lundy
S&P	Sanderson & Porter Inc.
S&S	Stewart & Stevenson Services Inc. (GT packager)
S&T	Sanders & Thomas Inc.
S&W	Stone & Webster Engineering Corp.
S-F	Sigoure-Freres Inc. (France)

S-H	Standard Havens Inc.
S-W	S. M. Smith/Worthington
S/ACEC	Siemens/ACEC
S/B	Subbituminous and bituminous coal
S/BCM	Sanders-BCM Engineering
SA	Siemens-Allis
SAAR	Saarberg
SAAT	G.W. Saathoff
SABCON	Balfour Beatty/Boving/Skansa/Sweco/ASEA
SADEL	SADELMI Cogepi Construction (Italy)
SAE	SAE Sadelmi SpA (Italy)
SAGP	Sir Alexander Gibb & Partners (England)
SAND	Sandwell Inc.
SAS	Saskmont Engineering
SAVI	Savigliano (Italy)
SAW	Sawdust
SBB	Schlup Becker & Brennan
SC	Sierra Constructors (JV of Guy F. Atkinson Co. and Harrison Western Corp.)
SC	Simon Carves (England)
SC	Single-cylinder steam turbine
SC	Spray canal
SCBP	Single-cylinder back-pressure steam turbine
SCDX	Single-cylinder double extraction steam turbine
SCDXBP	Single-cylinder double extraction back pressure steam turbine
SCE	Southeastern Consulting Engineers
SCEX	Single-cylinder single extraction steam turbine
SCEXBP	Single-cylinder single extraction back pressure steam turbine
SCEXMP	Single-cylinder single extraction multiple pressure steam turbine
SCGC	Southern California Gas Co.
SCH	Schuchart & Associates
SCHN	Schneider Inc.
SCMK	Schoonmaker
SCMP	Single-cylinder multiple pressure steam turbine
SCOTT	Scott
SCR	Selective catalytic reduction
SCRB	Particulate scrubber
SCRC	SCR cold (after FGD system)
SCRH	SCR hot (between economizer & air preheater)
SCS	Southern Company Services Inc.
SCSEN	SCS Engineering
SD	Spray dry FGD scrubber
SDA	Spray dry FGD scrubber system
SDNA	Spray dry FGD scrubber with sodium injection
SDXBP	Single-cylinder double extraction back pressure steam turbine
SEA	Seaward Construction Co.
SEIC	Southern Engineering International (subsidiary of Southern Co.)
SELF	Company did own work
SELMER	F. Selmer Engineering Pty Ltd. (Australia)
SEMT	SEMT Pielstick (IC manufacturer) [France]
SENA	Societe d'Energie Nucleaire Franco-Belge des Ardennes
SENIOR	Senior Foster Wheeler (HRSG manufacturer)
SERE	Sereland SA (Spain)
SFL	Standard Fasel Lentjes (HRSG manufacturer)
SFLASH	Single flash geothermal units
SFT	Solid Fuels Technology
SFTINC	SFT Consulting Engineers Inc.
SG	Sloped-grate furnace
SGAS	Biogasification gas (chicken droppings etc.)
SGE	Societe General d'Enterprises (France)
SGP	Simmering Graz Paukering (Austria)
SH	Oil shale
SHALE	Shawinigan Lavalin Inc. (Canada)
SHAW	Stepped hearth burner
SHB	Shanghai Boiler Works (China)
SHBW	Shanghai Generator Works (China)
SHGW	Shimco
SHIM	Shimizu (Japan)
SHIN	Shin Nippon
SHINK	Shinko
SHL	Saarberg-Hoelter-Lurgi
SHTW	Shanghai Turbine Works (China)
SI	Steuler Industriewerk (Germany)
SICOM	Sicom SpA (Italy)
SIE	Siemens AG (Germany)
SIG	Signal Environmental Systems Inc. (subsidiary of Signal Co.)
SILT	Anthracite silt
SIMONS	H.A. Simons Ltd. (Canada)
SIPHON	Siphon-type hydroelectric turbine
SJ	Stanley Jones
SJG	S.J. Groves & Sons
SKANSA	Skansa International AB (Sweden)
SKAP	Semi-kaplan hydroelectric turbine
SKI	Solar Kinetics Inc.
SKIN	Skinner Engineering
SKNR	Skinner (T/G manufacturer)
SLAT	Slattery Group Inc.
SLIP	Slipform Engineering (Hong Kong)
SM	Schuler Meyer Engineering
SMP	SMP Engineers Inc.
SMS	S. Morgan Smith (hydraulic turbine manufacturer)

SMS/LF	S. Morgan Smith/Leffel	
SNC	SNC Inc. (Canada)	
SNCR	Selective non-catalytic reduction	
SO	Sorefame (Neypric and Alsthom licensee) (Portugal)	
SOCIA	Societe pour l'Industrie Atomique (France)	
SOD	Sod peat	
SOFT	Soft Energy Associates	
SOGEA	SOGEA (France)	
SOGENE	Societa Generale per Lavorie Pubbliche Utilita (Italy)	
SOLAR	Solar Turbines Inc. (subsidiary of Caterpillar)	
SOLARX	Solarex Corp.	
SOLTRN	Solar Transition of the Desert	
SOR	Sorumsand (hydraulic turbine manufacturer) (Norway)	
SOREN	Sorenson Engineering	
SOYAK	Soyak Trading & Construction (Turkey)	
SP	Spray pond	
SP	Springfield Boiler Works	
SPIE	Spie-Batignolles (France)	
SPOND	Solar pond	
SPR	Sierra Pacific Resources	
SPRNG	Springhouse Energy Systems	
SR	Stearns-Roger Engineering Corp.	
SS	Siemens Schuckertwerke (Germany)	
SSR	Smith Seckman Reid Inc.	
SSW	Japanese KWU affiliate	
ST	Steam turbine	
ST	Sterling Energy Systems	
ST/BO	Steinmuller/Borsig	
ST/C	Steam turbine in combined-cycle	
ST/D	Steam turbine and desalinization	
ST/DU	Steinmuller/Durr	
ST/H	Steam turbine with hot water sendout	
ST/HZ	Stetson-Harza	
ST/S	Steam turbine with steam sendout	
ST/TOS	Stein/Tosi	
STAGED	Staged combustion	
STAL	Stal-Laval (now ABB Stal)	
STAN	Stanley Consultants Inc.	
STATE	State Contractors Inc. (Canada)	
STCAT	Stearns Catalytic (subsidiary of United Engineers & Constructors)	
STEAG	STEAG AG (Germany)	
STEAM	Steamboat Springs Inc.	
STEIN	Stein Industrie (boiler manufacturing subsidiary of GEC Alsthom) (France)	
STEN	Stephens	
STER	Sterling (boiler manufacturer)	
STEU	Steuler Industre Iewerke (Germany)	
STIL	Alden E. Stilson Associates	
STIR	Stirling (England)	
STK	Stoker	
STK/TG	Traveling-grate stoker	
STLY	Stanley ** USE STAN **	
STM	L&C Steinmuller (boiler manufacturer) (Germany)	
STM	Stamford/Newage International (generator manufacturer)	
STM IJ	Steam injection	
STN	Shutdown or standby	
STORK	Koninklijke Machinefabriek Gebr. Stork & Co. N.V. (Netherlands)	
STOWE	Stowe Engineering Co.	
STROM	Stromberg (Finland)	
STS	STS Hydro Turbine Co.	
STURT	Sturtevant	
STW	Struthers Wells Corp.	
SUB	Subbituminous coal	
SUBCR	Sub-critical steam conditions	
SULFUR	Sulfur	
SUM	Summit Construction	
SUMITO	Sumitomo (Japan)	
SUN	Solar power	
SUN/WS	Solar and MSW system	
SUNDT	M.M. Sundt	
SUP	Superior Division of Cooper Industries (IC manufacturer)	
SUPER	Supersystems Inc.	
SUSP	Suspension combustion	
SUTIN	G.L. Sutin & Assoc. (Canada)	
SV	Sverdrup Inc.	
SV&P	Sverdrup & Parcel	
SW	Southwire	
SW&W	Swinerton & Walberg	
SWANSN	Swanson Rink	
SWE	Southwestern Engineering	
SYN	Synergics Inc.	
SYNOIL	Synthetic oil (H2 enhanced)	
SZ	Sulzer (Switzerland)	
SZEW	Sulzer-Escher Wyss (Switzerland)	
T&B	Townsend & Bottom Inc.	
T&G	Tippett & Gee Inc.	
T/A	TIBB/Ansaldo	
T/B	Tubular hydro turbine/bulb type	
T/G	Turbine-generator set	
T/H/GE	Toshiba/Hitachi/GE	
T/K	Takenaka/Kajima (Japan)	
T/M	Franco Tosi/Ercole Marelli (Italy)	
T/O/S	Taisei/Obay/Shimizu (Japan)	

Revised December 1992

C-19

T/R	Truck and rail fuel delivery
T/S	Takenaka/Shimizu (Japan)
T/T	Franco Tosi/TIBB
T/T/K	Taisei/Takenaka/Kim (Japan)
TAKE	Takenaka (Japan)
TAKUMA	Takuma Company Ltd. (GT packager) [Japan]
TAL	Talon Construction Co.
TAMP	Tampella (boilers) [Finland] also Tampella Keeler (USA)
TAMPER	Tamper
TAMS	Tippetts-Abbett-McCarthy-Stratton
TB	Tompkins Beckwith
TBGC	JV of Townsend Bottum & Gilbert Commonwealth
TC	Terry Corp. ** USE TERR **
TC/NM	Thomson Construction Inc./Natt McDougal
TCCED	Thiess-Codelfa-Cogefar-Evans-Deakin
TCEX	Tandem compound single extraction steam turbine
TCMP	Tandem compound multiple pressure steam turbine
TD	Turbodyne Division of Dresser Industries
TE	Traction-Electrique (Belgium)
TEC	Tecogen
TECH	Techint (Argentina)
TEI	Turbonetics
TEL	Teledyne
TERM	Terminal Construction
TERR	Terry Corp. (hydro turbine manufacturer)
TERRA	Terra Engineering
TFC	Thermo-Flood Corp.
TGAS	Top gas
TGT	Topping gas turbine (in combined-cycle)
TH/CEA	Thyssen/CEA
THER	Thermo Electron Corp.
THERM	Thermal DeNox system
THERMO	Thermodyn Division of Framatome (steam turbine manufacturer) [France]
THERMX	Thermex
THOM	Thomassen International bv (Netherlands)
TIANJ	Tianjin (China)
TIBB	Tecnomasio Italiano Brown Boveri SpA (Italy)
TICR	TIC Raymond
TIRE	Scrap tires
TIRU	Traitement Industriel des Residus Urgains (France)
TIW	Toronto Iron Works
TKW	Tavernkraftwerke AG (Austria)
TLG	TLG Engineering Inc.
TM	Trane Murray
TMEC	Turbomeca
TMSI	TMSI (subsidiary of Daniel Mann Johnson and Mendenhall)
TNPG	The Nuclear Power Group (England)
TONEY	Ray Toney and Associates
TORNO	Groupe Torno (Italy) or Torno America Inc.
TOS/CG	Tosi/CGE (Italy)
TOS/EM	Tosi/Ercole Marelli
TOSI	Franco Tosi Industriale SpA (Italy)
TOSI/A	Tosi/Ansaldo
TOWN	Townsend & Associates
TOYO	Toyo Engineering Corp. (Japan)
TPL	Tampella Ltd. (Finland)
TPM	Turbo Power & Marine formerly Pratt & Whitney (division of United Technologies)
TPT	Tandem pump turbine
TR	Thermal Reduction Co.
TRACY	Tracy Constructors
TRAIN	Train Johnson Power Development
TRANS	TransAmerica Delaval
TRE	TRE (T/G manufactuer) [USSR]
TRGN	Trigon Engineering
TRI	TRIO
TRICL	Tricil Ltd. (Canada)
TRIGEN	TriGeneration Inc.
TRIHAS	PT Trihajra Sarma (Indonesia)
TRP	Trump
TRUCK	Truck fuel delivery
TS	Toshiba Manufacturing Co. [AC licensee] or Toshiba International Corp. (Japan)
TSC	Two-stage combustion
TSC/FR	Two-stage combustion/flue-gas recirculation
TSC/LN	Two-stage combustion/lo-NOX burners
TSI	Turbo Systems International
TSSI	Thomassen Stewart & Stevenson Inc. (Netherlands)
TT	Turbo Tecnica (gas turbine manufacturing unit of Nuovo Pignone (Italy)
TTD	Tampella Turbine Division (Finland)
TUBE	Tube-type hydroelectric turbine
TUDR	Tudor Engineering
TUMAC	Tumac Industries
TUR	Tuma Turbomach (GT packager) [Switzerland]
TURBO	Turbo Machinery
TURBO	Turbo expander (small power)
TURGO	Turgo (hydraulic turbine manufacturer)
TURN	Turner Corp.
TVBB	Tijdelijke Vereniging Burgelijke Bouwkude (Belgium)
TW	Taylor Woodrow Construction Ltd. (England)
TWO	Twombly Partners
TY	Terry Corp.

TYGER	Tyger Construction Co. Inc.
U&A	Uerling & Associates
UAE	United American Energy Corp. (unit of Guy F. Atkinson Co.)
UCP	United Centrifugal Pump Co.
UE&C	United Engineers & Constructors Inc. (subsidiary of Raytheon)
UENG	Utility Engineers Inc.
UES	United Environmental Services Inc.
UHR	Uhl Hall & Rich
UKAEA	U.K. Atomic Energy Agency
ULEC	ULEC Services Ltd. (England)
ULTRA	Ultrasystems Engineers & Constructors
UM	United McGill Corp.
UMA	UMA Engineering Lt. (Canada)
UN	Union Iron Works (boiler manufacturer)
UNICO	Unico Engineering Associates
UNION	Union Carbide Corp.
UNISYN	Unisyn (subsidiary of Washington Energy Co.)
UNIV	Universal Engineering Co.
UNK	Unknown
UOP	Air Pollution Divison of Universal Oil Products Co. (unit of Signal Industries)
UOP/LC	UOP/Lodge Cottrell
UP	Union Power Co.
UPC	Utility Power Corp. (subsidiary of Kraftwerk Union)
UR	Uranium
USACE	U.S. Army Corps of Engineers
USAEC	U.S. Alternate Energy Corp.
USBR	U.S. Bureau of Reclamation ** USE BUREC **
USM	U.S. Motors
USS	U.S. Fabricating
USSR	Unspecified Soviet vendor
UST	U.S. Turbine Corp.
USW	U.S. Windpower Inc.
UTC	United Technologies Corp.
UTEC	Utility Engineering Co. (subsidiary of Southwestern Public Service Co.)
UTHEA	Hernando/Entrecanales/Agroman (Spain)
UTIL	Utility
UTILCO	Utility Contractors Inc.
V/AAB	Voith/ABB
V/BBC	Voith/BBC
V/N/M	Voith/Neypric/MEP
V/SIE	Voith/Siemens
VA	Voest-Alpine International Corp.
VAC	Frank A. Vaccaro
VAL	Valmont Industries Inc.
VAN	Vanguard (WTG)

VAP	Vapor Corp.
VAR	Various
VAW	Verle A. Williams & Associates Inc.
VAWT	Vertical axis wind turbine
VDDA	Valley Detroit Diesel Allison
VEA	Vern E. Alden Co.
VEL	Velck
VENG	Vivian Engines
VENT	Ventech Engineers
VENT	Venturi particulate scrubber
VENZ	Venezuela
VERT	Vertical axis hydro turbine
VES	Vanguard Energy Systems
VEST	Vestas (WTG manufacturer)
VEW	Vancouver Engineering Works (Canada)
VIAN	Vianini Lavori SpA (Italy)
VICK	Vickers
VICON	Vicon Recovery Systems Inc.
VILL	Villares (hydro generators) [Brazil]
VIW	Vancouver Iron Works (Canada)
VK	Von Kalk (generator manufacturer)
VKW	Vereinigte Kesselwerke AG (boiler manufacturer) [Germany]
VL	Vianini Lavori SpA (Italy)
VL	Villares (Brazil) ** USE VILL **
VLD	Volund
VM/AJM	Vevey Meoka/Alsthom Jeumont
VOGEL	Vogel Brothers
VOI	J.M. Voith GmbH (also purchased A-C Hydro-Turbine operations) [Germany]
VOI/BD	Voith/Bardella (Brazil)
VOI/SS	Voith/SS
VOLC	Volcano Inc. (Canada)
VOLV	Volvo (Sweden)
VOLVO	Volvo (Sweden)
VR	Von Roll
VS	Vulcan Stirling
VS/CE	Vulcan Stirling/Combustion Engineering
VT	Henry Vogt Machine Co. (HRSG manufacturer)
VULCAN	Vulcan Chemical Co.
VVIW	Vulcan Iron Works (Canada)
W&A	Williamson & Associates
W&E	Wiamer & Ernst Umweltschnik AG (Switzerland) (subsidiary of Blount Inc.)
W&K	Winzler & Kelly
W&L	Williams & Lane Inc.
W&M	Westcott & Mapes

Revised December 1992

C-21

W-S	White Superior
W/G	Wood and natural gas
W/O	Wood and oil
WADE	Wade-Lupe
WALSH	Walsh Construction Co. (unit of Guy F. Atkinson Co.)
WALT	Walther & Cie. (Germany) ** USE WALTH **
WALTH	Walther & Cie. (now GEC Alsthom subsidiary) [Germany]
WART	Wartsila Diesel (Finland)
WARZN	Warzyn Engineering
WAT	Water
WAT	Watson
WAT IJ	Water injection
WAU	Waukeshaw Engine Division of Dresser Industries Inc.
WAVE	Wave energy
WAYNE	Wayne Energy Recovery Inc.
WB	Waagner Biro (Austria)
WBIT	Waste bituminous coal (coal refuse)
WCAR	Wet sodium carbonate scrubber
WDGAS	Wood gas (from wood gasifier)
WE	Waste Energy Inc.
WECRES	Resource Energy Systems Division of Westinghouse Electric Corp.
WEE	Western Energy Engineers
WEG	Wind Energy Group (British Aerospace/GEC/Taylor Woodrow) [England]
WEHRAN	Wehran Enviro Tech
WEL	Wellons Inc. (boiler manufacturer)
WELEC	Western Electric
WELL	Wellhead Electric Co.
WELL	Wellman-Lord FGD scrubber
WELLS	Wells (air-driven turbine)
WES	Western Engine Systems Inc.
WESP	Wet ESP
WEST	Western Precipitation Division of Joy Industrial Equipment
WEST/L	Western/Lodge Cottrell
WEST/U	Western/UOP
WEST/W	Western/Wheelabrator
WEU	W&E Umwelttechnick (Switzerland)
WF	Wheelabrator-Frye Inc.
WF/AI	Wheelabrator/Atomics International
WF/RC	Wheelabrator/Research Cottrell
WFC	William F. Cosulich Associates
WH	Westinghouse Electric Corp.
WH/GE	Westinghouse/General Electric
WH/ID	Westinghouse/Ideal
WHA	W.H. Allen -- now NEI Allen (steam turbine manufacturer) [England]
WHAM	William Hamilton
WHC	Westinghouse Canada
WHDC	Westinghouse Development Corp.
WICK	Wickes Boiler Corp. ** USE WK **
WILL	Williamette Iron & Steel (hydraulic turbine manufacturer)
WIMPEY	G. Wimpey (England)
WIND	Wind-powered turbines
WING	Wing
WISC	Wisconsin Associates
WK	Wickes Boiler Corp.
WL	Wet lime FGD scrubber
WL/A	Wet lime-alkaline fly ash FGD scrubber
WLGAS	Wellhead gas
WLST	Wet limestone FGD scrubber
WM	WindMatic
WMC	Windsor Machinery Co.
WMI	Waste Management Inc.
WMST	Windmaster Corp.
WOOD	Wood or wood-waste fuel
WORT	Worthington Division of Dresser Industries Inc.
WPS	Wind Power Systems
WRM	Wormser Engineering Inc.
WS	Wellman-Seaver
WSA	Wet soda ash FGD scrubber
WSC	Wet sodium carbonate FGD scrubber
WSCRB	Wet scrubber (unspecified)
WSM	Wellman-Seaver-Morgan
WST/LC	Western/Lodge Cottrell
WST/WF	Western/Wheelabrator
WSTGAS	Waste gas
WSTH	Waste heat
WSTOIL	Waste oil
WT	Water-tube boiler
WT	Wind Tech
WT/FIX	Fixed-grate water-tube boiler
WT/HZ	Water tube/horizontal boiler
WT/SPR	Spreader stoker water-tube boiler
WT/SUS	Suspension grate water-tube boiler
WT/TG	Travelling grate water-tube boiler
WTG	Wind turbine generator
WUHAN	Wuhan Boiler Works (China)
WV	Wet venturi particulate scrubber
WW	Waterwall boilers
WW/FIX	Fixed grate waterwall boiler
WW/RCP	Reciprocating grate waterwall boiler
WW/ROT	Rotary combustor waterwall boiler
WW/SG	Sloped grate waterwall boiler
WW/SPR	Waterwall/speader stoker boiler

WW/SUS	Suspension grate waterwall boiler
WW/TG	Waterwall/traveling grate boilers
WWSL	Waste-water sludge
WWT	Westwind Turbines (WTG manufacturer) [Australia]
WWT/TI	Wicker Water Tube/Toronto Iron Works
WY/LY	Weyher/Livsey
WYAT	Wyatt & Lipper Engineers Inc.
YANK	Yanke Energy
YAR	Yarrow (boiler manufacturer) [England]
YAS	Yaskawa (generator manufacturer)
YEARG	Yeargin Inc. (subsidiary of United Engineers & Constructors)
YK	Yokoyama (boiler manufacturer) [Japan]
YU	Yuba Heat Transfer Co. (division of Avondale Industries Inc.)
ZAM	Zamech (now ABB Zamech) [Poland]
ZIMP	Zimpro/Passavant Inc.
ZINK	John Zink Co.
ZN/EPI	Zurn/Energy Products of Idaho
ZN/RS	Zurn/Riley Stoker
ZOND	Zond Cogeneration Systems
ZURN	Zurn Industries Inc.

Revised December 1992

APPENDIX D

CONVERSION FACTORS AND TABLES

FOR ELECTRIC POWER PLANT DATA

UDI-033-92   December 1992

APPENDIX D

Conversion Factors and Tables for Electric Power Plant Data

As a general matter, primary source documents used in the development of UDI's international power plant data bases show data in metric units. The following conversion factors and values are provided to assist users of steam conditions data in UDI's files of international power plant statistics.

English to Metric Conversions

Steam Pressure

$(kg/cm^2)(0.9807) = BAR$

$PSI/14.5 = BAR$

$(Atmospheres)(1.013) = BAR$

Steam Flow

$(1000\ lbs/hr)(0.126) = kg/sec$

$(tons/hour)(0.252) = kg/sec$

Temperature

$[(degrees\ F) - 32](.56) = degrees\ C$

On the following pages are tables illustrating typical pressure, temperature and flow conversions.

D-1

TYPICAL PRESSURE CONVERSIONS

PSI	BAR
850	59
938	65
950	66
1450	100
1600	110
1900	131
2320	160
2400	166
2410	166
2474	174
2503	176
2600	179
2702	190
3500	246
3626	255
4494	316

TYPICAL TEMPERATURE CONVERSIONS

°F	°C
510	268
514	270
537	283
540	284
850	458
900	486
944	511
950	514
1000	542
1004	544
1005	545
1050	566

REPRESENTATIVE STEAM FLOW CONVERSIONS

10^3 LBS/HR	KG/SEC
1160	146
1662	209
1850	233
2180	275
2250	284
2270	286
2320	292
4100	517
4600	580